SHARED RISK:
COMPLEX SYSTEMS IN SEISMIC RESPONSE

Elsevier Science Internet Homepage — http://www.elsevier.nl

Full catalogue information on all book, journals and electronic products.

Related Journals

Global Environmental Change
Editor: M. Parry

Global Environmental Change Part B: Environmental Hazards
Editors: J.K. Mitchell and S.L. Cutter

Applied Geography
Editor: R. Rogerson

Free specimen copies of journals gladly sent on request: Elsevier Science Ltd, The Boulevard, Langford Lane, Kidlington, Oxford OX5 1GB, UK.

SHARED RISK: COMPLEX SYSTEMS IN SEISMIC RESPONSE

Louise K. Comfort
University of Pittsburgh
Pittsburgh, PA 15260, USA

1999

PERGAMON
An imprint of Elsevier Science

Amsterdam – Lausanne – New York – Oxford – Shannon – Singapore – Tokyo

ELSEVIER SCIENCE Ltd.
The Boulevard, Langford Lane
Kidlington, Oxford OX5 1GB, UK

© 1999 Elsevier Science Ltd. All rights reserved.

This work is protected under copyright by Elsevier Science, and the following terms and conditions apply to its use:

Photocopying
Single photocopies of single chapters may be made for personal use as allowed by national copyright laws. Permission of the Publisher and payment of a fee is required for all other photocopying, including multiple or systematic copying, copying for advertising or promotional purposes, resale, and all forms of document delivery. Special rates are available for educational institutions that wish to make photocopies for nonprofit educational classroom use.

Permissions may be sought directly from Elsevier Science Rights & Permissions Department, PO Box 800, Oxford OX5 1DX, UK; phone: (+44) 1865 843830, fax: (+44) 1865 853333, e-mail: permissions@elsevier.co.uk. You may also contact Rights & Permissions directly through Elsevier's home page (http://www.elsevier.nl), selecting first 'Customer Support', then 'General Information', then 'Permissions Query Form'.

In the USA, users may clear permissions and make payments through the Copyright Clearance Center, Inc., 222 Rosewood Drive, Danvers, MA 01923, USA; phone: (978) 7508400, fax: (978) 7504744, and in the UK through the Copyright Licensing Agency Rapid Clearance Service (CLARCS), 90 Tottenham Court Road, London W1P 0LP, UK; phone: (+44) 171 631 5555; fax: (+44) 171 631 5500. Other countries may have a local reprographic rights agency for payments.

Derivative Works
Tables of contents may be reproduced for internal circulation, but permission of Elsevier Science is required for external resale or distribution of such material.
Permission of the Publisher is required for all other derivative works, including compilations and translations.

Electronic Storage or Usage
Permission of the Publisher is required to store or use electronically any material contained in this work, including any chapter or part of a chapter.

Except as outlined above, no part of this work may be reproduced, stored in a retrieval system or transmitted in any form or by any means, electronic, mechanical, photocopying, recording or otherwise, without prior written permission of the Publisher.
Address permissions requests to: Elsevier Science Rights & Permissions Department, at the mail, fax and e-mail addresses noted above.

Notice
No responsibility is assumed by the Publisher for any injury and/or damage to persons or property as a matter of products liability, negligence or otherwise, or from any use or operation of any methods, products, instructions or ideas contained in the material herein. Because of rapid advances in the medical sciences, in particular, independent verification of diagnoses and drug dosages should be made.

First edition 1999

Library of Congress Cataloging in Publication Data
Comfort, Louise K. (Louise Kloos), 1935–
    Shared risk : complex systems in seismic response / Louise K. Comfort.
        p.   cm.
    Includes bibliographical references and index.
    ISBN 0-08-043211-5 (hc)
    1. Emergency management. 2. Risk management.   I. Title.
HV551.2.C63 1999
363.34'8–dc21   99-29391
                                                                           CIP

British Library Cataloguing in Publication Data
A catalogue record from the British Library has been applied for.

ISBN: 008 043211 5

∞ The paper used in this publication meets the requirements of ANSI/NISO Z39.48-1992 (Permanence of Paper).

Printed in the Netherlands

To Nathaniel and Honore

.... and the next generation.

Preface

The intellectual and emotional journeys involved in writing this book far surpass the physical distances traveled. I began this study by seeking to understand the decision processes involved in uncertain, rapidly changing environments, a problem that confronts practicing public managers with an extraordinarily difficult set of tasks. Living and teaching in California, where disasters regularly disrupt the fragile balance created between human communities and a beautiful but often threatening environment, I turned to emergency response. Teaching in the Master's Program in Public Administration at San Jose State University and California State University, Hayward, I had students in my seminars from police, fire, public works and emergency medical services who were grappling with real problems of decision making under uncertainty. They needed to balance existing laws, rules and procedures against a dynamic set of conditions for which the existing decision rules did not necessarily apply. They were legally compelled to act under conditions which many of them had not seen before, but yet had to determine the most appropriate action to protect their communities. Further, they often had to make these decisions under conditions of urgent stress and tight time constraints.

My first approach was to draw upon the theoretical literature of systems theory, which offered a framework for the study of dynamic decision processes. But listening to my students in emergency services worry about the problems they faced in daily practice, I realized that I needed to observe this process in actual field environments. Earthquakes represent a serious and recurring threat in California and carry the potential for creating severe, community-wide damage in terms of loss of life and property. I chose to focus on rapidly evolving response systems following earthquakes, because of the significance and complexity of the decision processes involved. Yet, this choice presents a difficult problem for a researcher. Moderate earthquakes occur every 40 to 60 years, severe earthquakes every 90 to 150 years, in any one location, but both occur with relative frequency around the globe. The study of rapidly evolving response systems following earthquakes necessarily became an international project, as I sought to examine this problem in actual field environments.

When I undertook my first study of the evolving response system followed the Mexico City Earthquake in September, 1985, I did not anticipate that it would lead to a series of nearly annual studies of seismic response over the next decade, ending with the Hanshin, Japan earthquake in 1995. The problem itself governed the search for additional data, appropriate measures, and a theoretical framework that provided a closer fit to the phenomena under study. Each new earthquake demonstrated a different aspect of the evolution of disaster response as the same type of hazard interacted with different local conditions to produce a distinct response system that reflected the characteristics of the economic, political, social, and cultural contexts in which it occurred. Once undertaken, the study demanded a reasoned explanation of the dynamic phenomena I was observing. Discovering the driving forces of this evolving process of response and recovery following earthquakes became a compelling assignment that I could not refuse. This book offers my interpretation of this dynamic process, and each of the eleven cases included in the book added important dimensions to the argument. The interpretation evolved as the data documented more fully the recurring patterns of interactions among multiple participants in the complex process of interorganizational response following earthquakes.

While this book represents an analytical study of rapidly evolving response systems following disaster, it was written with deepening awareness of the intense emotional trauma that such events create for all members of a stricken community. The differences in culture among the nations only accentuate the commonality of grief shared by those left behind. No words can capture the pain in a father's eyes, as he waits for word of his missing daughter. No amount of money can rebuild the memories associated with a destroyed home. No quantity of handkerchiefs can dry the tears of an entire village that has lost its children in the collapse of a school. In tribute to those memories, shared by tens of thousands of people across eleven stricken communities, I offer a poem, written in the wake of searing recollections from the earthquake of San Salvador, October 10, 1986.

Salvador

Holy sentinels guard the Valley of the Hammocks
    where Spaniards once camped

Sultry, seductive peaks draw mortals
    toward intermittent peril,
    hidden beneath lush earth

Gambling souls defy the odds of danger.

Morning sunlight masks the harsh friction
    of ordinary life

Preface

Women, balancing baskets on their heads, weave
    gracefully through meshing traffic,
    impassable to smoking machines

Men, smiling over carbines, nod "Buenos Dias" as they
    check rights of passage with glinting steel

Violence, mixed with daily courtesy, dulls the
    gnawing edge of risk.

In the center of the city, a desperate search for life
    presses past the edge of logical choice

Men in uniform stretch courage
    over collapsed concrete

Electric saws keen to high tension

Grit clouds the faces of those waiting, creating hope
    out of bleak shreds of uncertainty

Grumbling, the crane stops. Silence thickens
    as the signal for assistance drops

Soberly, paramedics untie green bags while
    judges prepare finality. Aching
    in disbelief, a father mourns alone

In guttural roar, the crane backs
    toward the misshapen wall

In this crucible of death, life is discovered anew
    for only compassion can absolve
    the sound and stench of horror.

Returning, Pennsylvania hills beckon
    with rich October colors

University halls echo the mundane
    "Faculty meeting starts in five minutes...
    Did your research grant come through? Is my
    stipend secure?"

Preface

Banalities wash over unspeakable trauma,
    blurring tragedy into acceptable conversation.

Etched indelibly in memory is the image
    of mountains, spilling danger
    over the valley

For those who know the bitter reek of death,
    there is no choice but action.

Louise Comfort
October, 1986

Acknowledgments

This study of dynamic response systems following eleven earthquakes in nine different countries over more than a decade represents much more than my efforts alone. This study could not have been done without extensive support and cooperation from governmental and nongovernmental organizations in each of these countries. It has been my privilege to work with, and learn from, an extraordinary group of participants in each of these response systems. I acknowledge, with warm thanks and appreciation, the institutions that have supported this research and the efforts of many, many people who have assisted me in its conduct.

First, I acknowledge the support of the National Science Foundation through the Quick Response Grant Program administered by the Natural Hazards Center of the University of Colorado, Boulder. Initiated in 1984 shortly before I began this study, the Quick Response program enabled me to leave almost immediately after an earthquake occurred to study the rapidly evolving response process in the field. Six of the eleven cases included in this study were supported, in part, through the Quick Response Program. I also acknowledge the National Research Council for its support of the 1987 Ecuadorian Earthquakes Study, the National Science Foundation for grants to support studies of the 1987 Whittier Narrows, CA Earthquake #CES 88-04285,1989 and the 1994 Northridge Earthquake, #BCS 94-10896, and the United Nations Department of Development Support and Management Services which facilitated the 1993 Maharashtra Earthquake Study.

I acknowledge, with thanks and appreciation, the several research units at the University of Pittsburgh, that have contributed resources, equipment, administrative and student support to the continuing program of research. These units include the University Center for Social and Urban Research, the University Center for International Studies, the Office of Research and the Graduate School of Public and International Affairs (GSPIA). I especially appreciate the flexibility of GSPIA, through several deans, in allowing me to interrupt my teaching schedule on several occasions in order to go immediately to the scene of an earthquake disaster.

With special thanks and warm affection, I acknowledge the continuing support of the Institute of Governmental Studies, University of California, Berkeley to this project. I was fortunate to spend a research year at the Institute in 1988–1989, during which the 1988 Armenian Earthquake occurred. With the support of Nelson Polsby, Director of the Institute, I returned for seven consecutive summers to do the analysis of successive earthquakes and to write the draft manuscript for the book. I acknowledge not only Nelson, but the entire staff of the Institute, who were consistently helpful in tracking down source materials and welcoming visitors from many nations who shared my research interests. This generous institutional support made a major contribution to the successful completion of this study.

In each country, while I benefited from the advice and assistance of many people, there were several persons who made especially valuable contributions of time, effort and knowledge to facilitate this research. For the Mexico City study, I gratefully acknowledge the assistance of the late John Funari, former dean of GSPIA, who had served in Mexico City with the Ford Foundation; Fernando Estrada and Federico Estevez, both of the Instituto Tecnologico Autonomo de Mexico (ITAM); Rosario Molinero, who served both as my translator and driver in the formidable traffic of Mexico City; George Natanson, CBS News; Paul Bell, US Office of Foreign Disaster Assistance; Linda Wallace, National Association of Search and Rescue; and Victoria Funari, who introduced me to the student groups in Colonia Morelos that were actively engaged in community response. Paul Bell proved to be a valued mentor and model of dynamic leadership in disaster response through each of the four Latin studies.

In San Salvador, I acknowledge with thanks and gratitude the assistance of Paul Bell, Director of Operations, OFDA Team in San Salvador; Alan Swan, US OFDA, Washington, DC; Douglas Jewett, Metro Dade Search and Rescue Team; Antonio Godoy, Cuerpo de Bomberos, Ciudad de San Salvador. I thank also Hector Maldonado, San Salvador, who conducted the survey of citizens affected by the earthquake; and Lynn Whitlock and Jesus Garcia, graduate students at the University of Pittsburgh, who provided thoughtful assistance with the data analysis. Oliver Davidson, formerly with the US Office of Foreign Disaster Assistance, graciously made his personal files available to me to verify events from disaster operations.

In Ecuador, I thank especially Blasco Penaherrera, then Vice President of Ecuador; Renan Herrera, Instituto Ecuatoriano de Mineria (INEMIN); Raul Montalvo, INEMIN, Maj. Luis Aguas, Ejercito de Ecuador, Cmdr., Battalon de Selva, Lago Agrio; Maj. Howard Mayhew, US Army, Southern Command, Quito; Neil Meriwether, US Office of Foreign Disaster Assistance, Quito; Mario Venegas, Corporacion de Apoyo a la Tecnologia y la Comunicacion (CATEC), Quito; Eduardo Lopez, Ford Motor Co, Quito. I am grateful to Alvaro Saenz, Director, INFOC and Departamento de Sociologia, Universidad Catolica, Quito, who assisted me with the survey of citizens affected by the earthquake.

For assistance following the 1987 Whittier Narrows, California Earthquake, I thank Shirley Mattingly, then with the Emergency Operations Bureau, City of Los Angeles; Lt. Gary Schoeller, Emergency Operations Bureau, County of Los

Acknowledgments xiii

Angeles; Theodore Anagnoson, then Chair, Department of Political Science, California State University, Los Angeles; George Lidtke, Department of Political Science, California State University, Los Angeles, and Patricia Campbell, a very able graduate student in the Department of Political Science, California State University, Los Angeles, who assisted me with data collection. I also acknowledge the excellent work of Bernadette Palumbo, a graduate student at Claremont University, who assisted me with interviews of directors of nonprofit organizations; and Keun Namkoong, University of Pittsburgh, who assisted me with the data analysis. George Hart, California State University, Pomona and David Ringsmuth, California State University, Northridge, offered advice and insight into the local administrative networks in emergency response. Paul Flores, then Executive Director of Southern California Earthquake Preparedness Program, provided a careful assessment of background information on emergency preparedness in the region.

Many people facilitated the study of the response system in Armenia. I am grateful to Peter Safar, M.D., International Resuscitation Research Group, University of Pittsburgh, for including me in the international, interdisciplinary research team that he coordinated with Victor N. Semenov, M.D., Director, Institute of Reanimatology, USSR Academy of Medical Sciences and Organizing Leader for the International US-USSR Team. I thank Miroslav Klain, M.D., Team Leader for the U.S. Team; Julia Anoshkina, Institute of Reanimatology, translator and staff assistant to the international team; Bishop Aris Shirvanian, Armenian Church of San Francisco; Stella Gregorian, Yerevan State University; Vladislav Teriaev, M.D., Director, Boris Gazetov, M.D., and Timothy Gorchaumelidze, M.D., all of the Institute of Emergency Medicine, Moscow; Boris Karapetian, Ph.D., Department of Engineering, Yerevan Polytechnic University; and Arthur Melkonian, M.D., Armenian Ministry of Health, who did the content analysis of the Armenian newspapers.

In the Loma Prieta, California Earthquake, I was essentially conducting research in my home region, which made the data collection easier, as I was familiar with the organizations and persons involved in disaster response operations. But conducting the analysis proved more difficult, as it required all the professional training I could muster to be objective about the damaging event and the consequences for the region. I am especially grateful to Nelson Polsby, Director, Institute of Governmental Studies, who gave support to the study in terms of space and communications. In addition, a number of people were very helpful to my data collection efforts. They include: Dr. Vitelmo Bertero, Director, Earthquake Engineering Research Center, University of California, Berkeley; Henry Renteria, then Director, Office of Emergency Services, City of Oakland; Charles Scawthorn, Team Leader for Emergency Response, National Research Council Investigating Team; Bob Olson, VSP Associates, Sacramento, J. Thomas Cooke, Sedway and Cooke, San Francisco.

In Costa Rica, I am again grateful to Paul Bell, Director of the US Office of Foreign Disaster Assistance team; Bernardo Mendez, Deputy Director of the

Comision Nacional de Emergencia (CNE), Luis Diego Morales, Director of Planning, CNE, Teo Sarkis M.D., Costa Rica Red Cross, and Lt. Col. Richard Price, U.S. Army Corps of Engineers, Southern Command.

In Erzincan, Turkey, several people greatly facilitated the conduct of this research. They include: Dr. Abdulkadir Ates, then Minister of Tourism, Turkey; Dr. Rusen Keles, Department of Political Science, Ankara University, and Oktay Ergunay, then General Director, General Directorate of Disaster Affairs, Ministry of Public Works and Settlements. I am also grateful to colleagues from the University of Pittsburgh, Dr. Ernesto Pretto and Dr. Bulent Kirimli, who led the US component of the International, Interdisciplinary Research Team for the study of the Erzincan Earthquake, and to Dr. Mehmet Sungur, Department of Psychiatry, Ankara University, who assisted me with interviews of practicing managers in Erzincan. Ali Tekin and Ayse Caner, then graduate students at the University of Pittsburgh, provided excellent support for the translation and analysis of the interviews.

In India, I am very grateful to Dr. Sharayu Anantaram, Department of Sociology, SNDT University Bombay, for her careful planning and arrangements for the study in the Marathwada region, her translation services for interviews conducted in the Districts of Latur, Osmanabad, and Solapur, and her conduct of the content analysis of Indian newspapers in both English and Marathi. Dr. Anantaram was a gracious colleague and traveling companion, and proved central to the success of the study in every way. I also am grateful to the families of Hari Godbole in Solapur and Nalin Sheth in Latur. Both graciously invited us to their homes, and shared with us their observations and experiences of the impact of the earthquake on the region. In addition, Praveensingh Pardeshi, District Collector of Latur, and Dineshkumar Jain, District Collector of Solapur, both members of the Indian Administrative Service, provided access to essential records and information regarding the disaster operations in the earthquake-stricken districts. Rajeev Anantaram, graduate student at the University of Pittsburgh, consistently provided information and voluntary support to the study.

In Northridge, CA, I was again working in the familiar context of California disaster operations, but several people greatly facilitated my research. I thank Lacey Suiter, Federal Emergency Management Agency (FEMA), for his support and assistance to this research. James Lee Witt, Director, FEMA, granted me access to the daily staff meetings of federal managers. Richard Andrews, Director, California Office of Emergency Services, granted me access to the daily staff meetings of state managers. William J. Petak, University of Southern California, Institute for Systems Safety and Management, provided thoughtful observations on the organization of response to the event, and managers at the City of Los Angeles and County of Los Angeles gave time, information and candid observations regarding the event. I am especially grateful to Dr. David Ringsmuth, Department of Political Science, California State University, Northridge, for his detailed knowledge of the earthquake-affected region and his tour of the area.

Acknowledgments

In the study of the Hanshin, Japan Earthquake, I am grateful to many people for their efforts to facilitate the study. First, my warm thanks and appreciation go to Dr. Shunjii Fuji, Taisei Corporation, Tokyo, for his considerable efforts in arranging interviews for me in both Tokyo and Kobe. I especially am grateful to Drs. Muneo Ohta and Tatsuro Kai of the Senri Critical Care Medical Center in Osaka, who arranged interviews for me in both Osaka and Kobe. Dr. Kunio Funahashi, Department of Architecture, Osaka University, gave considerable effort and time to assisting my study, including traveling to Tokyo for the interview with Kiyoshi Nishimura, Director, Ambulance and Rescue Service Division, Fire Defense Agency, Ministry of Home Affairs. Dr. Kentaro Serita, then Dean of the Graduate School of International Cooperation Studies at Kobe University facilitated both the arrangement of interviews and the use of materials at Kobe University's Library. I am grateful to Juntaro Ashikari, Osaka University, who served as translator, guide, and thoughtful observer of the disaster response operations, especially from the perspective on voluntary organizations. I also acknowledge Dr. Ernesto Pretto, University of Pittsburgh, for facilitating contacts in Japan, and my colleagues, Dr. Sharlene Adamson and Nancy Bowen from the University of Pittsburgh for their support and good will in the conduct of this study. I warmly thank Ms. Laurie Lofgren, Northwest Airlines, Narita, for her hospitality and assistance in arranging an interview with a private sector manager in the Osaka region. I am grateful to Takehiko Serai, MPIA, University of Pittsburgh, who served as my translator and guide in Tokyo, and who translated key reports following the field research.

Throughout the years of this study, from 1985 to 1995, I have had the benefit of support from a remarkable group of graduate students at the Graduate School of Public and International Affairs. It is one of the great privileges of teaching at GSPIA to have the opportunity to work with students from all over the world. In the conduct of this study, I turned repeatedly to students from the countries in which I was working. They offered much more than language skills, which were greatly valued, but also knowledge of their home culture and institutions, references, recommendations and practical advice on travel and study arrangements. US students also played a central role in the long and arduous data analysis for this study. Some are listed above, but I list here the full set of students who have worked with me in the conduct of the content analysis, coding and preparation of this manuscript. They have now completed their degrees and continued to develop their own professional careers. They include: Lynn Whitlock, Jesus Garcia, Jan Jernigan, Harry Dai, Keun Namkoong, Noel Benson, Sarah Nicolic, Ali Tekin, Lesley Mohr, Elizabeth Robedeau, Tamara Taylor, Susan Wade, Jeanne Schroeder, Andrea Beck, Julie Stetz, Rajeev Anantaram, Leonardo Alvarez, Joseph Narkevic, and Holly DeFoe. I am grateful to each of them for the hard work, dedication and effort they contributed to this study. It has been my privilege to work with, and learn from, each of them.

Several colleagues have played major, invaluable roles in assisting me to think through the difficult conceptual and methodological issues involved in this analysis. First, I acknowledge the thoughtful advice of my son, Nathaniel Comfort, who referred me to the literature on complex, adaptive systems. I thank Frederick Balderston, who read Chapters 1–4, offering perceptive suggestions on the conceptual model of complex systems. Revan Tranter read Chapters 1–3 from the perspective of a reflective practitioner, offering candid advice for this audience. I am especially grateful to H. Richard Priesmeyer, who demonstrated his Chaos! software, and patiently worked with me to apply it to the initial analysis of the data from the content analysis of newspapers for the set of eleven response systems. I appreciate immensely the thoughtful comments of Sidney Verba, who read the entire manuscript. Emery Roe and Walter Hays offered careful, invaluable advice that made the manuscript stronger, clearer and more carefully reasoned. Any errors in the manuscript are mine alone. And very importantly, I am grateful to my son, Nathaniel, and my daughter, Honore, for their unwavering encouragement, good humor and good will over a decade of work and travel that enabled me to keep my perspective and conclude the study, strengthened by their optimism and focus on the future.

List of Tables

1. Losses of Life and Property in Eleven Recent Earthquakes — 13
2. Assessment Indicators for Disaster Response Systems — 65
3. Preliminary Assessment of Eleven Disaster Response Systems following Major Earthquakes, 1985-1995, on Technical, Organizational, and Cultural Dimensions — 67
4. Types of Emergency Response Systems — 68
5. Characteristics of Nonadaptive Systems — 69
6. Characteristics of Emergent Adaptive Systems — 70
7. Characteristics of Operative Adaptive Systems — 72
8. Characteristics of Auto-Adaptive Systems — 74
9. Summary, Organizational Response System by Source of Funding and Jurisdiction, San Salvador, October 14 — November 3, 1986 — 85
10. Frequency Distribution: Types of Transactions in Disaster Response by Funding Source, San Salvador Earthquake, October 14 — November 3, 1986 — 86
11. Frequency Distribution: Types of Interactions in Organizational Disaster Response, San Salvador Earthquake, October 14 — November 3, 1986 — 88
12. Summary, Organizational Response System by Funding Source and Jurisdiction, Ecuador, March 5–6, 1987 — 98

List of Tables

13. Frequency Distribution: Types of Transactions in Disaster Response by Funding Source and Jurisdiction, Ecuador Earthquake, March 7–31, 1987   100

14. Frequency Distribution: Interactions by Funding Source and Jurisdiction, Ecuador Earthquake, March 7–31, 1987   100

15. Summary, Organizational Response System by Funding Source and Jurisdiction, Armenia, December 8–28, 1988   107

16. Response Operations, 1988 Armenia Earthquake   109

17. Frequency Distribution: Types of Transactions in Disaster Response by Funding Source and Jurisdiction, Armenia Earthquake, December 8–29, 1988   110

18. Frequency Distribution: Types of Interactions in Organizational Disaster Response, Armenia Earthquake, December 8–29, 1988   112

19. Summary, Frequency Distributions, Organizational Disaster Response Systems by Funding Source and Jurisdiction   115

20. Perceived Benefit of International Aid, 1985 Mexico City Earthquake   123

21. Perceived Availability of Information Regarding International Aid in District   123

22. Perceived Sufficiency of Information Regarding International Aid in District   124

23. Reception of International Aid, 1985 Mexico City Earthquake   124

24. The Relationship Between Availability and Sufficiency of Information Regarding International Aid, 1985 Mexico City Earthquake   124

25. The Relationship Between Availability of Information and Reception of International Aid, 1985 Mexico City Earthquake   125

26. The Relationship Between Reception of International Aid and Sufficiency of Information, 1985 Mexico City Earthquake   125

27. Frequency Distribution: Disaster Response System by Funding Source and Jurisdiction, 1985 Mexico City Earthquake   126

List of Tables

28. Frequency Distribution: Types of Transactions in Disaster Response by Funding Source and Jurisdiction, 1985 Mexico City Earthquake   128

29. Frequency Distribution: Types of Interactions in Disaster Response by Funding Source and Jurisdiction, 1985 Mexico City Earthquake   130

30. Summary, Logistic Regression Analysis, Public Organizations, 1985 Mexico City Earthquake   134

31. Frequency Distribution: Disaster Response System by Funding Source and Jurisdiction, 1991 Costa Rica Earthquake   139

32. Frequency Distribution: Types of Transactions in Disaster Response by Funding Source and Jurisdiction, 1991 Costa Rica Earthquake   140

33. Frequency Distribution: Types of Interactions by Funding Source and Jurisdiction, 1991 Costa Rica Earthquake   142

34. Logistic Regression Analysis, Public Organizations, 1991 Costa Rica Earthquake   145

35. Frequency Distribution: Exposure to Traumatic Stress, 1993 Erzincan Earthquake   149

36. Frequency Distribution: Information Sought as a Basis for Action, 1993 Erzincan Earthquake   150

37. Cross-Tabulation: Capacity for Action by Exposure to Traumatic Stress   150

38. Frequency Distribution: Disaster Response System by Funding Source and Jurisdiction, 1992 Erzincan, Turkey Earthquake   151

39. Frequency Distribution: Types of Transactions in Disaster Response by Funding Source and Jurisdiction, 1992 Erzincan Earthquake   152

40. Frequency Distribution: Types of Interactions in Organizational Disaster Response, 1992 Erzincan Earthquake   153

41. Summary, Frequency Distributions, Emergent Adaptive Systems, by Funding Source and Jurisdiction   157

42. Frequency Distribution: Disaster Response System by Funding Source and Jurisdiction 1987 Whittier Narrows, CA Earthquake   165

List of Tables

43. Frequency Distribution: Types of Transactions in Disaster Response by Funding Source, 1987 Whittier Narrows, CA Earthquake — 166

44. Frequency Distribution: Types of Interactions in Disaster Response by Funding Source and Jurisdiction, 1987 Whittier Narrows, CA Earthquake — 167

45. Frequency Distribution: Disaster Response System by Funding Source and Jurisdiction, 1989 Loma Prieta Earthquake — 175

46. Frequency Distribution: Types of Transactions in Disaster Response by Funding Source, 1989 Loma Prieta Earthquake — 176

47. Frequency Distribution: Types of Organizational Interactions in Disaster Response, 1989 Loma Prieta Earthquake — 176

48. Log: Disaster Operations, Latur and Osmanabad Districts, 1993 Marathwada, India Earthquake — 183

49. Frequency Distribution: Disaster Response System by Funding Source and Jurisdiction, 1993 Marathwada, India Earthquake — 187

50. Frequency Distribution: Types of Transactions in Disaster Response by Funding Source and Jurisdiction, 1993 Marathwada, India Earthquake — 188

51. Frequency Distribution: Types of Organizational Interactions in Disaster Response, 1993 Marathwada, India Earthquake — 190

52. Summary, Frequency Distributions, Operative Adaptive Systems by Funding Source and Jurisdiction — 194

53. Log: Multi-jurisdictional Operations, 1994 Northridge, CA Earthquake — 202

54. Frequency Distribution: Disaster Response System by Funding Source and Jurisdiction, 1994 Northridge Earthquake — 204

55. Frequency Distribution: Types of Transactions in Disaster Response by Funding Source and Jurisdiction 1994 Northridge Earthquake — 206

56. Frequency Distribution: Types of Interactions in Disaster Response by Funding Source and Jurisdiction, 1994 Northridge Earthquake — 208

List of Tables

57. Summary, Logistic Regression Analysis of Disaster Response Functions, 1994 Northridge Earthquake — 212

58. Log: Multi-jurisdictional Operations, 1995 Hanshin, Japan Earthquake — 215

59. Frequency Distribution: Disaster Response System by Funding Source and Jurisdiction, 1995 Hanshin, Japan Earthquake — 218

60. Frequency Distribution: Types of Transactions in Disaster Response by Funding Source and Jurisdiction, 1995 Hanshin, Japan Earthquake — 218

61. Frequency Distribution: Types of Interactions in Disaster Response by Funding Source and Jurisdiction, 1995 Hanshin, Japan Earthquake — 220

62. Summary, Logistic Regression Analysis of Disaster Response Functions, 1995 Hanshin, Japan Earthquake — 223

63. Summary, Frequency Distributions, Auto-Adaptive Systems, by Funding Source and Jurisdiction — 223

64. Summary, Frequency Distributions, Organizational Disaster Response Systems by Type, Funding Source and Jurisdiction — 233

65. Summary, Communication and Coordination Functions by Type of Response System, Jurisdiction and Funding Source — 237

66. Summary, Frequency Distribution: Types of Interactions in Disaster Response by Funding Source and Jurisdiction, Non-Adaptive Systems — 240

67. Summary, Frequency Distribution: Types of Interactions in Disaster Response by Funding Source and Jurisdictions, Emergent Adaptive Systems — 242

68. Summary, Frequency Distribution: Types of Interactions in Disaster Response by Funding Source and Jurisdiction, Operative Adaptive Systems — 244

69. Summary, Frequency Distribution: Types of Interactions in Disaster Response by Funding Source, Auto Adaptive Systems — 246

70. Summary, Logistic Regression Analysis, Nonadaptive Systems — 255

71. Summary, Logistic Regression Analysis, Emergent Adaptive Systems — 257

72. Summary, Logistic Regression Analysis, Operative Adaptive Systems   259

73. Summary, Logistic Regression Analysis, Auto-Adaptive Systems   261

List of Figures

1. 1986 San Salvador Earthquake: Phase Plane Plot, Public Organizations, Emergency Response by Communication/Coordination ... 90

2. 1986 San Salvador Earthquake: Marginal History, Public Organizations, Emergency Response by Communication/Coordination ... 90

3. 1986 San Salvador Earthquake: Logistic Regression, Public Organizations, Emergency Response ... 91

4. 1986 San Salvador Earthquake: Logistic Regression, Nonprofit Organizations, Emergency Response ... 91

5. 1986 San Salvador Earthquake: Logistic Regression, Nonprofit Organizations, Disaster Relief ... 92

6. 1986 San Salvador Earthquake: Logistic Regression, Private Organizations, Damage Assessment ... 92

7. 1987 Ecuadorian Earthquakes: Marginal History, Public Organizations, Emergency Response by Communication/Coordination ... 102

8. 1987 Ecuadorian Earthquakes: Logistic Regression, Public Organizations, Emergency Response ... 102

9. 1987 Ecuadorian Earthquakes: Logistic Regression, Public Organizations, Damage Assessment ... 102

10. 1987 Ecuadorian Earthquakes: Logistic Regression, Private Organizations, Damage Assessment ... 103

11. 1987 Ecuadorian Earthquakes: Logistic Regression, Private Organizations, Recovery/Reconstruction ... 103

xxiv *List of Figures*

12. 1988 Armenia Earthquake: Phase Plane Plot, Public Organizations, Emergency Response by Communication/Coordination 113

13. 1988 Armenia Earthquake: Marginal History, Public Organizations, Emergency Response by Communication/Coordination 113

14. 1988 Armenia Earthquake: Logistic Regression, Public Organizations, Communication/Coordination 114

15. 1985 Mexico City Earthquake: Phase Plane Plot, Public Organizations, Emergency Response by Communication/Coordination 130

16. 1985 Mexico City Earthquake: Marginal History, Public Organizations, Emergency Response by Communication/Coordination 131

17. 1985 Mexico City Earthquake: Marginal History, Public Organizations, Disaster Relief by Communication/Coordination 132

18. 1985 Mexico City Earthquake: Logistic Regression, Public Organizations, Emergency Response 132

19. 1985 Mexico City Earthquake: Logistic Regression, Public Organizations, Damage Assessment 132

20. 1985 Mexico City Earthquake: Logistic Regression, Public Organizations, Communication/Coordination 133

21. 1991 Costa Rica Earthquake: Logistic Regression, Public Organizations, Communication/Coordination 144

22. 1991 Costa Rica Earthquake: Logistic Regression, Nonprofit Organizations, Disaster Relief 145

23. 1992 Erzincan, Turkey Earthquake: Phase Plane Plot, Public Organizations, Emergency Response by Communication/Coordination 154

24. 1992 Erzincan, Turkey Earthquake: Marginal History, Public Organizations, Emergency Response by Communication/Coordination 154

25. 1992 Erzincan, Turkey Earthquake: Phase Plane Plot: Public Organizations, Emergency Response by Communications/Coordination 155

26. 1992 Erzincan, Turkey Earthquake: Marginal History, Public Organizations, Disaster Relief by Communication/Coordination 155

List of Figures

27. 1992 Erzincan, Turkey Earthquake: Nonlinear Logistic Regression, Public Organizations, Communication/Coordination — 156

28. 1987 Whittier Narrows, CA Earthquake: Phase Plane Plot, Public Organizations, Emergency Response by Communication/Coordination — 168

29. 1987 Whittier Narrows, CA Earthquake: Phase Plane Plot, Public Organizations, Disaster Relief by Communication/Coordination — 168

30. 1987 Whittier Narrows, CA Earthquake: Nonlinear Logistic Regression, Public Organizations, Damage Assessment — 169

31. 1987 Whittier Narrows, CA Earthquake: Nonlinear Logistic Regression, Public Organizations, Emergency Response — 169

32. 1987 Whittier Narrows, CA Earthquake: Nonlinear Logistic Regression, Private Organizations, Communication/Coordination — 169

33. 1989 Loma Prieta, CA Earthquake: Phase Plane Plot, Public Organizations, Emergency Response by Communication/Coordination — 178

34. 1989 Loma Prieta, CA Earthquake: Marginal History, Public Organizations, Disaster Relief by Communication/Coordination — 178

35. 1989 Loma Prieta, CA Earthquake: Nonlinear Logistic Regression, Public Organizations, Emergency Response — 179

36. 1989 Loma Prieta, CA Earthquake: Nonlinear Logistic Regression, Public Organizations, Damage Assessment — 179

37. 1989 Loma Prieta, CA Earthquake: Nonlinear Logistic Regression, Public Organizations, Communication/Coordination — 180

38. 1989 Loma Prieta, CA Earthquake: Nonlinear Logistic Regression, Private Organizations, Communication/Coordination — 180

39. 1993 Marathwada, India Earthquake: Phase Plane Plot, Public Organizations, Disaster Relief by Communication/Coordination — 190

40. 1993 Marathwada, India Earthquake: Marginal History, Public Organizations, Disaster Relief by Communication/Coordination — 191

41. 1993 Marathwada, India Earthquake: Nonlinear Logistic Regression, Public Organizations, Emergency Response — 192

xxvi *List of Figures*

42. 1993 Marathwada, India Earthquake: Nonlinear Logistic Regression, Public Organizations, Damage Assessment 192

43. 1993 Marathwada, India Earthquake: Nonlinear Logistic Regression, Public Organizations, Communication/Coordination 192

44. 1993 Marathwada, India Earthquake: Nonlinear Logistic Regression, Public Organizations, Disaster Relief 193

45. 1994 Northridge, CA Earthquake: Phase Plane Plot, Public Organizations, Emergency Response by Communication/Coordination 210

46. 1994 Northridge, CA Earthquake: Marginal History, Public Organizations, Emergency Response by Communication/Coordination 210

47. 1994 Northridge, CA Earthquake: Logistic Regression Analysis, Public Organizations, Emergency Response 211

48. 1995 Hanshin, Japan Earthquake: Phase Plane Plot, Public Organizations, Emergency Response by Communication/Coordination 221

49. 1995 Hanshin, Japan Earthquake: Marginal History, Public Organizations, Emergency Response by Communication/Coordination 222

50. 1987 San Salvador Earthquake: Marginal History, Public Organizations, Emergency Response by Communication/Coordination 246

51. 1987 Ecuadorian Earthquakes: Marginal History, Public Organizations, Emergency Response by Communication/Coordination 247

52. 1988 Armenia Earthquake: Marginal History, Public Organizations, Emergency Response by Communication/Coordination 248

53. 1985 Mexico City Earthquake: Marginal History, Public Organizations, Emergency Response by Communication/Coordination 249

54. 1991 Costa Rica Earthquake: Marginal History, Public Organizations, Emergency Response by Communication/Coordination 249

55. 1992 Erzincan, Turkey Earthquake: Marginal History, Public Organizations, Emergency Response by Communication/Coordination 249

56. 1987 Whittier Narrows, CA Earthquake: Marginal History, Public Organizations, Emergency Response by Communication/Coordination 251

57. 1989 Loma Prieta, CA Earthquake: Marginal History, Public
    Organizations, Emergency Response by Communication/Coordination 251

58. 1993 Marathwada, India Earthquake: Marginal History, Public
    Organizations, Emergency Response by Communication/Coordination 251

59. 1994 Northridge, CA Earthquake: Marginal History, Public
    Organizations, Emergency Response by Communication/Coordination 252

60. 1995 Hanshin, Japan Earthquake: Marginal History, Public
    Organizations, Emergency Response by Communication/Coordination 252

Contents

Preface — vii

Acknowledgments — xi

List of Tables — xvii

List of Figures — xxiii

Part I: Shared Risk in Theory: Context, Concept and Methods of Analysis

1. Shared Risk and Self-Organizing Processes — 3
2. Models of Transition in Complex, Dynamic Environments — 18
3. Measuring Change in Nonlinear Social Systems — 38
4. The 'Edge of Chaos': Creative Response in Dynamic Environments — 56

Part II: Shared Risk in Practice: The Evolution of Response Systems

5. Nonadaptive Systems: San Salvador, Ecuador and Armenia — 81
6. Emergent Adaptive Systems: Mexico City, Costa Rica, and Erzincan, Turkey — 119
7. Operative Adaptive Systems: Whittier Narrows, California; Loma Prieta, California; and Maharashtra, India — 159
8. Auto-Adaptive Systems: Self Organization or Dysfunction in Northridge, California and Hanshin, Japan — 197

Part III. Future Strategies: Managing Risk in Complex, Adaptive Systems

9. Adaptation to Disaster: Evolving Response Systems   231

10. Sociotechnical Systems and the Reduction of Global Risk   263

Bibliography   277

Appendices   293

Index   311

Part I
Shared Risk in Theory:
Context, Concept and Methods of Analysis

CHAPTER ONE

SHARED RISK AND SELF-ORGANIZING PROCESSES

Shared Risk: A Nonlinear Policy Problem

As the Northridge Meadows Apartments crumbled in the pre-dawn hours of January 17, 1994 in Northridge, California, neighbors first determined their own safety, then looked to see if others escaped danger. Vestiges of earlier training in earthquake preparedness guided startled residents as they rushed to door frames or slid under tables. Half-forgotten lessons of self-protection and collective safety re-emerged into conscious acts, as they sought to determine if family members were safe and whether neighbors had emerged unscathed. Timely action brought themselves and others out of danger, as shared risk turned to shared responsibility.

Shared risk represents public risk, one which affects all residents of a risk-prone community, whether or not they have contributed to the conditions producing the threat. Shared risk, consequently, invokes public response to mitigate the threat of danger to a specific community. But public organizations cannot meet the needs of the community alone. All organizations in the community are affected to some degree, and private and nonprofit organizations become engaged in the response process as well. Further, shared risk includes a class of policy problems that have defied solution by traditional means of analysis and planning. Such problems are interdependent, dynamic, and unpredictable. They require collective action for resolution, which is extraordinarily difficult to achieve under stress.

Mancur Olson (1965, 2), in his classic argument regarding *The Logic of Collective Action*, asserted that "...unless the number of individuals in a group is quite small, or unless there is coercion or some other special device to make individuals act in their common interest, rational, self-interested individuals will not act to achieve their common or group interests". We see this problem in everyday examples of communities living with long-term threats to their welfare, but which are unable or unwilling to address the problem collectively. These communities are neither able to mobilize action to reduce the threat measurably nor to increase their capacity to respond to danger. Members of the community are usually well aware of the danger, and may take individual actions to reduce risk, but the community as a whole remains vulnerable.

This study examines the question of collective behavior under threat and proposes that, under certain conditions, self-organizing processes emerge to enable the community to act voluntarily for its own welfare. However, these conditions are not well understood nor are the dynamic processes always effective. Under what conditions does a community acknowledge a threat, accept responsibility for reducing the perceived risk, and become engaged in constructive action for its own welfare? This study systematically explores the relationship between shared risk and the spontaneous emergence of self-organizing processes to reduce that risk.

Mobilizing action for problems of shared risk historically has occurred after a destructive event has generated massive disaster. Examples include the U.S. response to natural disasters, such as Hurricane Andrew, a Category 4 hurricane that struck South Florida and Louisiana on August 24, 1992; the Mississippi Flood of 1993, or the Northridge, California Earthquake of 1994; incidents of civil strife, such as the Los Angeles Riots of April, 1992 or the more intense hostilities in Haiti, Rwanda, and Bosnia. After each destructive event, citizens' groups mobilized volunteer action to address the needs of those stricken by the disaster, with many people contributing time, goods, and money toward restoring the community's basic welfare. The cost of response after a disaster has occurred, however, is usually many times higher than the cost of taking mitigating action prior to the threatened event. The need for both greater efficiency and more humane resolution compels a re-examination of problems of shared risk.

Traditional means of policy analysis and planning have been used to address problems of shared risk with little effect. Communities are resistant to solutions imposed upon them externally, yet unable to reallocate their own energies and action toward a more constructive and efficient strategy. Although much effort has been spent in developing policy and plans for reducing risk shared by whole communities,[1] decision makers have not been noticeably successful in reducing this type of risk, or have done so only with rare exceptions and for brief periods.

The problem of shared risk requires a new level of understanding and action by the community that incorporates risk reduction into management of daily operations at the local level as well as facilitates capacity for response at the macro or global level, should a destructive event occur. The transition needs to occur on at least two levels of operation and action within the community, often more. This multi-level set of needs generates a complex system of continuing anticipation and response to the problem.

Shared risk is nonlinear in that small differences in initial conditions, repeated in actions over time, lead to unpredictable outcomes. It is also dynamic, in that a change in the performance of one sub-unit of the affected system may directly influence that of other units in its immediate vicinity, creating a ripple effect of failure throughout the system. Problems of shared risk are not easily amenable to control strategies, particularly if the control is externally imposed. They appear to be more problems of collective learning, involving multiple groups at different levels of understanding, commitment, and skill, as well as requiring different types of knowledge, authority, and action for effective resolution.

Once seemingly intractable, problems of shared risk can now be reconsidered in light of advances in information technology. The technical capacity to order, store, retrieve, and disseminate information to multiple users simultaneously, to represent knowledge visually, and to monitor different types of functions at different levels of performance has created potential new approaches to problems of shared risk that involve collective learning and self-organization. By linking information technology to organizational capacity for framing, reviewing, and revising policies that affect the community as a whole, it may be possible to create an "information-rich environment" (Mohr 1982) that supports informed, voluntary action to reduce shared risk. This linkage creates a "sociotechnical system" in which the ability to exchange timely, accurate information among multiple participants facilitates a more open, responsive, creative approach to solving shared problems. This study explores the problem of shared risk, the modes of learning and action conducive to reducing this type of risk, and the potential for the development of "sociotechnical systems" that can facilitate transition and self-organizing processes.

Nonlinear vs. Linear Methods of Problem Solving

Methods for addressing problems of shared risk differ from those used in traditional policy analysis. Since the problems are nonlinear and dynamic, the assumptions underlying this inquiry are those related to discovery rather than control. Traditional means of bounding the problem are ineffective, since the boundaries between the environment and the participants are open. The threat may affect different groups within the subsystems differently, and these differences, iterated over time, lead to different consequences for the whole system.

Further, problems of shared or public risk differ from nonlinear dynamics directed toward controlling risk in closed systems, as in some business organizations (Priesmeyer 1992). In such organizations, the goal may be narrowly defined, such as profitability for the corporation. Different measures of profitability – each representing different levels of risk for the corporation – are used as the basis for projecting the next state of the corporation as a system. Using this information, corporate managers may intervene in the system's operation to bring about the desired change, seeking to keep the system's performance within a specified range of risk and profitability. The goal of the system remains control, although the basis of calculation has shifted to degree of change, rather than stationary status, in the system's performance.

In contrast, not only the problems, but also the decision-making responsibilities are shared in matters of public risk. Public managers are accountable to the citizens for the actions they take (or don't take) in the interests of public safety and welfare. Effectively resolving these problems becomes a continuing process of discovering common elements among different groups, clarifying issues for public understanding, and integrating different perspectives into a common base of understanding to support multiple types of action toward a system-wide goal. The methods needed to solve problems of shared risk require a continuing process of collective learning, rather than control, to support collective action.

The argument against collective action (Olson 1965) assumes a linear, rational model in which individuals act in their own best interest and in which no mechanism other than human reason is used to evaluate the available alternatives. Olson's model of rational choice does not acknowledge different concepts of time for different sub-sets within the system, nor does it factor in the degrees of interdependence among the actors which likely vary over time, nor does it admit feedback processes either within the system or between the system and its environment.

The rational choice model assumes that each individual makes an independent decision to join or not join in collective action, based upon rational consideration of her own interests and the best information available. However, that decision is constrained by the quantity and quality of information available to the individual, which in turn is limited by human cognitive capacity (Miller 1967; Simon 1981). If decision makers alter the amount, accuracy, and timing of information available to the full range of individuals who are involved in a situation of collective danger, and if they facilitate the free exchange of information among them, they alter significantly the collective decision process regarding action in response to the threat.

In practice, this change in information access has already occurred to a substantial degree, albeit unsystematically. Advanced information and satellite communications technology transmit news and update information to multiple locations simultaneously. This technical capacity to increase the amount of timely, accurate information that a system can hold and exchange fundamentally alters the decision process among individuals and groups regarding collective action. It also compels decision makers to reconsider concepts of time, distance, causality, and complexity in reference to collective action.

The linear model of rational choice that underlies Olson's argument has been a remarkably effective means of simplifying a very complex world to enable individual action. But, it is based upon the amount of information that one rational human being can hold and consider in reference to a specific decision for action. The simplifications that it requires are not consistent with practice in reference to the complex, interdependent problems of shared risk. In risk that endangers an entire community, interdependencies among technical, organizational, cultural and other types of systems affect a community's capacity to both mitigate and respond to disaster. The best interest of the individual is directly tied to the community's capacity to provide services that benefit the whole. There is no longer a single actor, but many actors, involved in interdependent decisions that increase or decrease the threat of danger to the community. Communities, by definition, are characterized by interdependencies which serve both as constraints upon, and facilitators of, action in regard to shared risk.

If decisions regarding collective action reflect the amount, accuracy, and timeliness of information that are available to the affected actors, then decision makers need a model that takes into account the system's capacity to hold and exchange information. To develop such a model, the whole community needs to be included in designing a system that can hold and exchange information among multiple actors with different levels of responsibility and vulnerability to risk.

Such a system is distinguished from the rational, linear, single actor model on the basic issues of causality, time, distance, and complexity. If many actors are involved in a collective decision, the reasoning process shifts from linear to nonlinear (Coveney and Highfield 1995). It is no longer certain that a given act will produce a specific result, if that act depends upon the commitment and action of many persons, who likely vary in their understanding and involvement with the problem. Many people involved in the process also means many different concepts of time, indicating the urgency with which they are willing to commit their energies and resources to solving a particular problem before other related problems. Finally, while the linear, rational actor model seeks simplification in order to clarify a course of action, a nonlinear system acknowledges the complex relationships involved in shared risk and integrates them into a functioning, operating system.

Causality, time, distance, and complexity are fundamental to framing courses for collective action in reference to problems of shared risk. These issues are deeply related to our human cognitive capacity and ability to generate action. Under certain circumstances complex, nonlinear systems generate self-organizing processes that reallocate energy and action in order to make a transition to a more stable, less risky, and more efficient state of operation. Discovering what those circumstances are, and how they can be recreated to address problems of shared risk, are the primary tasks of this study.

Self-Organization in Nonlinear, Dynamic Social Systems

Recent research on complex systems provides important insights on this problem of collective action in response to shared risk. Ilya Prigogine (1987, 102), Nobel laureate in chemistry and early researcher on complex systems, observed that "we are living in a world of unstable dynamical systems". Prigogine's insight presents a fundamental challenge to our assumptions of causality underlying the design of linear, ordered, administrative systems to govern collective action.

If shared risk reflects the world of unstable, dynamic systems asserted in Prigogine's assessment, then understanding the characteristics of these systems is vital to generating collective action to reduce risk. A significant literature describing the characteristics of nonlinear, dynamic systems has evolved over the last fifteen years, drawing on research in physics, chemistry, biology, and more recently economics and the social sciences.[2] A set of eight key concepts from this literature is central to understanding the conditions that lead to self-organization. These concepts, although identified at different times and by different authors, offer a coherent portrayal of dynamic, nonlinear systems that generate self-organization.

First, vulnerable communities demonstrate a "sensitive dependency upon initial conditions" (Prigogine and Stengers 1984; Kauffman 1993). That is, the capacity of a community to mobilize collective action in response to perceived risk depends directly upon the degree of awareness, level of skills, access to resources, and commitment to informed action among its members *prior* to the occurrence of a damaging event. Subsequent actions will depend upon the initial choices made or not made, the mix of strengths and weaknesses revealed in the community's capacity to coordinate its activities in response to threat, resulting in substantial variation in collective performance over time.

Second, random events which occur outside its normal operating routines have startling impacts upon the system's capacity to respond. Stochasticity, according to physicist Murray Gell-Mann (1994, 50), is a defining characteristic of complex, dynamic systems. Random events, in turn, generate effects that are irreversible within the system, a third characteristic of nonlinearity. The community retains a collective "memory" of the unplanned occurrence, which continues to influence subsequent choices and behavior by different actors and organizations at different times within the system. Prigogine and Stengers (1984) term this characteristic of irreversibility the "arrow of time," which moves relentlessly forward, never back.

For example, the Long Beach Earthquake of 1933 resulted in the collapse of hundreds of unreinforced masonry buildings and 120 deaths in this California beach city. The vivid memory of this earthquake and the deaths and destruction that it caused, shared not only by the community of Long Beach but by policy makers and professionals interested in seismic policy throughout the state, initiated a serious and sustained effort over sixty years in California to review and revise building codes against seismic risk (Petak and Atkinsson 1982). Consequently, subsequent earthquakes in California, such as the Whittier Narrows Earthquake of October 1, 1987 (M = 5.9, 8 deaths), the Loma Prieta Earthquake of October 17, 1989, (M = 7.1, 63 deaths) and the Northridge Earthquake of January 17, 1994 (M = 6.7, 59 deaths), resulted in lower death rates and degrees of damage to buildings. The Long Beach Earthquake, perceived as an unplanned, "random event", altered significantly the pattern of building construction in the seismic zones of California.

Fourth, iterative patterns of communication and coordination, or "feedback loops," transmit information and energy among subsets of actors within the community, allowing different actors to adapt to changing conditions via mutual adjustment (Nicolis and Prigogine 1989). Such feedback loops serve both to inform segments of the community of altered conditions and to anticipate alternative courses of action. They function as mechanisms of learning, control, and adaptation, vital to the operation of dynamic systems.

Fifth, multiple actors create constraints on action within the community, producing dependency relationships among actors and upon scarce resources. Such constraints contribute to the emergence of clusters of variant energy and action, or "strange attractors", that alter regular patterns of communication and coordination within the community (Kauffman 1993, 178–179). These irregular clusters of attraction and avoidance among community actors increase in influence over time, and eventually alter substantially the community's performance in reference to perceived risk. At a critical point in its operation, for example, the community may move suddenly in full transition to a different state, exhibiting the phenomenon of self-organization, a sixth characteristic of complex systems. Self-organization represents, essentially, a transition in the dominant orientation, communication, coordination, and action patterns within the community, where it both accepts new sources of energy and acknowledges other audiences or demands for attention.

Pursuing evolving patterns of choice, community organizations often produce unpredictable results in social action, a seventh characteristic of dynamic systems. Multiple actors, following internal patterns of selection, produce a distinctive, dynamic course of action (Kauffman 1993) for the community. Successive choices, combined with an iterative feedback process, produce cumulative alterations in performance within the community that yield often unpredictable results.

Finally, recurring patterns of communication and coordination allow nonlinear systems to reproduce self-similar patterns of behavior, the eighth characteristic. These repetitive patterns of organizing resources and action serve as a means of reproducing similar behavior to achieve the same goal in different contexts. The process creates "fractal organizations" (Peitgen, Jurgens, and Saupe 1992) or self-similar patterns of performance that cross boundaries of organizations, jurisdictions, disciplines, space, and time. Such fractal forms of reproducing its internal goal create an important link between the specific community and its wider environment.

Communities that share seismic risk exhibit these eight characteristics of nonlinearity to varying degrees. The presence of each characteristic may vary substantially among subsets of groups or organizations in micro level performance within the community. Yet, more important than micro level performance of any single characteristic by any given subset is the interaction generated among the set of eight characteristics that defines the macro level performance of the community as a whole system.

Given the dynamics generated separately and collectively by this set of characteristics within a vulnerable community, the emergence of self-organizing processes in a nonlinear, dynamic system minimally entails two levels of collective action: the micro level of community action involving households and organizations, and the macro levels of subnational, national and international action.

The processes involved in voluntary collective action to reduce shared risk, which include communication, selection, feedback, and self-organization, depend upon information. Managing the information processes needed to engage the number of households, organizations, communities, and jurisdictions involved in the reduction of risk quickly escalates beyond human capacity to produce or coordinate (Simon 1981). However, appropriate uses of advanced information technology can now provide the needed mechanisms to facilitate collective learning among individuals, organizations, and jurisdictions.

Advanced information technology provides a means of transition to a new level of communication, learning, and action in communities vulnerable to shared risk. It allows the design of sociotechnical systems that can address multilevel problems and support organizational learning and action in complex systems. Such systems enable communities to address problems of shared risk, previously considered only by individual actors.

Problem Solving Requirements for Shared Risk

Shared risk represents a class of policy problems that requires concepts and methods of facilitating information exchange and acknowledges differences in capacity and commitment among sub-groups within the system. This class of policy problems tends to generate a system of potential problem solving in which the boundaries are defined by the number of participants who share the risk of adverse impact. The principal function of such a system is to support the creative processes of integration and transition among the component units toward a common goal and to adapt the flow of information and feedback accordingly. Since the system is formed in part to respond to a threat from the immediate environment, it is open to a continual exchange of information, material, and demands from this environment.

Five basic requirements characterize problem-solving processes for issues of shared risk. First, problems of shared risk attract participants operating within different constructs of time. These constructs reflect the corresponding levels of autonomy, dependence, intensity, competence, and conflict among the participants in an evolving system. Different conceptions of time among the participants influence the potential for collective learning by the whole system in two ways. A system-wide schedule for development needs to acknowledge the different rates of learning within the system, and adapt both its information flow and requirements for monitoring performance and feedback accordingly. Further, the demands for adaptation and adjustment of the components to one another, as well as to the system-wide goal, require flexibility and a continuous process of learning and integration. The construction of time and its multiple effects on nonlinear processes of shared risk will be discussed in more detail in Chapter Two.

Second, analytic processes used to address problems of shared risk need to incorporate both negative and positive feedback. If only positive feedback is given, as is the prevailing practice in traditional management or problem-solving structures, the additional time required to discover and correct discrepancies in interactions among the participating units multiplies the dysfunctional consequences for the system, decreases the efficiency of the system's operation, and increases the costs and vulnerability to the affected community. Negative feedback has the important effect of refocusing the participants on the system-wide goal and, if necessary, compelling a redefinition of that goal in light of changed conditions.

Third, fundamental to solving problems of shared risk is the capacity to hold and exchange information among the different sub-groups affected by the risk, and between this system of participants in the problem-solving process and the environment that engenders the risk (Simon 1981; Comfort 1988). This requirement implies creating an information infrastructure for problem-solving that can be accessed, at least to some degree, by all participants in the process. Such an information infrastructure is beyond the cognitive capacity of human managers and requires the technical capacity to extend and adapt information processes both within the system and between the system and its environment in accordance with changing conditions.

Fourth, since shared risk problems are nonlinear, successful problem-solving requires the capacity to infer a global solution from local events. This requirement reverses the order of logical inference accepted in traditional policy making and poses new questions of measurement and design in policy analysis and implementation. This problem will be discussed more fully in the presentation of methodology in Chapter Three.

Finally, solving problems of shared risk requires the design of collective learning processes that can focus on long-term solutions as well as immediate, short-term needs. The ability to juxtapose short-term sacrifice for long-term gain – at different rates for different groups of participants in the process – is essential to finding a system-wide goal acceptable to the whole community. This concept means devising a balance in equity between needs and resources, and factoring in the ability of different groups to learn, to contribute, to fail, and to renew their contribution to the system, all at different rates. This requirement is more difficult than human managers alone can muster. Information technology offers promising potential for support.

Shared Risk in Practice

In this book, I will examine seismic risk as an example of an actual policy problem that illustrates the characteristics of shared risk. To capture the range and variation of responses to the same problem of seismic risk, I present a small-n comparative study of eleven disaster response systems that evolved following major earthquakes. Seismic risk provides a vivid illustration of the characteristics of shared risk for several reasons. Seismic risk involves different constructs of time and space in assessing the degree of vulnerability for specific communities.

Major earthquakes occur infrequently at any given location, but with relative frequency around the world. If we learn best from observing the patterns of response, interaction, and problem solving that evolve in actual practice, then we must conceptualize the problem in global terms, drawing evidence from common functions performed in a range of local conditions.

Seismic risk also represents a problem that involves very high stakes in the potential costs of lives and property for communities in vulnerable zones. For example, the recent earthquake that shook the Kansai Region of Japan, known as the Great Hanshin Earthquake of January 17, 1995, claimed over 6,000 lives and caused an estimated $200 billion in damages. Seismic risk is shared by vulnerable communities on both local and global scales. While every nation in the world has registered seismic events, 36 nations in modern times have experienced major, damaging earthquakes ranging in magnitude from 5.4 to 8.7 on the Richter scale.[3] Within known areas of seismic risk, major metropolitan areas and cities such as Los Angeles and San Francisco, California; Mexico City, Mexico; Quito, Ecuador; Leninakan, Armenia; Erzincan, Turkey, Tangshan, China and Tokyo, Japan are subject to recurring earthquakes. Over three billion people, or approximately three-fifths of the world's population, live in the thirty nations that report the highest levels of seismic risk.[4]

When a disaster does occur, it shatters existing patterns of power and practice, and requires creative response to meet the urgent, stressful needs of the whole community. This situation creates an opportunity for the emergence of self-organizing processes, if conditions conducive to their development are present prior to the event. Consequently, disaster environments represent a rare opportunity to observe the emergence of self-organizing processes in a relatively brief period of time.

Disaster environments provide a striking test of the community's existing policy, state of preparedness, and capacity for action in response to an adverse event. The event is public, and news and videotapes of public response are now transmitted instantaneously across the world. Discrepancies between policy and action are undeniable and incorporated into public record. Public managers must confront these discrepancies or lose their credibility and, often, positions of authority.

Finally, disaster provides a common focus for community response and collective learning. Earthquakes in particular are sudden, destructive events that arrest the attention and efforts of all members of the community. The severe losses compel a focus on action; the needs are immediate and cannot be ignored. The entire community is affected both physically and psychologically. Seismic risk, thus, represents the type of actual policy problem that illustrates the interdisciplinary, interorganizational, and interjurisdictional characteristics that have made problems of shared risk extraordinarily difficult to resolve.

In this study, I present comparative profiles of disaster response systems that evolved in response to eleven major earthquakes since 1985. The profiles capture the shared danger of earthquakes and the continuing threat of seismic risk to the affected communities. I examine patterns of communication and coordination among participating units and sub-systems of the set of response systems, and assess the presence and capacity for self-organization generated by the respective communities in response to needs created by the disaster event. The eleven earthquakes are listed in Table 1, with their respective Richter magnitudes and estimated losses in lives and property.

Research Premises

The following set of premises serves as an initial research framework for the study of shared risk in vulnerable communities:

Table 1. Losses in Life and Property in Eleven Recent Earthquakes[5]

Year	Month/Day	Location	Richter Magnitude	No. of Dead	Cost (US$)
1985:	Sept. 19,20	Mexico City, Mexico	8.1	10,000+[a]	10 billion+
1986:	Oct.10	San Salvador, El Salvador	5.4	1,200+	1.03 billion
1987:	Mar. 5	Napo Province, Ecuador	6.9	1,000+[b]	1 billion
1987:	Oct. 1	Whittier Narrows, California	5.9	8	352 million
1988:	Dec. 7	Northern Armenia	6.9	25,000+[c]	16 billion
1989:	Oct. 17	Loma Prieta, California	7.1	63	7.1 billion
1991:	Apr. 22	Limon Province, Costa Rica	7.4	55	965 million
1992:	Mar. 13	Erzincan, Turkey	6.8	800+[d]	3.4 billion
1993:	Sept. 30	Marathwada, India	6.4	7,582	36.6 billion
1994:	Jan. 17	Northridge, California	6.7	59	25.7 billion
1995:	Jan. 17	Hanshin, Japan	7.2	6,000+	200 billion+

[a] Official reports cite 10,000 dead; informed observers estimate 20,000 dead; actual figures may never be known
[b] Estimated figure, Ecuadorian Civil Defense
[c] Official reports cite 24,542 dead; informed observers estimate 55,000 dead; actual figures may never be known
[d] Official reports list over 800 dead; informed observers estimate 2,000 dead; actual figures may never be known

Sources: United Nations Economic Commission for Latin America and the Caribbean; US Office of Foreign Disaster Assistance; Ecuadorian Civil Defense; Armenian Civil Defense; US Geological Survey; California Office of Emergency Services; Comisión Nacional de Emergencia, Costa Rica; Earthquake Engineering Research Institute, Special Report on the Erzincan, Turkey Earthquake; Government of Maharashtra, India; US Federal Emergency Management Agency; National Land Agency, Tokyo, Japan

1. Shared risk represents a context of nonlinear, dynamic, interdependent relationships in which adverse conditions, or threats of same, rapidly escalate and de-escalate to affect the welfare of the entire community
2. Linear models and control strategies for reducing shared risk repeatedly fail in dynamic environments
3. Shared risk may be reduced substantially by designing processes of organizational and interorganizational inquiry and learning to inform collective action
4. Interorganizational processes of learning and action in dynamic environments depend upon timely, valid information processes that include search, representation, exchange, analysis, and feedback among multiple participants simultaneously
5. Information technology provides the technical means to support individual and organizational learning in dynamic environments
6. A learning model to mitigate shared risk involves integrating both technical and organizational components in a sociotechnical system to support timely, informed collective action.

This set of premises guides the inquiry and analysis of data from a comparative study of disaster response systems that evolved following eleven major earthquakes since 1985.

Significance of the Study

Three primary reasons justify the conduct of this study. First, the study of nonlinearity in social systems is relatively recent, but critical to our understanding of complex, interdependent problems of public policy. As Forrester (1987, 108) states, "...the real world is nonlinear; social science models, to be informative, need to capture these nonlinearities." This study seeks to identify characteristics of nonlinear systems as they are revealed in a set of rapidly evolving disaster response systems.

Disaster environments provide an extraordinary opportunity to observe organizations operating under conditions that are close to the "edge of chaos". Earthquakes generate heavy demands upon the organizational, technical, economic, and social infrastructure of a community. They occur suddenly and create devastating consequences for interdependent populations and infrastructure. The stricken communities, with support from the wider environment, mobilize a collective system of interorganizational action in response to the needs generated by the earthquake. These response systems exhibit known characteristics of nonlinear, dynamic systems that can be distinguished from the standard administrative procedures of the society.

The study, secondly, addresses an important methodological problem in the social sciences which has previously inhibited research on nonlinear social systems. This problem is the task of constructing valid aggregate models of social behavior from individual level data. Bartels and Brady, (1993, 130) in their assessment of quantitative methods in the discipline of political science, state:

"Serious theoretical research on deriving an aggregate level contextual model from empirically verified assumptions about individual level interactions should have a high priority for students of social context." While this study does not claim to specify fully an aggregate level model from individual level data, it clearly addresses a problem that moves from individual to organizational to systemic response in its levels of data collection and analysis. It, therefore, seeks to contribute preliminary insights to this difficult methodological problem.

Finally, seismic risk presents a major policy problem to metropolitan centers in thirty-six nations, including both industrialized and developing economies. The states and nations on the western coasts of the Americas – North, Central, and South – all lie on the "Ring of Fire," the system of earthquake faults that are a product of shifting tectonic plates of the earth's surface. Eight of the eleven earthquakes included in this study occurred in locations on the "Ring of Fire," resulting in tens of thousands of lives lost and billions of dollars in damage within the last decade (See Table 1 above). The "Ring of Fire" extends down the western coasts of the Americas, across the Pacific and up through populous Asian nations, creating a serious threat of seismic risk in Indonesia, The Philippines, Japan, and other island nations, finally linking again to the Americas through Alaska. The cases of Armenia, Turkey, and India included in this study document the threat as it is found in the Caucasus Region and South Asia. Seismic risk presents a serious problem that has not yet been addressed systematically in global perspective.

Scope of the Study

This book explores the potential for self-organizing processes as a means to address problems of shared risk, focusing specifically on seismic risk. In Chapter 2, I review briefly the extensive literature on self-organization and complex, adaptive systems and define a set of research questions and objectives for this study. In Chapter 3, I present the methodology for a comparative study of disaster response systems, adapting Stuart Kauffman's (1993) concept of an N-K system and its measures for this study. In Chapter 4, I examine the theoretical concept of the "edge of chaos" and use it to assess the evolution of response systems in a comparative review of eleven earthquake disasters. This assessment illustrates key functions and characteristics that are shared by all eleven cases, but also identifies unique features that distinguish each case from the others in important respects. This chapter establishes the range of variation and commonality among the set of eleven cases, and presents the categories for their subsequent analysis.

Since an important part of the analysis documents the emergence of self-organizing processes after disaster in very different technical, organizational, economic, political, and cultural conditions, the cases will be presented briefly in subsets of structure, flexibility, timeliness of information, and capacity for spontaneous action. Chapter 5 will compare response systems characterized by both low structure and low flexibility. These systems evolved following the earthquakes of

San Salvador (1986); Ecuador (March,1987); and Armenia (1988). Chapter 6 will compare response systems characterized by low structure and moderate flexibility. Such systems evolved following the earthquakes of Mexico City (1985), Costa Rica (1991), and Erzincan, Turkey (1992). The response systems that evolved following these earthquakes were characterized particularly by disjunctures between local level and national level response, which seriously limited performance in the dynamic disaster environments.

Chapter 7 will analyze a set of response systems characterized by moderate structure and moderate flexibility. This subset evolved following earthquakes in Whittier Narrows, California (October, 1987); Loma Prieta, California (1989); and Maharashtra, India (1993). Chapter 8 will examine a set of contrasting response systems, one with high structure and high flexibility, Northridge, California (1994); and one with high structure and low flexibility, Hanshin, Japan (1995).

The analysis will show how the response systems, grouped in categories of roughly similar structure and flexibility, nonetheless generated very different types of collective action that reflect in critical ways the initial conditions of the social, economic, organizational, and technical environments in which they occurred. Each type of response system, in turn, produced significantly different consequences for the affected populations.

Within the subsets classified by structure and flexibility, all cases will be analyzed in terms of five basic phases that are important for the emergence of self-organizing processes. They are: Initial Conditions, Information Search, Information Exchange, Organizational Learning, and Adaptive Behavior.

Chapter 9 will compare the four types of adaptation to disaster environments in terms of self-organizing processes across the full set of eleven cases. Chapter 10 will conclude the study with a summary of findings on rapidly evolving, complex systems in disaster environments, and also present a brief set of recommendations for the design of information strategies that facilitate the evolution of self-organizing systems in disaster environments. In its entirety, the study will contribute to our theoretical understanding of complex adaptive systems in the context of shared risks.

NOTES

1. The Model Cities Program of the mid 1960s is one of the most extensive efforts to examine the problems of urban communities in a holistic perspective.
2. Three recent books summarize this literature nicely, while contributing to it. Please see Stuart A. Kauffman. 1993. *The Origins of Order: Self-Organization and Selection in Evolution.* Oxford: Oxford University Press; Murray Gell-Mann. 994. *The Quark and the Jaguar: Adventures in the Simple and the Complex.* New York: W.H. Freeman & Co.; and L.D. Kiel. 1994. *Managing Chaos and Complexity in Government.* San Francisco: Jossey Bass Publishers.
3. United States Geological Survey. 1997. *Significant Earthquakes.* National Earthquake Information Center. Denver, Colorado.
4. Kurian, George Thomas. 1981. *The New Book of World Rankings.* New York: Facts on File, Inc.: 11–12.

5. In the summer of 1990, two major earthquakes occurred in Northern Iran (June 21, 1990, M = 7.7) and Baguio, The Philippines (July 16, 1990, M = 7.8). These earthquakes are not included in this study because the author was unable to conduct field research on site.

CHAPTER TWO

MODELS OF TRANSITION IN COMPLEX, DYNAMIC ENVIRONMENTS

Transition in Environments of Shared Risk

Environments of shared risk require organizations to adapt their existing performance to avert or minimize potential destruction. Determining how to bring about the desired change under uncertain conditions presents a recurring dilemma to practicing managers. Developing a strategy to mobilize complex, interdependent operations in response to a devastating event in any given community involves a conscious process of transition. Elements outside the managers' control affect possible courses of action in unanticipated ways. It is essential to engage the entire community as well as external sources of support in a process of collective action. Collective action requires many actors to perform different functions at separate locations in coordinated sequence, and involves a process of mutual adjustment and reciprocal learning. The degree of uncertainty involved in risk environments affects the decision process at each point of decision, with discretionary choices governed only by a common goal of protection for the community and influencing the potential capacity of the response system.

Time, distance, and complexity in environments of shared risk create serious challenges for collective learning and action. Each introduces elements of uncertainty that may impede learning and inhibit performance. For example, the distant rumble of an earthquake through Himalayan cities and towns sets in motion a complex chain of international disaster assistance. The disaster creates immediate, urgent needs for medical care, housing, basic supplies, and tools to rebuild communities suddenly shattered by natural forces. In the remote villages of Nepal and Bihar State, India, resources are scarce in a local economy that, for most of the population, is marginal at best. In other parts of the world, resources are mobilized to assist the residents of Nepal and Bihar, but the problem is to determine what is needed, who needs it, where it is needed, how to get it there, and how to establish a process that will utilize the assistance most effectively to meet the needs of people in the damaged communities.

The problem involves mobilizing multiple participants to act simultaneously in a long, but often tenuous chain of design and delivery of services to meet urgent human needs generated by the disaster. Creative action in such an uncertain, dynamic environment requires overcoming obstacles of time, distance, and function, while exercising discretionary choice at multiple decision points in order to produce coordinated response. Such a process necessarily requires each of the participants to adjust and adapt dynamically to shifting conditions in resources, constraints, and opportunities in order to attain a common goal: restoration of the damaged communities to normal activities. This process is essentially one of mutual adaptation, or voluntary coordination of individual or organizational actions, to achieve a shared goal.

While coordination of action among multiple participants to achieve a shared goal is frequently recommended as a solution to complex, interdependent problems, planners and policy makers have found it difficult to implement in practice.[1] Administrative theorists have not been able to define coordination in ways that do not imply coercion (Caiden and Wildavsky 1974, 277—279) or to devise means of facilitating coordination without compromising the shared goal (Wilson 1989, 268—274). Others have found the costs of seeking coordination to be higher than the actual benefits achieved (Chambers 1974, 25—26). Theorists critical of coordination have viewed it primarily as a structural problem. They defined it as a set of organizational procedures that would require multiple participants with different levels of understanding and degrees of responsibility to follow a common set of rules, most often externally imposed, to achieve a complex goal. Coordination in disaster response operations is largely achieved by different means. It is based primarily on the processes of information search, exchange and feedback that results in intra- and inter- organizational learning. The detail, timeliness and accuracy of the information shapes the process, and the participating organizations, individually and collectively, seek the most current and complete information as a basis for effective action. Coordinated action, or adaptive behavior, becomes the product of the information search, exchange and learning processes as the participants voluntarily seek the most appropriate and effective action to protect the community.

Voluntary coordination, or self-organization, depends upon the existence of an information infrastructure and common knowledge base prior to a threatening event. Policy makers and planners who focus on the design and implementation of a shared knowledge base are more likely to achieve the desired coordination through learning processes than through procedural rules and externally imposed requirements.

The set of interactions mobilized to meet a shared risk shapes a distinct response system (Luhmann 1989), made up of individuals and organizations that may not have worked together prior to the disaster. Through a shared goal of response to the needs of the earthquake-distressed communities, these individuals and organizations have generated new patterns of communication, cooperation, and coordinated action. The response system evolves through recurring interaction with, and adaptation to, the environment changed by the disaster. It expands and contracts, crossing disciplinary, organizational, and jurisdictional lines as it seeks to meet the needs of the distressed communities. The system is a network of relationships in process, rather than a fixed, physical structure.

Identifying and characterizing the rapidly evolving, interjurisdictional systems generated in response to major disasters gives us an opportunity to observe systematically the processes people use to cope with urgent, threatening conditions. These processes enable them to avert more serious consequences for the larger community. This problem has long caught the interest and attention of researchers in organizational theory (Barton 1969; Smart, and Vertinsky 1977; Shrivastava 1984) and sociology (Quarantelli 1978; Dynes 1969; Dynes and Tierney 1994; and Drabek 1990). The problem invokes stubborn dilemmas of organizational theory that include dynamic relations between micro and macro units within a system (March 1988). Such relations involve self-organizing patterns of communication, failure, and redesign that transform both the functions of social organizations and the missions of the institutions through which they operate.

In this chapter, I first identify briefly the general characteristics of complex, dynamic systems that emerge in response to disaster and the conditions that support them. Second, I review four distinct models of transition for mobilizing collective action, a major goal for communities facing shared risk. In this review, I examine the underlying assumptions regarding collective action, mechanisms of interaction, and the costs and consequences of change. Third, I discuss the requirements of effective transition for disaster response systems, and finally, I propose a preliminary model of self-organization that serves as an initial theoretical framework for assessing actual disaster response systems documented from field research following major earthquakes.

Characteristics of Dynamic Response Systems

Disaster response systems have defied characterization by jurisdictional boundaries because they are not limited to a specific geographic area. Rather, they represent a system of interactions, or "communicative acts" (Luhmann 1989), among an expanding and contracting set of participants focused on a particular problem: restoration of a damaged community. These interactions, in turn, occur with varying degrees of intensity, duration, and critical importance to the affected population, reflecting the changing conditions of the affected community. Viewing response systems as a set of communicative acts, however, enables us to conceive of the system as a structure that holds and exchanges information across organizations and jurisdictions.

The information capacity of the response system defines the limits and potential for coordinated action among its constituent units. While these limits may be increased, the capacity for coordinated action by the disaster response system cannot exceed its information base. In Herbert Simon's (1969, 1981) terms, we can only create what we already know. If the knowledge base to support coordinated action is not established prior to a devastating event, the participating actors are not likely to act efficiently or effectively in response to collective need. The heavy criticism by citizens of governmental response in South Florida following Hurricane Andrew in August, 1992 or in the Kansai Region following the Hanshin Earthquake in January, 1995 illustrate this point. Conversely, with an informed base for collective action, the participating organizations and jurisdictions are much more likely to respond creatively in coping with adverse events. For example, the multi-organizational, multi-jurisdictional task force created to serve housing needs following the Northridge Earthquake of January 17, 1994 represented an innovative response by Los Angeles City and County agencies that already had considerable experience with disaster and brought informed perspectives to the problem.

A disaster response system evolves in a parallel, nonlinear process. Local conditions govern the initial courses of action taken simultaneously by many actors at different locations (Ruelle 1991; Kauffman 1993). Subsequent developments of the system depend upon the first set of choices made by many diverse actors. Decisions made at critical points in the generation of the response system create the basis for the next sequence of interactions. The degree of resilience, or the capacity to adapt existing resources and skills to new situations and operating conditions (Wildavsky 1988), appears directly related to the degree of access to, and exchange of, information in systems seeking change.

In the context of disaster, linear assumptions of "cause and effect" are likely to fail. Prior rules do not hold when the basic parameters of the system change or collapse in a community suddenly altered by a major catastrophe, such as an earthquake. A single act may set off a series of subsequent reactions that produce very unexpected consequences (Prigogine and Stengers 1984; Kauffman 1993). A response system develops after every disaster to assist the injured, restore basic services, and return the community to normal operating conditions. The efficacy of that system, however, depends very much on the initial conditions of knowledge, training, communication, and economic resources available in the community prior to the event.

Transition in the operating context of a disaster response system takes different forms. Incoming information and events from the wider environment drive the system to adapt its performance in a continual dialogue of action, response, and adjustment to altered conditions (Luhmann 1989). This continuing dialogue between the system and its wider environment involves the search, analysis, and dissemination of information to support action. Incoming information enables the system to exercise discretionary choices, adapting its performance through internal choices based on capability, goals, preference, and opportunity rather than relying on external sources for direction or control.

Without the capability for transition, social systems lose effectiveness and stagnate, as the conditions in which they were designed to function change and other needs become more urgent. In the rapidly evolving context of disaster, systems unable to adapt inevitably falter and eventually fail, increasing the losses to communities they were designed to serve.[2]

Models of Transition in Complex Environments[3]

A distinct response system emerges following each disaster to mobilize collective action to meet urgent social needs. In Stuart Kauffman's (1993) terms, the emergence of the response system represents a move toward order from the chaos of disaster. Kauffman, a biologist, holds that all systems operate on a continuum ranging from chaos to order, and that systems at either end of the continuum tend to move toward the center. That is, in a chaotic situation, a system will move toward order. In an orderly situation, a system will move toward chaos.

In the center of the continuum between chaos and order lies a narrow region that Kauffman (1993, 174, 208–227) identifies as the 'edge of chaos,' where there is sufficient structure to hold and exchange information, and sufficient flexibility to adapt to changing conditions. It is in this narrow region, 'the edge of chaos,' that organizations and systems are able to make the most creative responses to conditions in their operating environments.

The requirements for action in complex social systems depend on the extent and effectiveness of information and communication processes operating within the system, and its ensuing capacity for collective learning and adaptation. Collective action requires, first, structure for the effective mobilization of resources. Response personnel need to know exactly what their mission is, where it is, what equipment and materials are available to them, what procedures will be observed by participants from different organizations engaging in joint action, and what likely conditions will be encountered. In complex environments, response personnel need to have sufficient information and resources, backed by a dependable organizational structure, to enable them to coordinate action among multiple actors operating simultaneously in different locations under rapidly changing conditions. For example, the Incident Command System, recently adopted by the California Office of Emergency Services as the statewide standard for emergency response, represents a systematic organizational plan for action in rapidly evolving environments.[4]

Effective response systems, second, exhibit adaptive flexibility in action. That is, they devise a process for making corrections in performance due to unexpected changes in operating conditions, incoming information, unanticipated events or constraints imposed by resources, time, geography, physical endurance or existing knowledge. This combination of structure and process to support action is essentially the balance of structure and flexibility identified by Kauffman as the 'edge of chaos.'

If the appropriate balance between structure and flexibility is attained for a given system, it allows that system to move easily between micro and macro levels of conception, operation, and assessment in the design and implementation of programs to reduce risk. This ability to move between levels of conceptualization and action in a complex sequence acknowledges the interdependence among stages of operations, and allows more accurate specification of the needs, equipment, skills, materials, and personnel required for successful completion of the mission. The process of review and redesign of action is continuous, and in turn depends upon the accurate assessment and evaluation of action at each stage and each level of operation in the process.

This continuous stream of incoming information and feedback from performance allows the response system to make frequent small adjustments in its operations in order to complete a given mission successfully (Staw 1991). Without these continuing small adjustments or adaptations, the system, inflexible in the midst of change, would become vulnerable to breakdown or failure.

Transition is facilitated, third, by a common knowledge base among participants in a community response system. Among professional emergency responders, this common knowledge base is achieved in large measure through training and experience. In a community vulnerable to seismic risk, creating such a knowledge base among the citizens about the mechanisms of mitigation, response, and recovery is a major part of the preparedness task.

Finally, evolving response systems include the operation of subsets within the larger system, and acknowledge that each subset has a clientele, with different incentives and different feedback loops operating among the subsystems and between the system and its environment. Coordinating the activities of the subsets and guiding their operations toward the same common goal becomes a primary function in achieving collective action. This function also depends upon the circulation of valid information through the distinct feedback loops within the system.

Dependent upon collective learning, transition occurs as a process that moves from (temporary) state to (temporary) state. Learning processes serve as the connecting links between the temporary states, and the specific states mark stages of development or a series of transformations of the system in interaction with its environment. During the (temporary) states, the participating organizations are able to coordinate their activities on the basis of shared knowledge of the existing conditions. When the conditions change, the participants must realign their actions in light of new information and move to the next (temporary) state. If the emergency response system includes sufficient access to information, feedback on performance, and resupply of effort, materials, and energy, it is able to overcome the obstacles of time, distance, diverse functions, and multiple disciplines that inhibit coordinated, collective action.

Historically, different models have evolved as mechanisms to produce collective action to reduce risk. Each of these models has strengths and weaknesses, and each has contributed to our understanding of the difficulty of producing collective action under uncertain conditions. None has proven entirely satisfactory in assisting communities to manage the problems of shared risk. Four models of collective action have contributed to our current understanding of mechanisms to reduce risk. Each will be reviewed briefly as a stage in the developmental process of learning to cope with risk.

Command and Control

Mechanistic models of systems in operation have been developed in the physical sciences and engineering. The basic assumption underlying these models is that if the problem is well defined and systems can be closed to outside interference and disturbance, they can function without error. Once the system is carefully designed and functioning, considerable effort is placed on the control of error, or any aberrant disruption or behavior. The system is assumed to be in command; all other disturbances need to be controlled or eliminated.

Military training and tradition exemplifies most aptly the concept of 'command and control' (Train 1986; Weissinger-Baylon 1986; Crecine 1986), but it is also reflected in the hierarchical design of authority structures and highly specified tasks of "tightly coupled systems" in the administrative science literature (Taylor 1911, 1967; Gilbreth and Gilbreth 1917; Perrow 1972). The principle of 'command and control' is clear specification of the authority relationships among subunits in order to increase control over performance of the whole organization. It is largely a deterministic model, and seeks to reduce uncertainty in organizational performance through detailed plans and training.[5] While this organizational design has proven functional and robust in well-structured, routine conditions, it is weakest in uncertain, dynamic conditions.

In complex, dynamic environments, models of control are vulnerable to "lock out" (Cohen and Levinthal 1990, 136), that is, the exclusion of relevant information from the decision process. In disaster management, organizations with legal responsibilities for first response — police, fire, emergency medical services — operate primarily with a command and control orientation in training and field action. Field commanders have recognized the weaknesses of this orientation in the complex, dynamic conditions of disaster. Extensive efforts have been made to adapt the strengths of command and control principles to disaster environments, where common training and skills enable multiple units to work readily in coordinated action, but flexibility is needed for rapid response. For example, the US Forest Service developed the "Incident Command System", a widely used adaptation accepted by emergency response services across the US.[6]

Models of control cluster on the order end of the continuum ranging from order to chaos. The major strength of these models is also their major weakness. In application to social systems, it becomes difficult, if not impossible, to close a system for any significant period of time without damaging a community's capacity to protect itself. In a democratic society, the problem of risk to a community is not well defined, nor is it possible to close the system without excluding information necessary to enable members of the community to act responsibly in event of danger.[7]

Anarchy

At the opposite end of the continuum are models that treat all organized behavior as anarchic (Cohen, March, and Olsen 1972; Kauffman 1993). These models, tending toward chaos, evolved largely in recognition of the failure of imposed order to control social behavior in uncertain conditions. In ill-defined contexts, narrowly construed or inapplicable models of order often confuse or paralyze action. When rules do not apply or do not exist in a given situation, members of a community habituated to governance by rules frequently become locked in conflict, paralyzed by fright, or indifferent to danger, and either do nothing or are unable to protect themselves against virtually certain harm.[8] Anarchic models encourage some action, even if vaguely defined, by relaxing controls on behavior and allowing chance and personal choice to influence individual action. The assumption is that intelligent individuals are able to recognize opportunities for constructive action, and will match possible solutions with timely opportunities to solve recurring problems.

This perspective views organizational decision making largely as "organized anarchy" (Cohen, March, and Olsen 1972). Under conditions of uncertainty and ambiguity, the authors assert that organizational decision making is not a rational process determined by careful planning, but rather is strongly influenced by the limitations of human capacity for attention, timing and the continual flux generated by a dynamic environment. The model reverses the assumption of organizations operating in static, well-structured environments and acknowledges demands placed upon organizational decision processes by conditions of "problematic preferences, unclear technology and fluid participation" (Cohen, March, and Olsen 1972, 1—2). In such environments, agreement on either goals or means is difficult to achieve. Without consensus on goals, acknowledgment of authority becomes largely irrelevant, and means of action are unrelated to specific objectives. Consequently, organizational decisions, when they occur, are largely accidents of timing and spontaneous recognition of an appropriate 'fit' between problem, solution, opportunity, and participants.

The capacity to focus attention on demands and resources from the external environment in an intelligent, timely manner is essential for organizational problem-solving (March 1988; Cohen 1986). Cohen (1986, 65—66) further proposes the criterion of "flexibility" in adapting to the demands and opportunities generated by the environment as a measure of organizational effectiveness. This flexibility allows participants to recognize novel solutions emerging from dynamic conditions that may fall outside the previously defined rules of "command and control". It also allows the community to shift its focus of attention to incoming problems, without wholly relaxing the structure needed for coordinated action.

Recognizing that decision processes in collective action may not be fully or consistently anarchic, several researchers have used the garbage can model to examine decision processes in more structured settings (Weiner 1976; Crecine 1986; Train 1986). Train's characterization of the stages of organizational decision in battle, acknowledging the constraints of heavy load and deadlines upon military commanders, offers a useful analogy to decision-making in disaster environments.[9] Disaster managers operate under similar constraints of heavy load and deadlines, but frequently without the common skills and orientation of military training among their personnel.

In comparing the performance characteristics of tightly coupled systems, such as those in command and control organizations, to those of loosely coupled systems, such as those characterized by organizational anarchy, Crecine (1986, 82–88) notes that both types confront conflicting goals among participants, cope with uncertainty under varying parameters of action, search for solutions to shifting problems, and engage in organizational learning. However, the two types of systems cope with uncertainty in fundamentally different ways, and this difference shapes their performance in other primary relationships.

Tightly coupled systems seek to establish greater control over their participants and operating conditions in response to unexpected or ambiguous demands from the environment, restricting organizational attention to specific problems and solutions. In contrast, loosely coupled systems acknowledge that previously defined problems may be inaccurate or their associated rules inappropriate under changing or ambiguous conditions. These systems relax structural controls on attention and allow the spontaneous matching of problems, solutions, and participants in creative response to perceived opportunities or needs. Neither type of organization, however, manages to achieve fully consistent, efficient performance in uncertain conditions. The difference between the two models lies in their assumptions about how organizations learn.

Redundancy

Acknowledging that even the most meticulously designed systems are subject to failure or exposure to uncertain conditions, organizational theorists searched for means to strengthen operating systems against sudden disruption or fluctuating performance. Solving this problem in mechanical systems is relatively simple through the principle of redundancy (Simon 1981). For every critical component, the system arranges a back-up component or replacement part to maintain operation of the system through sudden failure or unexpected demands (Landau 1991).

This method is used by the military and emergency services to reduce uncertainty. Common practice during a fire, for example, which represents a very dynamic, unstable environment, is to pile on resources in personnel, equipment, and technical strategies in order to overwhelm the flames and reduce danger. As the danger subsides and the fire comes under control, the chief will then release resources to serve other needs.[10] Redundancy keeps the system in operation, but often at great cost. Many public agencies simply cannot afford to operate on this principle, and the whole system is weakened to the extent that smaller organizations or municipalities are likely to be incapacitated under major threat. Further, resources invested in redundancy cannot be used to search for more economical means to reduce community risk. Used in practice by the military and emergency services to compensate for failures in control systems, action models that rely on redundancy prove too expensive for general use (Rossi, Wright, and Burden 1982; Sutphen and Bott 1990).

Inquiry

Models of inquiry represent a fourth method of coping with uncertainty that requires collective action. Most inquiring models rely on search and feedback processes, but they vary significantly in time allocated to inquiry, modes of integrating information from different sources, measures of performance, and methods of review and revision. Three types of inquiring models can be distinguished in the search for mechanisms of collective learning and action to reduce risk. They are: trial and error (Holland 1975; Axelrod 1984), professional social inquiry (Lindblom and Cohen 1979; Weiss 1977, 1998; Weiss and Bucuvalas 1984; Lindblom 1992) and cultural modes of "sense-making" (Weick 1990, 1993; Rochlin 1989; Roberts 1993). Each type represents a conscious need to reduce the discrepancy between perceived goal and actual performance, and each involves the processing of incoming information with existing knowledge of the context for action. Each also assumes that action occurs through some form of organizational structure, and that the inquiry process is conducted to support action.

Trial and error learning in social contexts (Holland 1995; Piaget 1980; Axelrod 1984; Comfort 1986; Haas 1990) draws on the theory of adaptation from evolutionary biology (Holland 1975). Holland observes that living organisms – human and nonhuman – consistently assess the consequences of their actions upon their immediate environment and, over time, will choose alternatives for action that strengthen their chances of survival. However, time is required for adaptation, and the feedback process driving organizational learning is indirect. In social settings, organizations may learn inappropriate lessons from limited experience and may seek to apply those inappropriate lessons in a similar context, but obtain very different results.

While trial and error learning offers insight into adaptive processes in organizational behavior over the long term, it does not meet the needs of disaster managers as they cope with sudden-onset events requiring immediate response. Organizations appear to learn in two ways. First, they assimilate information from experience and the external environment and store it in their repertoire of possible responses to apply in similar situations (Newell and Simon 1972; Argyris 1982; Piaget 1980). Second, and more directly related to problem solving, organizations create knowledge to fit novel situations (Simon 1983). Disaster managers urgently need this type of creative problem solving in the complex, time-dependent context of disaster, but it is difficult to achieve at the organizational level.

Professional social inquiry addresses the need for continuous learning under conditions of uncertainty through the design (or redesign) of systematic patterns of search, collection, evaluation, and exchange of information, leading to commitment to action (Deutsch 1963; Churchman 1971; Weiss 1977, 1998; Lindblom and Cohen 1979; Meltsner and Bellavita 1983). This perspective acknowledges information as the driving force of organizational action (Deutsch 1963), and considers skills in search, processing and utilization of information central to the design of structures for action (Churchman 1971; Habermas 1979; Burt 1982).

Professional social inquiry represents systematic methods used to examine, evaluate, investigate, and explain problems of interest or concern to a given audience or community. These methods, proposed initially by C.E. Lindblom and David Cohen (1979), and developed more fully by Lindblom (1992), differ from the standard methods of evaluation and policy analysis in their continuous probing of unfolding events, rather than defining controls for the investigation and conduct of rigorous analysis within a set of well-structured problems. The difference derives from the authors' awareness of continual change occurring in the social conditions under study, and the inadequacy of analysis directed toward definitions of problems that rapidly become obsolete.

Professional social inquiry assumes, instead of permanent laws governing social relationships, the presence of a system for organizing action toward a defined goal. The goal of the system governs the framework for collective action, and the organizational components of the system adjust their actions to one another, and to the wider environment, in terms of their respective understanding of the systemic goal and its requirements. This organizational system, in turn, becomes the vehicle for common discovery and transmission of information from the environment to its constituent parts and back again, enabling the members to reach a collective goal.

An inquiring system is fundamentally a means of organizing information and communications processes in order to solve problems for a specific group (Churchman 1971). It is goal-seeking, open-ended, enabling and fosters cooperation among the participating members (Churchman 1971, 200). This perspective allows organizations to draw from a wider array of knowledge and resources than is possible for individuals acting alone, enabling them to address larger, more difficult, and more complex problems.

The difficulty in designing an inquiring system to solve a particular set of problems lies in keeping that system sufficiently open for continuing, broad information search and processing, but sufficiently focused to carry out specific action. The system needs to operate simultaneously at macro and micro levels in the search for, processing and transmission of, information. At the macro level, decision makers need to view the entire organization's operation to maintain their focus on the basic goal. At the micro level, managers need detailed information to plan specific actions that, cumulatively, achieve the goal for the whole system. Information about specific units at the micro level is aggregated to inform macro level policies for the entire organization, while macro level monitoring of the environment, in turn, provides resources and sets limits on micro level action. The interaction between the micro and macro processes of inquiry is continuous.

A third model of inquiry derives from the innately human drive to interpret one's immediate surroundings in ways that "make sense" to the individual and his/her community (Argyris 1982; Weick 1990, 1993). In this model, cognitive processes integrate new information from the environment with existing knowledge and cultural norms in order to produce a coherent (and revised) interpretation of the immediate situation that informs and enables action. This model is vulnerable to the limitations of initial knowledge and experience of the participating actors which, under sudden, unfamiliar or dangerous conditions, can lead to the collapse of "sense-making" resulting in disaster (Weick 1993).

Related to the cognitive process of sense-making, Luhmann (1989) identifies a powerful, driving force for creative self-expression (autopoeisis) in individuals that, if extended to social groups and organizations through articulated communication processes, serves as a vital source of creativity, renewal, and regeneration in social systems undergoing change. When established patterns of performance and interaction no longer elicit the expected results, individuals and organizations seek creative ways to achieve their goals, often by combining old elements with new information and techniques to produce improved results.

The search for meaning, coupled with opportunities for creative expression, creates the basis for individual action (Frankl 1970; Simon 1981) that is extended through organizations to the community. Without meaning, action at the individual, organizational, or community level is either paralyzed or contentious.

Models of inquiry quickly encounter the limits of human information-processing capacity in addressing large, complex, social problems. Although the basic ideas have been present in the literature for nearly thirty years, researchers have not linked information processes with structural designs for action in consistently successful ways. Shifting the focus on inquiry from the individual to the organizational level presents problems of communication and coordination that are significantly more complex and interactive (Argyris and Schon 1974, 1978). Descriptive accounts of organizational networks (Meltsner and Bellavita 1983; Rochlin 1989; Weick and Roberts 1993) document successful organizational efforts to achieve specific goals, but these networks become vulnerable to changes in inputs of individual energy and commitment over time (Bardach 1977).

Current information technology extends human problem-solving capacity through the use of computers and telecommunications, resulting in a problem-solving approach that is strong both technically and organizationally. While the technical capacity is available to meet the information processing requirements for large, complex problems, the organizational designs to establish and utilize these systems productively are not yet defined or developed.

The three types of inquiring models — trial and error over time, professional social inquiry, and interpretive inquiry or "sense-making" — represent a progressive increase in conscious acceptance of responsibility for action in uncertain contexts. The set of models reveals, most importantly, the capacity of individuals and organizations to advance their modes of inquiry regarding a specific problem to more rigorous methods, while simultaneously broadening the strategy of action to include revision of cultural and organizational norms.

Information processes drive collective action in all four models of transition in uncertain conditions — control, anarchy, redundancy, and inquiry. All four models tend to reduce uncertainty by minimizing conflict, asserting common purpose and shared goals under the urgent conditions of disaster. The shared threat compels all parties to work toward the objectives of coping and recovery from danger, although they may interpret the facts differently and may bring different capacities to bear on the common effort. Under these conditions, divergence from the common objectives — that is, conflicts among the parties — detracts attention and energy from common action and violates the assumption of cooperation for a shared goal. Each of the models outlined above deals with divergence of interests differently, which also affects the processes of organizational learning. The command and control model, for example, characteristically copes with divergence among its members through suppression and discipline. The anarchy model, in contrast, deals with divergence through bargaining among the differing parties. In the redundancy model, divergent goals are treated by partitioning roles, assigning buffers, or reducing conflict through allocation of 'slack' resources. The inquiring system model uses methods of education and discovery through error over time to reduce conflict. The different methods of coping with divergent interests not only reflect different conceptions of authority, but also result in different patterns of communication and exchange of information within the system to support the respective models.

The results obtained from the actual exercise of information processes differ with each type of organizational structure. Improving the technology of information processes in disaster management alters a basic parameter in each model, and links information to action in ways that reduce uncertainty in the disaster environment. In models of control, information technology, adapted to perform routine tasks of monitoring performance as well as allocation and utilization of resources in well-structured situations, frees disaster managers to devote more time and thought to creative redesign of action in dynamic environments. In models of anarchy, advanced information technology employed to focus the attention of multiple participants on the same information for the same problem at the same time enables the participants to create a more efficient, informed basis for common action.

In models of redundancy, information technology, used to store valid data regarding vulnerable conditions, or to allow multiple managers to retrieve the same information easily and to project from it possible courses of action to reduce risk, allows a more efficient use of resources and a more economical allocation of personnel and equipment in mitigating risk. Models of inquiry depend upon ready access to current, valid, timely sources of information. Appropriate uses of information technology increase the timeliness, accuracy, and validity of information available to responsible managers in disaster environments, enabling them to track dynamically evolving events more accurately and to anticipate likely courses of action more successfully. The limitations of each model become more apparent as the urgency of the crisis eases, and conflicting goals are exposed through more accurate, timely, and comprehensive information processes.

Integrating information technology into organizational designs to reduce risk, however, requires reconceptualizing the structure and processes by organizations engaged in risk reduction. Organizations need to invest in technical assistance and training as well as equipment to enable participants to acquire the necessary skills to search for, and use, relevant information using this technology. Introducing information technology also creates opportunities to restructure the organization's work processes (Goodman, Sproull, and Associates 1990) and to reexamine the allocation of authority, attention, and resources in disaster operations on the basis of more timely, accurate, and comprehensive information. Such proposed changes are likely to encounter resistance in established organizations reluctant to yield accepted practice, even if unproductive, to unfamiliar technologies (Argyris 1990; Goodman, Sproull, and Associates 1990). An effective strategy for integrating information technology into organizational policies and practice to reduce risk requires a clear, carefully developed, and sequenced design.

Requirements of Transition in Complex Systems

Based upon the preceding discussion, five conditions appear to invite and support the type of transition that leads to collective action in reducing risk. These conditions include:
1. "Discovery" of a common threat (March 1988; Argyris 1982)
2. Common understanding among the affected group of both the problem confronting the community and the goal for action (Simon 1981)
3. Mechanisms of information exchange and feedback that support learning among the participating members (Argyris 1982, 1990; Schon 1983; Cohen 1986; Comfort 1993; Weick and Roberts 1993)
4. Means of integrating incoming information with existing knowledge to create a timely, informed basis for action at each operating node in the emerging response system (Hayes, Waterman, and Roth 1983; Comfort 1990a, 1991b)
5. Means of evaluating performance and incorporating this information into a common knowledge base that informs the next decision in the evolving process (Argyris 1993; Deming 1986)

Each of these conditions, and the interaction among them, contributes to collective learning and creates shared knowledge, the basis for collective action.

The transition between learning and action produces change in organizational performance. While many mechanisms of adaptation in dynamic conditions are discussed in the literature, two appear especially insightful in explaining this process under uncertain conditions. The substantial body of research on high reliability organizations (Rochlin, LaPorte, and Roberts 1987; Rochlin 1989, 1996; Laporte and Consolini 1991; Roberts 1993; Weick and Roberts 1993) identifies methods of assessing dynamic constraints on action as products of organizational training and culture. Weick and Roberts (1993) term this process "heedful interrelating", which is maintained by keeping a common focus of attention among a highly trained and acculturated crew. These methods produce an extraordinarily high rate of reliability under very dangerous conditions. Although data from this research derive from observations made of take-off and landing operations aboard an aircraft carrier, the same focus of attention can be observed among highly trained and acculturated members of emergency response organizations. The difficulty in maintaining heedfulness comes when disaster operations are extended to a wider scope, with a large number of participants who have little or no training and who hold a diverse set of perspectives regarding what to save, what to leave, or what to do.

Feedback loops among the component subsets of the disaster response system constitute a second mechanism that supports collective learning. These loops operate in recurring processes, incorporating new information about current conditions and consequences of response actions into the next iteration of risk reduction measures. In nonlinear, dynamic systems, such as those evolving in disaster response, feedback loops may produce a "strange attractor", that is, a slightly aberrant action which, under repeated occurrences, attracts energy in a different form that operates at a slightly different rate from the rest of the system.

For example, an earthquake in a metropolitan area may have damaged the electrical power lines, blacking out street lights and traffic lights. The lack of operating traffic signals then snarls traffic, and worried citizens chafe at the delays caused by traffic in an already stressful situation. At one intersection, a driver suddenly pushes his car off the roadway and begins to direct traffic. The flow of traffic resumes, heavy but at least moving. Relieved drivers wave in appreciation. At the next intersection, another citizen, who has seen the first impromptu action to reassert traffic control, does the same. At other intersections across the city, citizens spontaneously repeat the pattern, and finally traffic is moving regularly in the city again.

In such conditions, new clusters of energy increase in size and force, and finally transform the operation of the whole system (Prigogine and Stengers 1984; Ruelle 1991; Kauffman 1993), just as the citizen traffic cops reordered the jammed flow of traffic. The transformation of an operating system occurs when repeated aberrations succeed in shifting the dominant feedback loop in the organization to a new level of performance or a different organization for direction. Both cultural and feedback methods of collective learning involve a re-evaluation of present action in dynamic contexts and a reinterpretation of the goals for action in light of new information.

Conditions that support collective action are conditions that foster collective learning. When collective learning occurs, it has significant consequences for the performance of organizations engaged in risk reduction. First, the locus of authority shifts between levels of governmental jurisdiction from a centralized, hierarchical structure to a flatter, more interactive structure that encourages participants to take responsibility for hazard mitigating action at their respective positions in the intergovernmental system. Second, the focus of policy and practice shifts from measuring performance against a set of structured rules to monitoring the process by which the organization(s) reach their goals (Bateson 1980). Finally, extending human cognitive capacity through information technology facilitates management of the complex information and communication processes required to mobilize and coordinate action appropriately among multiple participants who seek a common goal, but have multiple perspectives on the means of attaining it.

Preliminary Model of a Dynamic Disaster Response System

The preceding discussion allows a preliminary characterization of a dynamic disaster response system. This characterization serves as the basis for an exploratory study of actual disaster response operations. The primary characteristics of such a system and its operations are:

1. Organizations — public, private, and nonprofit — vary in terms of authority, skills, knowledge, resources, and capacity to act in response to needs created by disaster; yet interactions among the set of organizations involved in a given event are governed by the local conditions and generate a disaster response system specific to that event
2. A disaster response system constitutes the set of relationships that enables coordinated action among local, county, state/provincial, national, and international organizations to meet the needs of the affected population
3. An articulated set of communication and information processes (Luhmann 1989) activate and enable relevant organizations in an affected community to participate in disaster response in accordance with their respective levels of responsibility and capacity
4. The efficacy of communication processes within a disaster response system depends upon the timeliness, accuracy, and validity of information circulating through them
5. Structuring information and communication processes to support action both creates and maintains the disaster response system
6. This interactive set of communication and information sub-processes generates a continuing process of individual and organizational learning and action to support dynamic response to, and management of, disaster

This study focuses on emergent systems which demonstrate creative response to dynamic disaster environments. These emergent systems cross organizational and jurisdictional lines to gain support and mobilize assistance from local, state, national, and international sources. They provide an important means of communicating social knowledge to the affected population, as well as the technical and organizational support to meet the urgent needs of disaster-affected communities.

A dynamic model of disaster management operates in direct contrast to the traditional model of command and control, which is usually invoked after a community is already devastated and no longer has the capacity to act. The traditional model has the short-term advantage of little or no investment by the community prior to the disaster, but the long-term disadvantage of higher economic, social, and humanitarian costs when a disaster does occur. In this research, I explore the information and communication requirements for establishing an interactive, interorganizational, interjurisdictional information system to support decision-making in this complex context and propose a dynamic model of disaster management.[11] This model includes mitigation and preparedness, but focuses specifically on response and recovery, which have been the most prevalent activities in disaster management. It outlines a set of state and process phases to guide the inquiry.

In this preliminary model, the initial conditions represent the existing state of the community at risk prior to a specific hazardous event. It includes the basic resources available for learning and action, as well as the current operating context of the community. These conditions shape the possible courses of development for coordinated response to an actual event. Given a distinctive set of initial conditions, three action phases characterize a sequential learning process among individuals and organizations that reflects their respective responses to a destructive event and produces the next (temporary) state for the vulnerable community. The set of interactive responses by participating organizations, repeated over time, constitutes the dynamic response system for that disaster. The model includes:

Initial Conditions. Initial Conditions are defined as those existing in a risk-prone community prior to an actual damaging event. These conditions characterize the context in which the disaster strikes and influence the subsequent development of response operations through the sequential processes of Information Search, Information Exchange, and Organizational Learning. Specifically, these conditions shape and limit the capacity for Information Search, upon which the subsequent processes depend. The set of processes lead to the outcome state of Adaptive Performance, which represents the next (temporary) state of community operations, altered by the experience of disaster.

Information Search. This set of functions involves actions taken to acquire information about the event, organize that information for presentation to other participants in the disaster response process, and utilize that information in conjunction with other knowledge extant in the community as a basis for action.

This set of functions involves technical, organizational, economic, and ethical capabilities and perspectives that influence different organizations in their search for information. These functions both shape and set the technical and organizational constraints for the process of information exchange.

Information Exchange. This set of functions produces the exchange of information among participants in the disaster response process.[12] The functions involve technical, organizational, economic, and ethical factors that affect the ability of organizations to initiate or receive an exchange of information during disaster operations. The actions performed in this process create the basis for shared understanding of the problems confronting the community.

Intra- and Inter-organizational Learning. This set of functions includes the actions taken by participating organizations to interpret the consequences of a destructive event for their continuing performance and that of the community. It involves the record kept of the event, efforts made to identify error, and the review and feedback processes used both within and among organizations in response to the event. It also includes the technical means used to create a distinctive 'organizational memory' for the response system. That is, it involves the extent to which a filing system, computers, or access to other knowledge bases were used to record the consequences of the event, and actions taken in response. These data are used to interpret the impact, or to 'make sense,' of the event for the affected community. It summarizes the community's current assessment of its vulnerability under the changed conditions and the extent to which it has sought to communicate its interpretation of the damaging event to other relevant actors in the evolving response system.

Collective Action. The outcome measures of this process are actual changes in performance to mitigate future risk or recommendations for change that will be carried out in the future. These measures represent the extent to which a community has integrated information from an actual event into its ongoing operations in accordance with its stated goal of risk reduction. The system-wide set of changes constitutes the new (temporary) state of the community in an evolving process of interaction with its environment.

Based on this model, five research propositions are offered to characterize the sequential process of organizational learning and action in complex environments. They are:
1. Disaster response systems are dependent upon the initial conditions of technical, organizational, economic, and social development in the stricken communities
2. Adaptive performance among individuals increases in an environment of substantial complexity when acting within organizations appropriately structured for information exchange
3. Adaptive performance among organizations, in turn, increases in an environment of substantial complexity when acting within interjurisdictional systems appropriately structured for information exchange
4. Collective action depends upon the form and degree of intra- and inter-

organizational learning that occurs within and between organizations participating in information exchange. Information exchange is facilitated by appropriate uses of information technology
5. Inadequate information flow decreases organizational performance in actual disaster environments and, further, inhibits intra- and inter-organizational learning and action in complex environments[13]

In this model, communication and information processes activate a system of inter-organizational learning and action following a disaster that engages all organizations in the community in dynamic response. The response system constitutes a mechanism for substantive change in the initial conditions, and in the process of its operations, contributes to an evolving new (temporary) state of the community. The next chapter will present the research design and methodology for the study.

NOTES

1. I am indebted to Emery Roe for this observation on the role of coordination in administrative practice.
2. The historical record documents that governments in power at the time of a major disaster frequently lose credibility and support from their populations. The collapse of the Somoza government in Nicaragua following its mishandling of international assistance following the earthquake of 1972 is a vivid example. The rapid devolution of power within the former Soviet Union following the Armenian earthquake of December 7, 1988 illustrates the same pattern. Although other decentralizing forces were at work in the former USSR in late 1988, the impact of the Armenian earthquake in accelerating these forces, especially in Armenia, was significant. Field report, Interdisciplinary Study Team, Moscow-Pittsburgh-Yerevan. March, 1989.
3. This section draws heavily upon an earlier article, L.K. Comfort. 1993. "Integrating Information Technology into International Crisis Management and Policy". *Journal of Contingencies and Crisis Management*, Vol. 1, No. 1: 17–29.
4. California Office of Emergency Services. 1989. *Introduction to the Incident Command System*. Sacramento, CA: State Board of Fire Services. This manual has since been replaced by the State Emergency Management System, adopted statewide in 1992.
5. For example, missions performed by the US Army in disaster operations require that all mission commanders prepare and submit a 'plan of operations' before action is taken. Such a plan includes five basic elements:
   1. Problem
   2. Mission
   3. Resources
   4. Administration and Logistics
   5. Command and Control

   Specifying the content of these elements for all personnel involved in the mission clarifies the conditions of the operation, establishes the parameters for action and reduces the uncertainty associated with the mission. Operating in disaster environments, military commanders apply their previous training and experience to increase the success of their missions. The trade-off, however, is the time required for planning in life-threatening events. Ambiguity returns when other organizations interpret the problem or mission differently, or resources, communication and coordination with external organizations that are part of the larger disaster response operations fail. US Military Commander, Interview, San Jose, Costa Rica, April 29, 1991.

6. For example, the California Office of Emergency Services has accepted the Incident Command System as its primary mode of training and operations in disaster response. This system was effectively used in disaster operations at the Cypress Structure collapse following the Loma Prieta Earthquake of October 17, 1989. Chief, Oakland Search and Rescue Team. Interview. Oakland, CA, Cypress Structure Command Post, October 21, 1989.
7. The tragedy of the Bhopal, India disaster of December 4, 1984, in which over 20,000 people were killed by a release of methyl-isocyanate from the nearby Union Carbide plant, was that the plant officials had not informed the population that wet washcloths placed over their noses and mouths would enable them to breathe if exposed to the gas, and would give them sufficient time to leave the area. In hearings held following the disaster, plant officials said they had not informed the people of these simple methods of protection because they did not want to alarm them about the potential danger in their midst. Please see Paul Shrivastava. 1992. *Bhopal: The Anatomy of a Crisis.* 2nd Ed. London: P. Chapman Pub.
8. During the Pennsylvania tornadoes of May 31, 1985, residents of the northwestern part of the state were unfamiliar with the phenomenon of tornadoes. There was no adequate tornado watch and warning system in place. Some individuals stood and watched the tornadoes, as they roared towards them. Others got into their cars and tried to drive away, getting caught in the storm. Please see Louise Comfort and Anthony G. Cahill. Increasing Problem Solving Capacity between Organizations: The Role of Information in Managing the 31 May, 1985 Tornado Disaster in Western Pennsylvania. In Louise K. Comfort ed. 1988. *Managing Disaster: Strategies and Policy Perspectives.* Durham, NC: Duke University Press. 280–314. The collapse of the Heizel Stadium in Brussels illustrates another type of destructive chaos that can occur when a model of control is inadequate to the demands placed on it. See Paul t'Hart and Bert Pijenburg, "The Heizel Stadium Tragedy". In Rosenthal, Uriel, Michael T. Charles, and Paul t'Hart, eds. 1989. *Coping with Crisis: The Management of Disasters, Riots, and Terrorism.* Springfield, IL: Charles C. Thomas: 197–224.
9. Train (1986, 299–307) identified a set of phases in a recurring decision cycle that characterizes military organizations in battle. These phases are instructive for disaster managers, as they often face similar constraints. These phases are: a) problem recognition; b) crisis avoidance; c) crisis denial; d) problem definition; e) allocation of resources: time, attention, material; f) transition from preparedness to response operations; g) relation between field and command during operations; h) recognizing crisis termination; i) accountability and authority. The sequence of these phases is altered for disaster managers in sudden onset events, when crisis avoidance and denial are eliminated by the obvious manifestations of the event.
10. This perspective was offered by Paul Blackburn, Deputy Chief, Fire Department. Los Angeles County, in an interview following the Whittier Narrows Earthquake, Los Angeles, CA, June 3, 1988.
11. This model was originally presented in the research design submitted as the proposal for funding this study. L. Comfort. 1989c. "Interorganizational Coordination in Disaster Management: A Model for an Interactive Information System", National Science Foundation Grant #CES 88-04285:6–18. It is restated briefly here as a background reference for the presentation of research findings.
12. Initially termed "information transfer", the descriptive term for this set of functions was changed to "exchange", a term that more accurately describes the two-way or n-way dissemination of information characteristic of community response to disaster.
13. These hypotheses build on a set of hypotheses presented by Michael Cohen (1984) in his study of performance in local/global relationships in his article, "Conflict and Complexity: Goal Diversity and Organizational Effectiveness". *American Political Science Review.* 78(2): 435–451.

CHAPTER THREE

MEASURING CHANGE IN NONLINEAR SOCIAL SYSTEMS

Measurement in Nonlinear Systems

Dynamic systems that evolve among organizations operating in conditions close to the 'edge of chaos' require methods of study and measurement that accurately assess their nonlinear characteristics and processes. If we accept the premise that the fundamental structure of social systems is nonlinear, we need to reconsider our concepts and techniques of social science measurement on four primary issues: causality, time, spatiality, and complexity.

The requirements for determining causality advanced in linear models of social behavior cannot be met for rapidly evolving nonlinear systems. The assumption of independence among rational actors that underlies linear models of cause and effect (Blalock 1972, 128–134; Kerlinger 1986, 98–102) is not valid in evolving, dynamic, interdependent contexts. Rather, the rate and quality of performance of nonlinear systems vary with the context and initial conditions of their operations (Nicolis and Prigogine 1989, 181; Kauffman 1993, 174, 178; Gell-Mann 1994, 259–260). Acknowledging Heisenberg's "uncertainty" principle that it is impossible to determine both the location and the speed of a moving object at the same time (Peitgen, Jurgens, and Saupe 1992, 12–13), variation is always present in living systems. Small variations in initial conditions lead to large differences in results when the process is repeated over time (Peitgen, Jurgens, and Saupe 1992, 48).

The difficulty of imputing causality in nonlinear systems distinguishes them from linear models of cause-and-effect that have been used extensively, but with limited success, in social science. Pietgen, Jurgens, and Saupe (1992, 15) assert that: "...the validity of the causality principle is narrowed by the uncertainty principle from one end as well as by the intrinsic instability properties of the underlying natural laws from the other end..." There are too many variables outside the control of the policy initiator, too many factors contingent upon conditions that are subject to change, too many differences in individual reactions and responses that generate, or limit, different possibilities of collective action.

Instead of seeking to identify causal relationships among social variables to predict specific outcomes in this study, I characterize the operation of a set of disaster response systems in relation to their operating environments. The measurement tasks begin with identifying the components of a disaster response system and discovering the relationships and degrees of interdependency among them. It also includes observing the events that activate the operation of the system, and tracing the sources of energy and incentives that drive its action. Measurement tasks also include identifying the vulnerabilities, constraints, and boundaries of the system's performance. These steps lead to descriptive inference, or the "process of understanding an unobserved phenomenon on the basis of a set of observations" (King, Keohane, and Verba 1994, 55). In nonlinear systems, the whole is truly more than the sum of its parts, and our task is to devise forms of measurement that document the varying relationships between sub-unit and system-wide actions that lead to a given goal.

Measurement of nonlinear systems needs to account for random events and their potential influence on the direction and performance of existing systems (Gell-Mann 1994, 58). An earthquake, without warning, creates the startling effect of a random event in a community. Once it has occurred, the response system seeks to mobilize action to minimize the damage. Nonlinear systems are characterized by unpredictability in outcomes (Prigogine and Stengers 1984; Kauffman 1993, 178; Gell-Mann 1994, 259–260), even starting from similar initial conditions. The interactions among component units, and between the system and its environment, vary with the degree of discretion, energy, and attention available for active engagement among its component parts and with its environment.

Nonlinear systems also acknowledge the irreversibility of time and its consequences for the development of stable, causal relationships. Nicolis and Prigogine (1989, 14) state:

Far from equilibrium, that is, when a constraint is sufficiently strong, the system can adjust to its environment in several different ways. Stated more formally, several solutions are possible for the same parameter values. Chance alone will decide which of these solutions will be realized. The fact that only one among many possibilities occurred gives the system a historical dimension, some sort of "memory" of a past event that took place at a critical moment and which will affect its further evolution.

Nonlinear systems are continually in process, only temporarily in steady state. Incorporating the concept of process into a working model of a nonlinear system acknowledges an evolving exchange of energy and information between the system and its environment that transforms prior relationships. A nonlinear system reflects change, not necessarily progress, but at least the possibility of producing better performance toward an evolving goal.

Measuring the movement of an evolving nonlinear system implies the concept of spatiality. A nonlinear system involves the intersection of multiple dimensions, which can be identified at a single point in space for an array of variables in dynamic interaction. As interaction among the dimensions shifts the point of intersection in space, the trajectory of the evolving system can be mapped over time. This evolving record of the system's dynamic operations in both space and time provides a profile of its complex interactions at a given point that can serve to guide the next steps in action.

Nonlinear, dynamic systems thus generate distinctive characteristics that set them apart from linear systems. Most importantly, all nonlinear systems exhibit the capacity for self-organization, but not all nonlinear systems achieve this level of function. Self-organization depends upon a viable set of communication and information processes that enable the participants in a system to make informed choices. Self-organizing systems spontaneously reallocate resources and rearrange their activities to create a better 'fit' between their internal operations and their immediate external environment. The search for a better fit often leads to more complex relationships as different actors adjust their performance to one another as well as to the environment, but without a network of multi-way communication, these mutual adjustments cannot occur. While adaptation does not always result in improved performance, nonlinear systems reveal energies directed to change their structure through internal dynamics. Linear systems, relying on external direction, are not able to generate spontaneous, endogenous re-organization. In the study of nonlinear systems, measures are needed to assess the degree to which a system is able to generate, maintain, conserve, and redirect energy in order to achieve its desired goal. Policy implementation in nonlinear systems is intensely interactive and often increasingly complex.

This set of assumptions implies a fundamental shift in the purpose of social systems from designed mechanisms of control to vital processes of inquiry and collective learning leading to change. Creating a carefully developed and documented profile or 'map' of their primary characteristics and functions (Yin 1993, 1994) constitutes the first step in the systematic study of dynamic systems that evolve in response to 'shared risk'. This chapter presents the design and methods of observation and measurement used to characterize dynamic, interorganizational systems that evolved in response to urgent community needs following earthquakes. The response systems, each seeking to aid and facilitate recovery in their respective communities, reflect the current state of awareness, preparation, and capacity to cope with seismic risk at the time of the event.

Requirements of Response Systems Following Disaster

In order to facilitate the development of a response system to meet demands generated by the earthquake, disaster managers, in each case, needed to shift their focus of attention and action from the micro needs of their specific organizations to the macro needs of the community as a whole. This process requires interorganizational communication and coordination which creates a systemic capacity for collective response. Although interorganizational communication and coordination are assumed in every community's emergency plan, if one exists, the question is whether participating organizations are actually able to produce this shift in behavior under emergency conditions. This study will explore the conditions that facilitated or inhibited this interactive shift between micro and macro perception and action to enable coordinated interorganizational response to disaster in eleven different technical, organizational, and cultural settings.

As stated in Chapter 2, I propose that disaster response systems are generated through a sequential process of information search, exchange, and organizational learning that leads to a new (temporary) state of performance. The initial conditions characterizing a given organization's operating context as it enters the search process influence that organization's passage through the subsequent states to its integration into the disaster response system. Initial conditions are defined as the existing state of technical, organizational, and cultural practice in the community prior to the occurrence of a given disaster event.

Based on these premises, emergency response operations constitute a "sociotechnical system" that relies upon human organizations to deploy technical and organizational response for the protection of communities under threat. A primary function of such a system is its capacity to engage the wider community in collective action to protect life and property under threat of disaster. This capacity depends upon the type of communications and information processes that are functioning in the community prior to the event and the mobilization of additional resources to supplement communication and coordination to support the response process.

Measurement of Dynamic Disaster Response Systems

Developing a research design to identify, describe, and plot the evolution of dynamic disaster response systems requires a mix of qualitative and quantitative measures. Each type of measure has its limitations in practice, but both seek to explain the operation of disaster response systems from observable data. Although they differ in style, quantitative and qualitative measures adhere to the same logic of scientific inference (King, Keohane, and Verba 1994, 6–9). That is, they use systematic procedures of inquiry to collect and analyze empirical data in order to understand the social world. The differences in style, in turn, vary according to the clarity of the problem under study, access to data relevant to the problem, and limitations of time and resources. In this study, I use both quantitative and qualitative measures where appropriate, and integrate data gathered from the set of eleven case studies into a common knowledge base to further our understanding of rapidly evolving, disaster response systems.

Measurement in this study of dynamic response systems does not seek to identify a sequence of causal relations that reliably produce specified outcomes. Rather, it identifies what steps are taken through a field of many possible courses of action and obstacles in an evolving process that is guided by a specific goal. This process is analogous to the play of soccer on a field, in which the evolving course of action needs to take into account the varying direction of the wind, the unexpected plays of the opposing team, and the mix of strengths and vulnerabilities of the home team. Measurement of performance, winning the game, is perceived as much the product of interaction with the opposing team as performance in accordance with the preset plays of the home team. The coach's primary task is to maintain and focus the energy of her team against the strengths and vulnerabilities of the opposing team, recognizing the possibilities for discretionary judgment within the rules of the game.

Similarly, increasing the capacity for collective action in a community following a destructive earthquake requires maintaining both order and creative energy. Rules are used to focus the symbiotic energy of the group toward the common goal — recovery and reconstruction from disaster — not to limit it, a frequent mistake in most administrative organizations. In evolving disaster response systems, quantitative measures of the current state of different organizations at different locations and at different times are essential to inform qualitative decisions regarding priorities for action. Such measures also help to define alternative courses of possible action, and estimate likely costs and consequences of selected actions upon the community.

In terms of scientific method, the concept of causality shifts to a second order of abstraction in the study of nonlinear systems. The policy problem becomes not how to achieve a specific outcome, but rather how to generate and sustain a process of iterative inquiry and action that will, through its operation, lead its members to create new and more appropriate policies and practices in response to needs from its environment.

A Small-N Comparative Case Study of Disaster Response Systems

To explore the operations of nonlinear, dynamic systems in an actual field environment, I undertook a small-n comparative study of eleven rapidly evolving response and recovery systems following earthquake disasters. Using an exploratory case study method (Yin 1993, 1994), I have examined the dynamic systems of interorganizational response and recovery during disaster operations from the perspective of complex, adaptive systems. According to Yin, the case study method is appropriate when the research problem is not yet clearly structured, and it is necessary to construct a detailed profile of the components and their interactions that constitute the problem in order to design more rigorous research. Dietrich Rueschemeyer (1991, 32), writing on the case study method in comparative research, identifies four guidelines for its conduct. They are:

1. Work with a consistent conceptual grid for analysis of all cases, and use a basic framework of ideas as orientation for all case analyses
2. Ask of each case a set of theoretically grounded questions about the relative performance of major functions in the system
3. Deploy certain hypotheses in analyzing the set of cases
4. Maintain focus of analysis on the main task

He further stresses the relationship of a theoretical framework to the interpretation of results from data collection and analysis of case studies. Rueschemeyer (1991, 34) states:

> The theoretical framework, once developed on the basis of earlier research and argument, then informs the comparative case investigations, and it will in turn be specified and modified through these analyses. The result is, on the one hand, a set of historical cases accounted for with a coherent theory and, on the other, a set of propositions about the conditions of rapidly evolving organizational systems that have been progressively modified and are consistent with the facts of the cases examined as well as with the preceding research taken into account.

Given the exploratory stage of the study of nonlinear, dynamic systems in social science research, a small-n comparative research design appears most appropriate for this study of actual disaster response systems.

Research Objectives

The study has four initial research objectives. They are:
1. To identify the boundaries, size, and types of interaction that characterize the systems of public, private, and nonprofit organizations engaged in disaster response following eleven major earthquakes
2. To identify the role of information and communication processes in facilitating or hindering the emergence of these response systems
3. To document the dynamic evolution of interorganizational disaster response systems, identifying critical states in the process
4. To evaluate the role of self-organizing processes for increasing local capacity to reduce vulnerability to future disaster

Assumptions and Research Questions

In this study, I sought observable data to document self-organizing processes; that is, to determine whether communication and information processes activate the organizational learning and adaptive behavior involved in the evolution of disaster response systems. In the context of the disaster operations, the primary assumptions underlying this study are:

1. Initial conditions of education, training, prior experience, and primary responsibilities influence the development of information and communication processes that are used by organizations as they make their first assessment of the disaster situation and its requirements for action
2. These information processes include what types of information are sought, what sources are used, how incoming information is organized, integrated with existing information, and formulated into alternatives for action
3. Information processes, in turn, shape the content, frequency, direction, and form of information exchange, including transmission of information, interaction with other organizations, and means of transmitting information
4. The functions of information exchange activate organizational learning, which is measured by the following indicators: types of information retained; degree of identification of error; number of review and feedback processes; extent of perceived vulnerabilities in the community information system; and types of information technology used, ranging from paper and pencil to computers
5. Interorganizational learning leads to reallocation of resources, energy and time to produce recommendations for action and actual change in performance of the response and recovery system
6. Actions that demonstrate change in performance represent learning, and lead to substantive reduction of future risk for the community, altering initial conditions for the next seismic event

Three basic research questions have guided the study. They are:
1. How do the content and exchange of information affect the decision-making capacity of public/private/nonprofit managers engaged in disaster operations?
2. What kinds of information are required by public/private/ nonprofit managers to coordinate actions of their respective organizations appropriately in a disaster environment?
3. In what ways can information content and exchange be structured to maximize adaptive performance within and between organizations in a disaster environment?

Data are collected and analyzed at two levels in this study. The comprehensive unit of analysis is the disaster response system with its component sub-systems and transactions with the wider community it serves. Each disaster response system constitutes a case, and the number of cases (n) equals 11, the number of disaster response systems that are included in the study. Within each case, the sub-unit of analysis is the organization, and the n of organizations participating in a specific disaster response system varies by case. The unit of observation for both case and organizational levels of analysis is the individual participant in disaster response operations. The study traces action from the individual to the organizational to the systemic level of disaster response.

Selection of Cases

Selecting dynamic response systems that evolved following earthquake disasters as the unit of field study for complex, nonlinear systems has both advantages and disadvantages. The advantages for research are many. The sudden urgency and traumatic experience of an earthquake focuses the attention of the population, business and community organizations, and government upon the same event with powerful demands for immediate action. A destructive earthquake in a populated area generates quickly a complex system of interacting individuals, organizations, and jurisdictions engaged in the common task of response and recovery for the affected community. This response system, although composed of individuals, organizations, and jurisdictions that have other obligations on a routine basis, nonetheless can be distinguished from ordinary government, business or community operations by its specific goal of response and recovery from disaster.

Responsible collective action both generates and requires communication and information sharing across old boundaries in new ways to produce the desired goal of coordinated performance under urgent conditions of need. The recurring phenomenon of rapidly evolving systems of response and recovery following major earthquakes provides an unusual opportunity to study the mechanisms and processes of complex, nonlinear systems in practice.

Earthquakes, however, present the major disadvantage of unpredictability in location and time. Consequently, it is not possible to select cases *before* the event occurs. Rather, it is necessary to design a study to capture certain types of organizational characteristics, considering the phenomenon of seismic risk over time and in global perspective.[1] In this study, selection of cases for study was based upon four criteria: (1) severity of the earthquake's impact upon the community; (2) sufficient interaction among multiple organizations and jurisdictions to produce a distinct response system; (3) access to the disaster site and operations personnel; and (4) occurrence within a given period of study.[2] The eleven earthquakes, in chronological order, include:

Mexico City, September 19, 1985	(M = 8.1)[3]
San Salvador, October 10, 1986	(M = 5.4)
Napo Province, Ecuador, March 5, 1987	(M = 6.1, 6.9)
Whittier Narrows, California, October 1, 1987	(M = 5.9)
Northern Armenia, December 7, 1988	(M = 6.9)
Loma Prieta, California, October 17, 1989	(M = 7.1)
Limon, Costa Rica, April 22, 1991	(M = 7.4)
Erzincan, Turkey, March 13, 1992	(M = 6.8)
Marathwada, India, September 30, 1993	(M = 6.4)
Northridge, California, January 17, 1994	(M = 6.7)
Hanshin, Japan, January 17, 1995	(M = 7.2)

This set of eleven earthquakes presented a range of intensity and severity of seismic events and consequent impact upon their respective communities. After each seismic event, a disaster response system evolved, but with varying characteristics, rates of speed, and degrees of efficacy and efficiency among the local contexts. This comparative study examines the set of eleven cases to identify the primary characteristics and processes of the response systems, the ensuing consequences for their respective communities, and the balance between order and flexibility that enabled participating organizations to respond collectively to the destructive trauma of an earthquake.

Earthquakes of different seismic intensities generate different types of response systems in interaction with the local contexts in which they occur. The response systems reflect different patterns of structure and flexibility in the initial conditions of the communities in which the earthquakes occurred. These initial conditions affect the immediate choices that are made by local decision-makers, which in turn affect the subsequent choices and evolution of the respective response systems.

Using the basic measures that define the creative 'edge of chaos', I classified the 11 cases into four subsets that reflected different states of balance between structure and flexibility. Using data derived from documentary reports and interviews, I developed ordinal measures for structure and flexibility, and coded each response system on scales of high, medium, and low. Out of a possible 3×3 matrix of structure and flexibility, four subsets of cases emerged. This classification represented how close the response system was to the creative 'edge of chaos' that indicates its capacity to make the transition from a micro-level state of organizational operation to a macro-level state of a self-organizing, community-wide response system. The measurement and classification matrix for the set of 11 earthquakes is included in Chapter 4. The discussion of the initial conditions for each response system, as coded by this schema, is presented more fully in Chapter 4.

The four sub-sets of response systems represent different states of balance between structure and flexibility. The first group included those response systems that were low on both structure and flexibility: San Salvador (1986), Ecuador (1987), and Armenia (1988). The second group included those response systems that were low on structure, but medium on flexibility: Mexico City (1985); Costa Rica (1991); and Erzincan, Turkey (1992). The third set included those response systems that were medium on both structure and flexibility: Whittier Narrows, California (1987); Loma Prieta, California (1989); and Maharashtra, India (1993). The final set includes two contrasting systems. Both are high on structure, but one is high on flexibility, Northridge (1994), while the other is low on flexibility, Hanshin, Japan, (1995). The different proportions of structure and flexibility lead to very different patterns of performance in response operations, despite similar investments in earthquake engineering for physical infrastructure.

Of these eleven cases, there were no response systems that were low on structure and high on flexibility, or medium on structure and either low or high on flexibility. Nor were there any cases high on structure and medium on flexibility.

In all eleven cases, the earthquakes provided sobering tests of existing seismic awareness, public policies, and practice.

Qualitative and Quantitative Methods

The methodology used in this field research applied Kauffman's (1993, 175–209) concept of an N-K system to actual disaster environments to determine if these concepts would yield more consistent, accurate measures of disaster response and recovery systems. Kauffman suggests that increasing complexity in these systems can be tracked if we consider them as N-K systems, where N = the number of components in a system and K = the number of interactions among the components. This measurement is similar to the methodology of network analysis that offers a means of measuring the density and direction of evolving social patterns (Krackhardt 1992; Knoke and Kuklinski 1982).

In an N-K system, the process of co-evolution depends on the number of interacting parts and the frequency, content, and direction of information exchange among them. Further, it recognizes the element of choice exercised by the specific system components in accepting or rejecting the information exchanged, and in adapting their behavior accordingly in reciprocal actions. Change thus requires both order and flexibility, reflection and action.

This exploratory study seeks to identify the following measures for the interorganizational response and recovery systems that evolved during and after the eleven earthquakes named above:
1. N = number of organizations participating in disaster response
2. S = source of support for organization's response and recovery activities – public, private, nonprofit
3. T = number and type of transactions/exchange among participating organizations
4. K = estimated number of interactions among participating organizations
5. P = shared goal of organizations, 'bias for choice' in actions
6. D = duration of interactions among organizations

These measures allow me to examine specifically the number, types, extent, and influence of interactions among public, private, and nonprofit organizations as they engage in collective action and community learning in disaster response and recovery. While the search for data focused on the most specific information available, the precision of the data varied from case to case and involved both quantitative and qualitative methods.

Data Collection

Data collection included four types of data for each set of disaster operations. These types of data are:
1. On-site field observations of the disaster area, operating environment, technical and organizational capacity for response and recovery

2. Review of archival records, official reports, news analysis and videotapes, after-action critiques by participating agencies, and in-depth analyses by professional organizations
3. Semi-structured interviews with key decision-makers from public, private, and nonprofit organizations participating in disaster response
4. News reports of disaster operations from local newspapers for three weeks following each earthquake

In all eleven cases, I conducted interviews with disaster managers and operations personnel. with the assistance of local researchers. In seven of the eleven cases – Mexico City; San Salvador; Napo Province, Ecuador; Whittier Narrows, California; Erzincan, Turkey; Armenia; Marathwada, India — I also collected data directly from disaster-affected residents with the assistance of local researchers. In Turkey and Armenia, I participated as a member of an interdisciplinary research team in systematic surveys of response to the disaster, focusing on communication and coordination in medical response. In Mexico City, San Salvador and Ecuador, I collaborated with local researchers to conduct surveys of citizens following the disasters. In the Whittier Narrows Earthquake, I conducted a systematic survey of managers of public, private, and nonprofit organizations, as well as public operations personnel, with a carefully designed sample that represented the city, county, and state jurisdictions involved in response operations. In Maharashtra, India, I conducted interviews with selected community, district and state leaders with the assistance of a local researcher. In the Loma Prieta and Northridge Earthquakes, I have also relied upon secondary data gathered by other researchers who addressed similar questions.

For each of the eleven cases, I reviewed archival documents and used data from them to supplement the analysis of initial conditions prevailing in the communities before the earthquakes occurred. Archival records included a documentary review of the jurisdiction's response to the earthquake, agency reports, and in some cases transcripts of agency records and official hearings, media accounts, and other written analyses of the events and community response to them.

Data for this analysis were collected in field studies conducted over more than a decade, from September, 1985 to May, 1995. While a decade is a lengthy period in terms of social science research, it is extremely brief in terms of geologic time. Considering the problem of seismic risk in global perspective has allowed me to collect data from eleven richly varied cases, each of which offers a valuable perspective on the problem.

In collecting the data for each of the eleven cases, I followed six basic steps:
1. Identify the primary organizations that were involved in the response system, using news reports or professional reports
2. Construct a representative sample of key public, private, and nonprofit organizations involved in the interjurisdictional disaster response system
3. Conduct a set of semi-structured interviews with key response personnel and policy-makers engaged in disaster operations on site

4. Where possible, conduct a survey of disaster-affected citizens with the assistance of local researchers
5. Collect archival data regarding the organizations and response process, including professional reports and official documents, much of which is only available locally
6. Collect a full set of local newspapers that reported news of disaster operations for three weeks following the earthquakes

These steps in data collection produced comparable datasets for the eleven cases. The quantity and quality of data available varied to some degree over the set of 11 cases, but there is sufficient commonality among the set to allow comparative analysis. Differences in data or analysis will be noted in the presentation of findings.

Data Analysis

Within each case, I carried out the following steps in data analysis of the disaster response systems, using a common format for the set of eleven cases:

1. Review the record of archival data, interview data, and news accounts of response operations to determine the 'initial conditions' of the community in terms of its technical, organizational, economic and cultural capacity to mobilize action in disaster response and recovery
2. Identify, using content analysis of local newspapers, a system list of all organizations — public, private, and nonprofit — that were reported to be engaged in disaster response operations for a period of 21 days following the event
3. For each organization identified as participating in the response system (N), identify a set of primary characteristics: K (number of interactions); S (source of funding; if public, indicate jurisdictional level); T (number and types of transaction), P (goal), and D (duration)
4. Identify the boundaries of the response and recovery system that organized to meet the needs created by the disaster, specifically noting points of intersection across organizational and jurisdictional lines, as well as constraints on performance
5. Using local newspaper reports, construct a daily record of performance for the disaster response system, noting the frequency of mentions for each organization by day for 21 days
6. Code each organization participating in the system according to its dominant function in the disaster response system: emergency response, damage assessment, communication/coordination, disaster relief, recovery/reconstruction, and financial assistance
7. Enter the frequency data for each of 21 days by disaster response function into a nonlinear software program[4] to analyze the rate of change and degree of coordination in performance among key functions in the response system
8. Analyze the frequency data, using nonlinear techniques[5] to assess the extent of coordination among emergency functions within each system, and com-

pare the patterns of performance on emergency functions between sub-sets of each response system and among the set of response systems included in the study
9. Compare the characteristics and performance of the disaster response systems to the set of research propositions presented above (p. 63)
10. Assess the extent to which the actual performance differs from the proposed model and explain the difference in a revised model of emergent response systems, determining whether the systems produced adaptive behavior in disaster response practices

In carrying out these steps, this study seeks to identify the mechanisms (King, Keohane, and Verba 1994, 84–85) that generate collective response from a stricken community and its wider environment following a specific event, a major earthquake. These mechanisms document the evolution of the disaster response process as it occurred in each case. The response function moves necessarily into recovery operations, and it is often difficult to determine exactly when response ends and recovery begins. Since this study focuses on the first three weeks following an earthquake, the emphasis is on response, with recovery functions emerging during this period. Reports of organizational activity in disaster operations were then grouped by six major emergency functions for each disaster response system to provide an overall profile of disaster operations following that earthquake. The set of profiles represent, as accurately as reported in the news accounts, the evolution of the disaster response systems over time.

In conducting this analysis, I sought to identify the points and timing of the interaction among organizations as sub-units of the disaster response system, as well as the points and timing of interaction between the whole system and the environment in which it operated (Schneider 1992, 54–56). The disaster response system, thus, is perceived as a product of dynamic interaction both among the organizations that participate in disaster response activities and between the evolving system and its external environment. This perception reflects an action science orientation (Argyris 1982, 1985; Comfort 1985) that considers engagement in an actual situation an important part of the discovery and learning process in individual and organizational behavior.

Data Interpretation

In analyzing the data from the content analysis of news reports, I have used a method of nonlinear analysis developed by H. Richard Priesmeyer (1992). This method analyzes the change in organizational performance over a twenty-one day period following the earthquake, as reported in local newspapers. This period represents the first three weeks of emergency response, and documents the rapid evolution of the emergency response system. This analysis is central to the use of the case study method, as newspaper reports proved to be the most consistent form of data across all eleven cases of disaster response systems. Each of the eleven communities had newspapers that gave extensive coverage to disaster operations. While this form of data collection is imperfect and limited by journalistic practice, it nonetheless provides a common source of data for each of the eleven disaster response systems. The system lists of participating organizations derived from newspaper sources were, in each case, cross-checked against other data sources and verified by knowledgeable in-country experts.

Several types of analysis can be done using nonlinear techniques, but in this study, it will be used primarily to plot change in organizational performance over time. These data will allow comparison of system performance on individual emergency response functions within a given system, as well as comparison among a subset of systems. Data will be presented in phase plane plots with accompanying marginal history charts and logistic regression equations.

Phase planes describe the evolving "state of the system". Richard Priesmeyer (1994b, 2), describing the use of phase planes as a nonlinear method, states:

Specifically, a phase plane describes the changing condition of any system by identifying the "state of the system" on a common two-dimensional Cartesian plane. The two dimensions are simply two selected measures of the system... Phase planes plot *changes* in the two measures underlying the ratios rather than the actual values of the variables. That is, they plot marginal values, the differences between each measure's value and its value in the previous period. Obviously, changes can be either positive or negative from period to period. The center of the phase plane is coordinate 0,0 representing no change in either measure. Quadrants are traditionally numbered counter-clockwise, starting in the upper right quadrant.

In the content analysis of news reports on disaster operations, the phase plane plots the change in the number of references to a specific response function, e.g. emergency response, in relation to the change in the number of references to a second response function, e.g. communication/coordination over a specific period of time. For example, if the number of reports of emergency response activities following an earthquake increases from Day 1 to Day 2, the phase plane plots the corresponding change — increase or decrease — in communication/coordination activities. If emergency response activities increase at the same time that communication/coordination activities are increasing, one can observe the evolution of coordination in the response system. If communication/coordination activities increase, but emergency response decreases, the system's functions are likely disordered, indicating instability. This method assumes that correlation between the two functions is essential for effective system performance.

Priesmeyer and Sharp (1995, 3) explain the results produced by this analysis in the following terms:

Each axis on the phase plane measures changes in one of the two measures. The center of the phase plane is coordinate 0,0, which represents no change in either measure. Positions to the right and upward represent increases, those to the left and downward represent decreases. Any point on the phase plane represents the "state of the system" at a moment in time describing precisely how the relationship between the two measures is changing. The evolving state of the system is plotted as a trajectory on the phase plane.

The trajectory, which is essentially a line showing some movement or change in the system, is produced by plotting points at the intersections of the changes in each of the two measures over time, and then tracing a line from each point to the next. The dynamic behavior of the two measures in interaction is revealed by the changing position of the trajectory on the phase plane.

The four quadrants of the phase plane plot represent different dynamic states of the interaction between the two functions over the specified period. Plotting the change in the relationship between the two variables over time reveals the direction and pattern of change, as well as the degree of coordination between the variables for that period. Because the data in this analysis represent daily frequencies of reported organizational activities, each point on the trajectory represents a daily change in the relationship between the two response functions.

The logistic regression equation provides a measure of the stability of the whole system and estimates its point of transition to a new state of order in performance. R. Priesmeyer and W.T. Andrews (1994, 4) use logistic regression to identify an underlying equation, $X_{n+1} = kx_n(1-X_n)$, that can be applied at the beginning of the data set to generate all subsequent values.[6]

This method requires standardization of the observed values, and then repeatedly generating several series of logistic values by incrementing k and X within their relevant ranges. Second, the series of logistic values are then standardized and compared to the standardized observed values, using a traditional sum of squares regression (SSR) calculation. The values of k and X which generate the highest R^2 are kept as the optimal logistic parameters. Fitted and forecasted values are computed by scaling the optimal standardized logistic values, using the mean and standard deviation of the target values. The steps taken to estimate k and X in any data set are presented in Appendix A.

Specifically, the value of X describes the current state of the system, while the value of the constant parameter, k, describes the whole set of characteristics of the system which causes it to be either stable (period 1-type), oscillating (period 2-type), oscillating in a complex manner (period 4-type) or chaotic. The measure, k, produced by this equation, represents the extent to which the system either absorbs and integrates changing conditions into its existing structure of performance, or slides into chaos, requiring a new form of order in relation to its operating environment. The value of k ranges from 1.0 to 4.0. The system bifurcates at approximately k = 3.0, and moves into a second bifurcation at k = 3.5, and into chaos at approximately k = 3.66. Systems with low k values are quite resilient to changes occurring in the operating environment. Systems with high k values are chaotic, and operate in unexpected and unpredictable ways. Information provided by the k-value in social systems offer fresh insight into the operation of these systems and the possibility for guidance and reduction of risk.[7]

Findings from this nonlinear analysis will be presented in terms of a model of transition that incorporates the concept of a complex, dynamic system moving from state to state through a set of learning processes. The model identifies the set of Initial Conditions as the existing state in which the hazardous event occurs. This set of conditions influences the system's subsequent evolution through sequential processes of Information Search, Information Exchange, and Intra- and/or Inter-organizational Learning to reach the (temporary) outcome state of Adaptive Performance. Although the initial conditions in which the eleven earthquakes occurred differ, the same phases of transition can be observed across the set of eleven response systems. Transition is viewed as an evolving sequence of interactions that track the exchange of information, resources, and energy between the vulnerable communities and their seismic environments.

Initial conditions are defined as the organizational conditions existing prior to the occurrence of a disaster event. Measures of these conditions include, for example, characteristics of field operations personnel: education, specialized training in disaster management, years of service, and primary responsibilities in disaster response. Measures also include the geographic location of the community, its physical terrain and lifeline systems, e.g. communications, transportation, water, and power facilities, as well as the level of emergency preparedness and awareness of the affected population. The initial conditions will vary among the local contexts of disaster. Therefore, their careful assessment is the first step in characterizing the subsequent processes of an evolving disaster response system in its local context. The rate and content of the sequential processes — information search, information exchange, and organizational learning — depend in turn upon the initial conditions and shape the outcome state of adaptive performance. To the extent that these processes support mutual interaction and adaptation among the participating organizations, the disaster response system can be considered a self-organizing process (Kauffman 1993).

This study also seeks to identify, document, and measure the processes of self-organization in the response systems evolving after disaster in multiple environments. It applies a methodology intended to capture the dynamics of complex systems composed of public, private, and nonprofit organizations involved in disaster response and recovery operations. It presents observations and assesses the evolution of these complex, organizational systems in actual disaster operations. The critical decisions that shape the evolving disaster response systems are made in the first hours, days, and weeks of response and recovery operations. It is important to capture multiple observations of this process in its initial phases, as the data may erode over time. The set of disaster response systems included in this study offers an unusual opportunity for evaluating the N-K methodology for analysis of organizational response to seismic risk in other locations, other technical and organizational conditions, or in relation to other types of hazards occurring within interdependent communities.

The case profiles, while presenting a detailed account of the specific characteristics of each case, also offer a basis for comparison of general characteristics across the set of cases that permits the preliminary development of a model of rapidly evolving disaster response and recovery systems. In this comparison, I seek a coherence of behavior (Schneider 1992) in disaster response systems to discover the characteristic ways in which different organizations with similar responsibilities respond to the challenge of an earthquake disaster in different policy contexts.

Conclusions

This study examines the selection, adaptation, and self-organization processes that evolve following major earthquakes and form disaster response and recovery systems. The study seeks, first, to compare the characteristics and processes of the evolving response and recovery systems that represent different patterns of structure and flexibility in terms of their capacity for timely response. Most cases included in the study are moderate to strong in terms of earthquake severity, yet serious because they occurred in heavily populated areas. This type of hazard event provides a useful test of existing mitigation policies and preparedness programs for communities vulnerable to seismic risk.

Second, the study focuses on interorganizational response and recovery systems that have a capacity for adaptation and creative response in disaster environments. These systems cross disciplinary, organizational and jurisdictional boundaries to gain sources of support and to mobilize assistance from local, state, national, and international sources. They provide an important means of communicating social knowledge to the affected population as well as providing the technical, organizational, and informational support to meet community needs.

Third, the study implements a methodology for identifying characteristics of nonlinear disaster response and recovery systems that cross multiple boundaries. It also builds a cumulative profile of data gathered on the organizational performance of disaster response and recovery systems in ten different seismic regions[8] of the world. These events have created a unique international fraternity of professional disaster managers who have observed one another in action and exchanged information and advice over the years. Consequently, this study provides valuable insight into the extent to which organizational learning from one earthquake experience is translated into improved performance in response to a similar event in different geographic, technical, organizational, and cultural contexts.

NOTES

1. This study has been greatly facilitated by the Quick Response Study program funded by the National Science Foundation and administered by the Natural Hazards Research and Applica-

tions Information Center at the University of Colorado, Boulder. Eight of the eleven studies of disaster response systems that evolved after earthquakes have been conducted with financial assistance through this program or regular grants from the National Science Foundation, Washington, DC. I also wish to acknowledge the University of Pittsburgh and its Office of Research, Centers of International Studies, Latin American Studies, Russian and East European Studies, Urban and Social Research, and the Graduate School of Public and International Affairs that have consistently supported my research on this study.

2. Two additional major earthquakes — Iran, June 19, 1990, M = 7.7; and The Philippines, July 16, 1990, M = 7.8 — occurred during this time period, but are not included in the study because I was unable to do field research on site. Iran, a Muslim nation, did not allow women researchers on site. In July, 1990, research funds were not available from the Quick Response Research Program to study the Philippine event, due to heavy expenditures committed to investigate the Loma Prieta Earthquake and Hurricane Hugo, events which occurred in September — October, 1989.
3. The magnitude of the earthquakes is reported according to the Richter scale.
4. I used *The Chaos System Software*, developed by H. Richard Priesmeyer. 1994. Fair Oaks Ranch, TX: Management Concepts, Inc.
5. This analysis uses nonlinear techniques developed by H. Richard Priesmeyer and presented in his book, *Organizations and Chaos: Defining the Methods of Nonlinear Management*. 1992. Westport, CT: Quorum Books.
6. The sections on phase plane plots and logistic regression draw heavily on the work of H. Richard Priesmeyer and his colleagues. I am indebted to them for their clear exposition of this methodology.
7. R. Priesmeyer has applied this methodology to the analysis of data on infant mortality rates, substance abuse, and crime in his 1994 paper, A Logistic Regression: "Method for Describing, Interpreting and Forecasting Social Phenomena with Nonlinear Equations". Presented at the Chaos and Society Workshop. Universite du Quebec a Hull, Hull, Quebec, CA, June, 1994.
8. Two of the eleven earthquakes occurred in the Los Angeles Metropolitan Area: the Whittier Narrows, California Earthquake of October 1, 1987 and the Northridge Earthquake of January 17, 1994.

CHAPTER FOUR

THE 'EDGE OF CHAOS': CREATIVE RESPONSE TO CHANGING ENVIRONMENTS

When women in the barrios of Morelos and Guerrero in Mexico City began to boil water before using it for drinking or cooking after the 1985 earthquake, they were participating in a city-wide process of changing their daily behavior to adapt to altered conditions for maintaining their families' health. These simple acts, taken at the household level, reflected an interactive effort by the Mexican Government to engage the citizenry in maintaining their own health following severe damage to large sections of the city's water and sewage distribution mains by the earthquake.

Sections of the city's water distribution system dated back to the 15th century. There were no maps, no plans of where the pipes were laid. Other sections, once opened, revealed that sewage pipes were laid in the same trench with water pipes, and that cracked water pipes were now being contaminated with raw sewage. The Mexican Government confronted a new, and perhaps more dangerous, crisis in the contamination of its water supply in this capital city of over 18 million people. It could not even hope to maintain reasonable standards of water quality without the active participation of the citizens. Government officials needed to devise some means of engaging the citizens in reducing the risk. Overnight, the Water Department developed, printed, and distributed throughout the City colorful brochures in simple language explaining how and why to boil water before drinking or cooking. A major epidemic was avoided, and the Government and citizens largely achieved the desired change in collective behavior.[1]

Rapid, simple, direct means of information exchange between the Government and the residents of Mexico City alerted citizens to the dangers of possible water contamination and provided them with a simple, workable framework for action. Informed citizens then acted on their own initiative to protect themselves and their families. This exchange reflected a community effort to take responsible collective action under conditions altered by the earthquake and represented an effective capacity for transition to a more ordered, stable state of public health practice.

Transition is an interdisciplinary problem. Every academic discipline has studied the mechanisms of change, yet no single discipline has definitively explained this process. In the social sciences, theorists such as Thompson (1967), March (1988), Argyris and Schon, (1974, 1978), Argyris (1982, 1991), and Schon (1987) have addressed change in the context of individual and organizational learning, exploring its characteristics and limits. Others, such as Holland (1975, 1995), Axelrod (1984), Haas (1990), Nelson and Winter (1982) have examined change as an evolutionary process, identifying its organizational interactions over time. Still others, including Simon (1981), Perrow (1972, 1984), Shrivastava (1992), Goodman, Sproull, and Associates (1990), have examined change as a function of the dynamic relationship between human systems, technology, and the wider environment. Dryzek (1987) observed change as a co-evolving process among many actors and environmental conditions in a wider social ecology, while Luhmann (1984) sees change as an integral part of interactive communication among human actors.

Recent literature on the theory of complex systems offers fresh insight into the process of initiating change. Prigogine and Stengers (1984) initiated this discussion of the process of change as it occurs in different types of systems. In closed, linear systems that operate with causal rationality, change is largely undesirable. For example, mechanical or physical systems, such as steam generators or electrical circuitry, are designed to operate with maximum dependability, reliability, and minimum change. Nonetheless, the authors observed that even in linear systems, small variations occur which, cumulatively, affect the operation of the system and may produce large variations in performance outcomes. That is, small differences in initial conditions create variations among the subsets of the system that then perform at different rates. The cumulative build-up of these differences in subunit performance then destabilizes the operation of the entire system, shifting it into "chaos", a transition phase, as it seeks to develop a new systemic pattern of operation that accommodates more appropriately the differences observed among the subsets of the system. Such systems are termed complex, adaptive systems (Gell-Mann 1994, 18–19). They are nonlinear in explaining relationships among components, dynamic in their operations, and unpredictable in their outcomes. Bak and Chen (1991) extend this concept of "chaos" by noting the capacity for 'self-organization' even in inanimate systems.

How systems change, why systems change, and when they do not change are questions that spill over from the physical to the social sciences, as social scientists have increasingly acknowledged the dynamic, nonlinear aspects of public policies and programs operating in open social and economic systems (Lindblom 1992; Rivlin 1993). The dynamics of change in complex, adaptive systems underlie the difference between policy and practice in multiple disciplines.

Stuart Kauffman addresses the problem of change directly by acknowledging that constraints are placed upon possible alternatives for action in complex systems. Kauffman (1993, 51) notes the increase in reciprocal actions generated among components of a system as that system increases in size when stating:

> As systems of many parts increase both the number of those parts and the richness of interactions among the parts, it is typical that the number of conflicting design constraints among the parts increases rapidly. Those conflicting constraints imply that optimization can attain only ever poorer compromises.

In living systems, Kauffman sees the states of order and chaos existing on opposite ends of a continuum of performance. In either state, entrenched order or rampant chaos, Kauffman (1993, 233) observes that creative change is unlikely. But the center region of the continuum, the 'edge of chaos', supports the exchange of information, enabling informed choice by multiple participants. Informed choice, in turn, encourages adaptive behavior consonant with changing conditions in the environment. To Argyris (1980, 1982, 1993) and other organizational theorists, informed choice fosters individual and organizational learning, the conditions most conducive to producing change in behavior.

In important ways, both Kauffman and Argyris see change as a product of information search and exchange among the component parts of a system and between a system and its external environment. The exchange of information activates separate learning processes among the system's components, each generating different responses which, in turn, lead to different forms of adaptive behavior in responding to the same set of events. With powerful insight, Kauffman (1993, 234–235) recognizes that evolution is a product both of self-organization and natural selection. Individual responses or choices play a critical role in defining alternatives for action, which are then subject to competition for survival. Successful responses may cumulate in an orderly way to form substantively different interpretations of the same environmental conditions and result in substantially altered performance for the organization.

Viewed in Kauffman's (1993, 237) terms, social systems are "massively parallel-processing, nonlinear, dynamical systems" which crystallize new forms of order. That is, organisms co-evolve with one another in a dynamic environment, responding differently to the same set of conditions or events. As one organism responds spontaneously to shifting conditions, it creates an altered circumstance for those in close proximity, stimulating new responses in their interactions with one another and the environment.

Maintaining the desired balance between order and flexibility suggests a dynamic organizational structure, one that acknowledges a subset of "essential functions" that are kept intact for the safety of the system, while other functions may be relaxed or eliminated temporarily. A large system can thus tolerate damage or fluctuations in performance in some of its units and still adapt to altered conditions successfully. Change represents a transition to a new order within the entire system. Discovering the optimum form of that emerging order is an important function of the change process. In this study, I propose that an adaptive model of organizational learning and action based upon information exchange leads to systemic change, or transition, in behavior.

The limitations of decision making for organizations operating under uncertainty have spurred researchers to explore means of reformulating the concept of transition to include technical as well as organizational infrastructure to increase problem solving capacity (Comfort 1991a; 1991b; Bruzewicz and McKim 1995). Decision making under conditions of uncertainty requires a dynamic model that allows the integration of information from diverse sources in a continuously adapting format to reflect accurately the interaction of events, actors and conditions in the changing environment. Within this environment, organizational transition represents a significant increase in complexity from behavioral change in an individual. It involves engaging multiple organizations simultaneously in many types of action to address a complex problem (Simon 1981; Comfort 1991b).

Three concepts drawn from Niklas Luhmann's (1989) recent work offer insight into important characteristics of a model of organizational transition under uncertain conditions. First is Luhmann's (1989, 7) concept of "autopoiesis" or the instinctual drive for continuing creative acts of self-expression. Different from Darwin's concept of "survival of the fittest", Luhmann identifies a more refined and intense drive for creative expression that can be observed in individuals, but more importantly, can also be observed in social organizations. It is the energy that drives an organization to search for new information, to innovate, to try new approaches when existing patterns fail to achieve their intended objectives. Autopoiesis necessarily involves interaction with the environment, and compels the organization to respond to changing conditions in order to maintain its own vitality. Once a threshold level of original performance has been achieved by an organization, it develops an internal drive to continue that level of self-expression. This energy, or "autopoiesis", recognizable in vital organizations, drives continuing social interaction with its environment in a process of evolving complexity.

The second concept, clarified by Luhmann (1989) but recognized by other researchers (Parsons 1951; Almond and Verba 1962; Burt 1982) is the concept of functional differentiation. Functional differentiation serves as a means of reducing complexity (Luhmann 1989) to facilitate organizational performance. This concept refers to the analytical task of identifying differences as a means of conceptualizing unity. That is, a set of actors may distinguish itself as a 'system' by identifying its boundaries as distinct from the environment. This task involves identifying which interactions are designed to be performed by the organization to achieve certain goals in the larger environment. This specific set of interactions is then redefined as an organizational entity distinct both from the parent organization and the environment. It enables the new organization to operate within its selected environment with more clarity and purpose, thereby temporarily resolving the paradox of excessive environmental or organizational influence that inhibits the achievement of specific goals.

The newly formed organization, with an established drive for creative expression and clearly differentiated functions to accomplish its valued goals, nonetheless is able to do so only by generating sufficient "resonance" with its environment to elicit the consensus and cooperation needed to support purposive action. Luhmann's (1989) third concept, that of resonance, occurs when an organization is able to activate supportive response from the society on the basis of articulating shared meanings for a wider, more diverse audience and offering an appropriate format for common action.

All three concepts operate through the communication of information. The process of discovering shared meanings with receptive social groups stimulates an organization to integrate new information with previous experience and extend its capacity for mobilizing action to a larger social audience. This process, which generates its own heady drive for renewal and continued expression, leads to the design of institutions and programs to protect the newly discovered values and accelerate the fulfillment of goals shared with the community.

The capacity for organizational transition necessarily creates a second level of inter-organizational transition, as the initiating organization is able to marshal sufficient understanding, support, and resources from other relevant organizations in its environment to sustain its own capacity for effective action in the larger arena. This new, second level of inter-organizational transition now develops its own autopoiesis, or self-generating capacity for creative action in the wider environment. The process continues to evolve in iterations of increasing complexity, as new systems become capable of addressing larger problems, and consequently create even more complex means of intervening in the environment to coordinate common action among (previously) separate organizations and mobilize resources to solve broader problems. The response systems in practice may be temporary systems (Meltsner and Bellavita 1983), decreasing in importance as the urgency of the disaster fades and member organizations return to routine operations.

To observe this process in a practical environment, it is necessary to identify the interdependent relationships that link organizations in action to assist a community exposed to seismic risk. For example, a major earthquake severely disrupts the physical infrastructure of a community, causing buildings to collapse, roads and bridges to fail, electrical power, communications, water and sewer lines to break. These failures trigger serious social consequences, as people dependent upon this infrastructure lose their housing, transportation, and basic necessities of life: water, power, communications. Programs initiated to meet urgent social needs spend scarce resources and energy that are, then, not available for rebuilding the physical infrastructure in ways that would withstand a future earthquake. Determining how to meet the range of immediate needs generated by an earthquake and how to allocate scarce resources between present needs and future prevention for the community vulnerable to recurring risk becomes the set of policy problems that responsible managers of public, private and nonprofit organizations confront.

In practice, these problems are addressed through a set of "communicative acts" within and between organizations that produce the actual response to the disaster. Information, transmitted through communicative processes, activates organizational performance in both internal and external processes. The critical goal is to develop a consciously self-organizing community that uses its available resources in creative ways to balance present need against future threat. The process depends upon the communication of information in ways that engage the creative capacity of individuals, organizations, and governmental entities to make needed transitions between states of risk, crisis, and redesign.

These observations, drawn from practice, serve as the basis for formulating a model of inter-organizational transition that characterizes the formation of disaster response systems following earthquakes. The model, which builds on the theoretical concepts presented earlier, is restated briefly below for the context of a major earthquake.

1. Initial Conditions: A set of four conditions appears to be necessary, but not sufficient, to initiate inter-organizational transition. The four conditions include:
    a. Articulation of commonly understood meanings within and between organizations that are seeking to initiate change and their relevant audiences in the environment, creating a sufficient degree of shared understanding to support collective action toward a common goal — protection of life and property in the community
    b. Sufficient trust among leaders, organizations, and citizens to enable participants to accept direction toward achieving this shared goal
    c. Sufficient resonance between the organizations seeking change and their relevant audiences to elicit resources and support for action
    d. Sufficient capacity and resources to sustain collective action among participating organizations under varying conditions in order to achieve the shared goal
2. Earthquake! An actual event disrupts the normal operation of interdependent organizations in a community — households, businesses, public agencies, nonprofit organizations — compelling action in response
    a. Goal of performance (P): Goal of the emerging response system — protection of life and property for the whole community — becomes vividly clear to all members of the community
3. Information Search: Organizations affected by the earthquake search for new information regarding the impact of the event on the community and the possible means of response
    a. Number of actors (N): Organizations — public, private, and nonprofit — that are affected by the earthquake, engage in the search for information regarding the impact of the event both within their own organizations

and among the set of organizations that share the damage and risk of further consequences
   b. Source of funding (S): identification of the types of organizations that enter the response system by source of funding — public, private or nonprofit
4. Information Exchange: connection of the organizational components of the evolving response system through multi-way communication patterns, with 'feedback loops' between individuals, organizations, and subsystems, producing a response system which, in turn, links to the wider environment
   a. Number and density of interactions (K): actions in which several organizations share responsibility and resources to perform a common set of tasks within a specific time period; this measure provides an indicator of the degree of coordination achieved in the response system
   b. Number and type of transactions (T): tasks performed by organizations in response to the earthquake, both separately and in coordination with other organizations
   c. Duration (D): Time period of an organization's involvement in disaster operations; for this study, data are collected for twenty-one days past the event
5. Intra- and Inter-organizational Learning: Through the process of information exchange, knowledge about the effects of the earthquake upon the community circulates through the system and accumulates as an evolving knowledge base for the system. Intra-organizational learning facilitates action at the micro level of individuals, households and organizations, and contributes to inter-organizational learning, which facilitates action at the macro level of sub-national, national, and international operations. The cumulative knowledge from these transactions and interactions serves as the basis for continuing mutual adjustment and adaptive behavior among the participating organizations and jurisdictions
6. Adaptive Performance: The transition from micro level actions to macro level operations occurs when the component organizations of the system are able to coordinate their actions sufficiently to form a recognizable level of integrated performance. The system then moves to a new and/or improved level of coordinated performance in interaction with the dynamic environment. The measure of adaptive performance is the extent of action taken by the community to mitigate future seismic risk
7. Obstacles: Inadequate or interrupted information processes among the components of the newly defined system impair its capacity for inter-organizational problem solving and coordinated action

The model characterizes the self-organizing drive shown by community organizations, focused on a common goal, to maintain their autonomy and capacity for innovative adaptation through collaborative interaction with the environment. Doing so means acknowledging perceived opportunities and/or threats, and responding appropriately to actual events in ways that minimize risk and enhance creative expression. Operating within constraints of limited time, attention and resources, organizations seek to reduce complexity in the environment through functional differentiation of tasks among participating organizations and a re-integration of sets of related tasks under basic functions of disaster response. The process of re-integration of tasks performed by different organizations into a coherent set of disaster operations creates a more stable, effective, macro response to demands from the disaster environment.

The set of organizations confronting the shared consequences of the earthquake defines a new inter-organizational system and effectively distinguishes a fresh set of broadly shared meanings from the boundaries of the previous organizational units, now considered subsystems of the larger system. Supported by a technical infrastructure for information search and exchange, the new inter-organizational system develops an autopoiesis of its own, with a distinctive energy, shared goals, and common understanding that drive its self-generating creative capacity. The component organizations have successfully made the transition to a new state of cooperative performance, and function both at intraorganizational and inter-organizational levels.

Searching for a context in which to observe the dynamics operating at the 'edge of chaos' in actual practice, I have employed an analytical method to identify the level of coordination within an evolving system, using nonlinear logistic regression. This method used data from a series of field studies to document rapidly evolving organizational response systems following major earthquakes. The balance between structure and flexibility in these response systems is crucial to their performance, and this balance is affected by many variables both internal and external to the system.

Accepting Kauffman's proposition that the most creative response to uncertain conditions occurs at the 'edge of chaos', where the system is operating with sufficient structure to hold and exchange information but sufficient flexibility to adapt to changing conditions, I identify instances from actual practice in which the response system moved toward order or reverted to chaos. In doing so, I characterize briefly the initial conditions in which the earthquakes occurred for each response system through 'thick description' (Geertz 1973, 6). My purpose is not to make judgments about the efficacy or efficiency of any system. Rather, I seek to identify the critical factors — and the interactions among them — that move a response system toward creative, timely, performance under urgent, uncertain conditions, or, equally important, toward chaos. In each case, the response personnel involved put their maximum effort into protecting their communities, and are acknowledged for their effort and courage under extremely difficult conditions. However, if we can identify the factors that facilitate response, we can ease the heavy physical, psychological, and cognitive burdens placed upon response personnel and citizens of communities at risk, and enable these communities both to reduce seismic risk to lives and property and to respond more effectively when earthquakes do occur.

Returning to the model of transition presented above, I searched for similarities and differences among the cases in reference to the evolution of response systems. In all eleven cases, three factors are essentially the same: the event triggering action is a significant earthquake; the goal is to protect the lives and property of the citizens of the community; and the actors are public, private, and nonprofit organizations at local, provincial/state, national, and international levels of jurisdiction. The primary differences appear to lie in the initial conditions of technical design and organizational infrastructure to support collective action in the suddenly altered circumstances of a major earthquake. These conditions affect the availability of means for multiway communication among the participating actors and groups, and access to these means, in turn, affects the direction, content, frequency and flow of information among the participants. It is the exchange of information through multiway communications processes that enables a set of organizations operating in a prior state of relative autonomy to make the transition to an effective response system, operating in a coordinated manner to meet the needs of the community generated by a specific urgent event. The two factors of technical design and organizational infrastructure are primary components in creating the balance between structure and flexibility that is attained in the evolving response systems.

While these variables reflect the definition of the 'edge of chaos' suggested by Kauffman, I found a third factor, cultural openness to new information, important in influencing both organizational flexibility and technical structure in practice. Adding this variable creates a preliminary assessment based on the interaction between technical, organizational, and cultural conditions (Linstone 1984) existing prior to each event. In each case, I assumed that the initial conditions, interacting with the occurrence of the earthquake, generated the particular form, direction, and rate of development for the evolving response system.

For each of the eleven cases included in this study, I reviewed the set of initial conditions that shaped the evolution of the disaster response system, using documentary sources, newspaper reports, and responses from semi-structured interviews. Based upon these data, I assigned preliminary rankings to each response system on the three dimensions of technical structure, organizational flexibility, and cultural openness. Since the data available to support these rankings were largely qualitative, I used ordinal categories of "high, medium, and low" to distinguish gross differences in the evolving response systems. Ordinal scales are imprecise, and some categories had noticeable variation within each cell. The rankings were based on the dominant characteristics of the response system on each dimension. With that caveat, Table 2 presents the indicators for the classification scheme, and Table 3 presents the preliminary rankings of the eleven disaster response systems.

Table 2. Assessment Indicators for Disaster Response Systems

Dimensions: I. *Technical Structure:*

Indicators:
  High:
  Existence of informed assessment of seismic risk
  Existence of building codes calibrated to seismic risk
  Existence of requirements for soils analysis and engineering surveys prior to construction of new buildings, transportation
  Existence of alternative communications capabilities in event of earthquake
  Existence of special Emergency Operations Centers and equipment for disaster response
  Identification of major facilities in community that are vulnerable to seismic risk

  Medium:
  Historic record of seismic events and their effects on technical structures
  Some seismic design features included in construction, such as steel rebar reinforcements, ceiling to wall connectors
  Existence of equipment and operations centers for emergency services: police, fire, emergency medical corps
  Existence of basic communications capacity among local organizations during normal operations
  Identification of possible sources of risk and assistance in seismic events

  Low:
  Little to no assessment of risk to community from possible seismic event
  Little to no change in building construction due to previous seismic events
  Little to no investment in emergency preparedness at the local level
  Little to no anticipation of future seismic events

II. *Organizational Flexibility*

  High:
  Existence of a national law establishing legal authority for emergency response
  Existence of a disaster response plan that integrates capacity from several jurisdictional levels to meet the needs of an earthquake-damaged community
  Existence of an interdisciplinary, inter-organizational, interjurisdictional knowledge base for seismic risk, response, and possible consequences for the community
  Existence of multiway patterns of information exchange within and between organizations and jurisdictions that participate in emergency response
  Existence of trained, professional managers with the experience and authority to adapt existing administrative plans to the demands of the event
  Existence of trained reserve personnel available on recall in disaster events

  Medium:
  Existence of a national law that establishes authority for disaster response
  Existence of general training programs for public agencies in disaster response
  Existence of disciplinary knowledge bases within specific organizations regarding possible consequences and criteria for action in disaster
  Presence of professional staff able to take innovative action under urgent conditions
  Capacity to establish communications patterns with 'feedback loops' with new organizations

Low:

Little or no national authority for disaster response

Little or no training or preparedness for disaster response

Little or no common knowledge regarding the risk, possible consequences, or criteria for action for seismic events

Little or no professional staff with authority to act in emergency conditions

Little or no communication within or between organizations and jurisdictions regarding seismic risk

III. *Cultural Values*

High

Shared values regarding humanitarian assistance to those in need

Commitment to goal of protecting life and property for all members of community

Ready acceptance of new information from valid sources

Openness to new methods of working and acting with other organizations and jurisdictions in order to achieve a shared goal

Information exchange between organizations and between jurisdictions

Willingness to review actions taken and to correct mistakes discovered between organizational and/or jurisdictional working groups

Continual search for relevant, accurate, timely, information to protect community

Medium:

Shared values regarding loyalty to organization, profession

Information exchange within organizations and within jurisdictions

Willingness to review actions taken and to correct mistakes discovered within organizational and/or jurisdictional working groups

Willingness to consider alternative modes of problem solving and action within organization or working groups

Reliance upon historical information to inform decisions regarding protection of community

Low:

Little to no value placed upon public and/or community interest

Little to no shared knowledge regarding risk or resources among organizations, jurisdictions, or citizens

Little to no information search or exchange among organizations and jurisdictions

Little to no trust or experience in solving common problems among organizations, jurisdictions, or citizens

Little to no willingness to review actions taken or to correct mistakes for future

---

Since all eleven cases were selected for their destructive impact upon communities, it is apparent that seismic events at similar levels of severity have very different effects upon the affected communities. For example, the Armenia Earthquake (M = 6.9) resulted in the greatest loss of life, with over 25,000 persons[2] reported dead and an estimated $16 billion in damage. The Northridge Earthquake (M = 6.7) claimed 59 lives, but resulted in an estimated $25 billion in damage. The Hanshin Earthquake (M = 7.2) resulted in the greatest economic loss, over $200 billion, in addition to a very high loss of life, over 6,300 dead, whereas the Costa Rica Earthquake (M = 7.4) took 55 lives and caused $965

Table 3. Preliminary Assessment of Eleven Disaster Response Systems following Major Earthquakes, 1985-1995, on Technical, Organizational, and Cultural Dimensions

Response System	Technical Structure	Organizational Flexibility	Cultural Openness
San Salvador, 1986	Low	Low	Low
Armenia, 1988	Low	Low	Low
Ecuador, 1987	Low	Low	Medium
Mexico City, 1985	Low	Medium	Low
Costa Rica, 1991	Low	Medium	Medium
Erzincan, 1992	Low/Medium	Medium	Low
Maharashtra, 1993	Low/Medium	High	Medium
Whittier, 1987	Medium	Medium	Medium
Loma Prieta, 1989	Medium	High	Medium
Hanshin, 1995	High	Low	Low
Northridge, 1994	High	High	High

million in damage. The San Salvador Earthquake (M = 5.4) registered the lowest magnitude on the Richter scale, but claimed more than 1,200 lives and caused more than $1 billion in property losses. While obviously associated, there is no clear linear relationship between severity of seismic shock and degree of destruction for the affected communities. The location of the epicenter, the quality of building construction, the presence of an informed public, the existence of an information infrastructure all contribute to the mix of factors that define the scale, scope, number of deaths, and amount of damage resulting from a given earthquake. Different capacities to reduce seismic risk, and respond to earthquakes when they do occur, are demonstrated by the eleven response systems included in this study. Determining what factors contribute to explaining these differences is a primary objective of this study.

The classification schema outlined above produces a three-dimensional model of transition to self-organization, that is, the capacity to reallocate resources and action in response to the demands created by the earthquake. This model is more complex than Kauffman's metaphor of the 'edge of chaos', because it seeks to capture the interaction among the dimensions of technical structure, organizational flexibility, and cultural openness that leads to successful transition for inter-organizational disaster response systems. It interprets the 'edge of chaos' as an arena for informed choice in an interdependent setting, in which multiple actors exchange information, make choices for action, and, observing the interaction between their choices, choices of others, and conditions in the environment, adapt their performance accordingly in a continual process of seeking to achieve a shared goal.

Based on these three-dimensional rankings, the 11 response systems are grouped into four sub-sets reflecting the dominant characteristics of their state of transition toward self-organization during the first three weeks following each

Table 4. Types of Emergency Response Systems

Nonadaptive Systems	Emergent Adaptive Systems	Operative Adaptive Systems	Auto-Adaptive Systems
San Salvador	Mexico City	Whitter Narrows	Northridge
Ecuador	Costa Rica	Loma Prieta	Hanshin, Japan
Armenia	Erzincan	Maharashtra	

earthquake. Table 4 presents a summary listing of the four subsets for the set of 11 response systems.

The four subsets, referring back to their rankings given in Table 3, represent the dominant patterns found in the actual cases, rather than an exact match with each indicator. These measures capture the state of each system in transition, for a particular period of three weeks immediately following an earthquake event. Each response system is unique to that event, and the same community may generate a different response system following a different earthquake. The four subsets, however, provide useful types for illustrating the process of transition, and the characteristics that both facilitate and hinder the emergence of self-organizing processes in an earthquake-stricken community. Each type will be characterized briefly below.

Nonadaptive Systems

Nonadaptive systems are those that are low on technical structure, low on organizational flexibility, and, for the most part, also low on cultural openness. Three response systems fall into this category, with some variation: San Salvador, 1986; Ecuador, 1987; and Armenia, 1988. Each of these communities was operating under significant political and economic stress prior to the earthquake, a condition that obviously affected the evolution of the response system immediately following the earthquake event.

The San Salvador response system evolved under the constraints of a bitterly-fought civil war, then in its sixth year. The Ecuadorian response system evolved under the double constraint of a tense relationship between the President and military chiefs prior to the earthquake, and the traumatic loss of 50% of the nation's income due to the rupture of the Andean oil pipeline, a consequence of the earthquake. The severe economic loss, in turn, triggered national strikes organized by the trade unions to protest administrative measures taken by the President to cope with the earthquake damage, and threatened the stability of the political system. The Armenian response system evolved under the tense situation prevailing between the republic and the Union in the last years of the USSR. These adverse contexts affected seriously the four initial conditions identified earlier as necessary, but not sufficient for the evolution of an effective response system. These conditions will be reviewed briefly here, to document the classification of the three response systems in this subset.

Although seismic risk was a known factor in each of the three communities, the meaning of this risk had not been integrated well into public practice and awareness in any of the three cases. In San Salvador, the epicenter of the M = 5.4 earthquake occurred on a known earthquake fault that runs directly beneath the city (Harlow, Rymer, and White 1986, 2). Although the record of seismic events in San Salvador goes back to the seventeenth century (Harlow et al. 1986), this historical experience had not actively informed public practice in terms of developing construction codes or programs for educating the citizenry regarding this risk.

For example, the Ruben Dario building which collapsed in the center of the city, trapping over 300 persons inside, had been damaged in a previous earthquake in 1969 (Comfort 1989a). Although the owner was informed of the damage and the need to retrofit the building, no official action was taken to enforce this recommendation. The collapsed building and the lives it claimed became a negative symbol for the next condition, trust between the leaders and the citizens. Lack of trust between public leaders and citizens inhibited the resonance between the response system and the environment, resulting in an inadequate supply of resources to mitigate seismic risk prior to the earthquake and an insufficient infrastructure — both technical and organizational — to respond effectively to the urgent social needs after the earthquake occurred. Similar conditions existed in Ecuador and Armenia, but with an emerging resonance between citizens and earthquake victims, if not the existing government. Table 5 presents the rankings on these four initial conditions for each of the three response systems in this subset.

Given these initial conditions, the response systems that did evolve following each of these earthquakes failed to achieve the internal characteristics of self-organization and sustainable performance that enabled a continuing adaptation to seismic risk. The response systems remained largely the products of external direction and control. The Ecuadorian and Armenian systems did show evidence of a medium degree of resonance between the evolving disaster response systems and their wider societies in terms of a wide range of organizations and citizens that contributed to disaster relief for the injured and displaced. But the resonance

Table 5. Characteristics of Nonadaptive Systems

	San Salvador	Ecuador	Armenia
1. Articulation of common meanings	Low	Low	Low
2. Trust between leaders and citizens	Low	Low	Low
3. Resonance between response system & environment	Low	Medium	Medium
4. Sufficient resources to sustain collective action	Low	Low	Low

achieved in each case was temporary, insufficient to sustain a continuing system for mitigating, or adapting to, the known seismic risk in their communities. Although the rubble has long since been cleared in San Salvador, Leninakan, and Napo, Pichincha, and Imbabura Provinces in Ecuador, relatively little adaptation to the long-term problem of seismic risk in these communities has occurred as a result of the respective earthquakes. The characteristics of each case and the comparisons among them will be discussed more fully in Chapter Five.

Emergent Adaptive Systems

Emergent adaptive systems are those characterized by low technical structure, medium organizational flexibility, and emerging openness to new cultural meanings of seismic risk in their respective communities. Using the same criteria for assessing initial conditions, three disaster response systems fall within the next category of Emergent Adaptive Systems. This subset is shown in Table 6.

In an emergent adaptive system, the earthquake event serves as the stimulus for the community to begin adapting its practices to fit an environment of seismic risk. Previous experience with earthquakes may or may not have initiated mitigating activities, but the technical structure for the community was clearly inadequate for the degree of its exposure to seismic risk. In each of the three cases – Mexico City, Costa Rica and Erzincan, Turkey – the responsible governments demonstrated a remarkable degree of organizational flexibility in responding to the damage once the earthquake occurred, and revealed a beginning degree of openness to new information about seismic practice and methods of risk reduction.

In Mexico City, for example, the functions of Government were temporarily disrupted when the communications towers of TELMEX (Teléfonos de Mexico) were disabled, and telephone communication was disrupted for several days. Government functions were further disrupted as the buildings of major ministries, e.g. Navy, Health, Communications and Transportation, Foreign Relations and others, were severely damaged, destroying records, offices, and work sched-

Table 6. Characteristics of Emergent Adaptive Systems

	Mexico City	Costa Rica	Erzincan
1. Articulation of common meanings	Medium	Medium	Low/Medium
2. Trust between leaders and citizens	Low	Low/Medium	Low/Medium
3. Resonance between response system & environment	Low/Medium	Low/Medium	Low/Medium
4. Sufficient resources to sustain collective action	Low/Medium	Low	Low/Medium

ules for personnel who had no physical base of operations.[3] This situation was exacerbated by the lack of a functioning national disaster plan. In July, 1985, two months prior to the September 19, 1985 earthquake, the Mexican Government had closed the offices of its federal disaster management agency in an austerity measure to cope with its worsening economic situation. At the time of the earthquake, there was no formal agency responsible for coordinating disaster response. The Mexican Government demonstrated remarkable flexibility in this chaotic situation, however, as then-President Miguel de la Madrid called upon twelve major departments of his Government with capacity for disaster response to form an Intersectoral Commission for Aid to the Metropolitan Zone.[4] This Commission would plan and direct the national response to this catastrophic event. The Mexican Government was able to reallocate resources within its own system of operation, just as community residents in the various colonias of Mexico City reorganized their time, energy and efforts to respond to needs generated by the earthquake in spontaneous ways.

In Costa Rica, located at the intersection of three tectonic plates, seismic risk is a known factor for most of the population. The capacity to translate this knowledge into informed practice, however, encountered other constraints. While the tiny nation had developed an extensive emergency plan on paper, the implementation of the plan focused primarily on the capital city of San Jose, with the largest concentration of population. When the earthquake occurred outside the San Jose area, the technical capacity for communication, transportation, and damage assessment was not in place, and the emergency plan faltered.[5] Then-President Rafael Calderon countered this situation by placing two of his Cabinet members, the Minister of Agricultura y Ganadería and the Minister of Vivienda y Asentamientos Humanos, in charge of disaster operations in the province of Limon.[6] He effectively reallocated the resources of major governmental agencies to meet the needs for rebuilding the damaged roads, bridges, and the Atlantic port of Limon, restoring vital economic production and trade.

Erzincan, Turkey, a city with an historic record of earthquakes documented over ten centuries, nonetheless had inadequate technical infrastructure to resist severe damage to buildings from the March 13, 1992 earthquake. Advanced communications between Erzincan and Ankara, the nation's capital, enabled the national government to receive virtually immediate reports from Erzincan regarding the size, location and magnitude of the earthquake. Entering these data into a computerized model, the national earthquake research center calculated an estimate of damage in Erzincan within 30 minutes of the actual event.[7] The technical structure for the relatively new building stock of Erzincan, rebuilt on the ruins of the 1939 earthquake, proved inadequate again in 1992. The national administration responded with a rapid mobilization of national resources, but without local capacity to receive and implement the resources in Erzincan, there was no basis for immediate response operations in the city.[8] The national level plan was severely limited by limitations at the local level, both technical and organizational. The requisite resources had not been invested in local level planning, construction and organization to reduce seismic risk.

In each of the three cases, the earthquake event served to initiate longer range planning and calculation of costs of inadequate infrastructure. Each nation was receptive to new methods of estimating and reducing seismic risk. Planning centers and organizations focusing on risk reduction, some already in existence prior to the actual earthquakes reported in this study, used the events to justify the relevance and necessity of their work. The disaster response systems, while not immediately self-organizing, showed strong potential for becoming so and attracting the necessary support from the wider society to sustain their efforts.

Operative Adaptive Systems

Operative adaptive systems are those in which the technical structure, organizational flexibility and cultural openness to new methods of perceiving and responding to risk are approximately medium. At this stage of development, response systems evolve that enable communities to mobilize a reasonably coherent response to an earthquake. In these communities, the capacity for response to a sudden, urgent event exists prior to the earthquake, and existing administrative systems are able to reallocate resources and mobilize response to meet the immediate needs of the event. The systems may not be optimal in all aspects of their operations, but they are able to function in a coherent manner. The subset of response systems that are classified in this category are shown in Table 7.

The two California earthquakes of Whittier Narrows, 1987 in Southern California and Loma Prieta, 1989 in Northern California fall into this category. Both earthquakes occurred in areas of known seismic risk. California has an active Office of Emergency Services that trains its personnel well, and regularly organizes simulated disaster operations in risk-prone areas. While these training exercises facilitate interaction between jurisdictional levels, they focus primarily on local level response for cities and counties.

In the actual events, both disaster response systems operated primarily at the local level, providing a test of local emergency planning and preparedness. Both

Table 7. Characteristics of Operative Adaptive Systems

	Whittier	Loma Prieta	Maharashtra
1. Articulation of common meanings	Medium	Medium	Medium
2. Trust between leaders and citizens	Medium	Medium	Medium
3. Resonance between response system & environment	Medium	Medium	Medium
4. Sufficient resources to sustain collective action	Medium	Medium	Low

response systems proved capable of coping with demands created by the respective earthquakes, but each revealed significant weaknesses in the communications and coordination processes.[9] These weaknesses indicate points at which the systems would likely break down in a larger, more destructive event. Weaknesses in communication and coordination were particularly apparent in the response system to the Loma Prieta Earthquake, a more complex event with centers of devastation and disaster operations spanning three counties, each, in turn, involving separate municipal agencies. The linkages between the three centers of emergency operations were not strong, resulting in differential timing of response operations and delayed treatment of certain groups of residents who suffered damage.

Maharashtra, India represents an unusual version of an operative, adaptive system. While the technical infrastructure of the villages of the Latur and Osmanabad Districts was very primitive indeed, the strong communications infrastructure provided by the Indian national satellite system enabled a very rapid response to the extensive damage caused by the earthquake.[10] Further, the professional Indian Administrative Service provided a common set of training and skills for district administrators that constituted a common knowledge base for disaster operations.[11] It also supported an organizational flexibility which allowed local, district, state and national administrators to use the technical communications structure to full advantage in conducting a coherent strategy of response. Consequently, despite the severe poverty of the districts in which the earthquake occurred, the responsible organizations were able to mobilize remarkable quantities of personnel and material support in timely response to needs of the damaged communities. They were aided in this process by a generous response from nongovernmental organizations which volunteered time, effort, money and material goods to assist in the recovery of the damaged area and its people. Combining the selected technical structure of communications with professional organizational flexibility and a humane response to the needs of the displaced villagers, the Indian response system operated with far greater effectiveness than expected, given its limited economic base.

Auto-Adaptive or Self-Organizing Systems

Systems operating at the 'edge of chaos' have the potential for creative response to meet suddenly altered conditions of operations in effective ways. Yet, the margin for choice is narrow. Systems that do not move toward creative new actions will slide back toward chaos as their old patterns of performance fracture under the stress. Systems that move toward creative new actions are termed auto-adaptive or self-organizing. Such systems are high on technical structure, high on organizational flexibility and high on cultural openness to new information and new methods of action. Systems unable to adapt may be high on technical structure, but lack the flexibility or openness to new information that would allow innovative performance in dynamic conditions. Two response systems in this set showed

Table 8. Characteristics of Auto-Adaptive Systems

	Hanshin	Northridge
1. Articulation of common meanings	Low	Medium/High
2. Trust between leaders and citizens	Medium	High
3. Resonance between response system & environment	Medium	High
4. Sufficient resources to sustain collective action	High	High

initial potential for self-organization, but revealed very different results in practice. These two systems are shown in Table 8.

The Northridge response system, evolving after the third California earthquake in seven years, tested the existing policy and practice of California seismic risk reduction and response. After the preceding Whittier (1987) and Loma Prieta (1989) earthquakes, the California Office of Emergency Services and the Federal Emergency Management Agency (FEMA) had made major adjustments in their risk reduction and response practices to improve coordination in disaster response. FEMA, reviewing its operations in a series of large and very costly disasters, had enacted a new Federal Emergency Response Plan in April, 1992.[12] In January, 1994, FEMA also had a new director, James Lee Witt, who assumed management of the previously troubled agency in 1993.

Mr. Witt had immediately set about improving the morale and performance of the agency, and focused on communication and coordination both within the agency and between the agency and its state and local counterparts.[13] For example, FEMA had funded the training and equipment for specialized urban search and rescue teams for major disasters to operate at local levels. There were five such teams located in California. The City and the County of Los Angeles each had developed specialized response teams with 56 personnel, trained in urban search and rescue procedures, engineering, emergency medicine, and communications/coordination.

Further, in January, 1994, California emergency personnel had experienced the extraordinary situation of working together in fourteen federally declared disasters over the previous thirty months.[14] Although all planned changes were not fully in place, both emergency personnel and residents of the Los Angeles community had current knowledge of seismic risk and procedures for mitigating same. Sobered by weaknesses in performance in prior earthquakes, the emergency response community as well as citizens in the region were aware of the vulnerability of their community to seismic risk and receptive to demands for urgent action. On nearly all measures of initial conditions, the Northridge response system ranked high, with the exception of a sizeable number of immigrant groups who were unfamiliar with the California procedures. Many — especially those from Latin America and Armenia — were all too familiar with seismic risk and frightened by previous experiences in earthquakes in their home countries.

Initial conditions for the Hanshin response system were quite different. The Hanshin metropolitan area ranked high on technical structure, but surprisingly low on both organizational flexibility and cultural openness to new methods of action under urgent conditions. This situation resulted in a delayed response to the immediate demands of the earthquake, allowing the dynamics of the event to evolve in destructive ways (Comfort 1996). Once a clear understanding of the damage and consequences of the earthquake was established, the strong technical structure facilitated the evolution of an effective response in the later period of recovery and reconstruction. This response, in turn, elicited needed resources from the wider society for recovery in the Hanshin region. The delayed response, however, likely contributed to the very high costs in both lives and property following the Hanshin Earthquake.

Conclusion

Reviewing the four sub-sets of response systems presented briefly above against the preliminary model of transition, I conclude that the response systems in each of the four sub-sets made a temporary transition to meet the immediate demands of the earthquake in their environments. The more critical issue is whether the transition represented lasting change in the patterns of performance of the organizations participating in response that could reduce the risk to their respective communities in subsequent earthquakes.

In the first subset of response systems, those termed nonadaptive, the experience of the earthquakes left little lasting effect upon the technical and organizational infrastructure for disaster response. In each case — San Salvador, Ecuador, and Armenia — groups of scientists and researchers have continued their inquiry into the phenomena surrounding seismic risk, and have increased cultural openness to new approaches to seismic policy. This small but significant change may prove the basis for later change in the technical and organizational dimensions.

In the subset of emergent adaptive systems — Mexico City, Costa Rica, Erzincan, Turkey — the degree of transition to lasting change is problematical. Once the recovery measures are fully in place, the event tends to be forgotten. Researchers continue their work, but policy-makers have other issues, termed more urgent, to address. The continued allocation of resources and time needed to maintain performance at higher technical, organizational and cultural levels is uncertain at best.

The third subset of operative adaptive systems — Whittier Narrows, Loma Prieta and Maharashtra, India — show a very interesting facet of the process of transition. The three California response systems included in this study offer an unusual opportunity to examine the effect of experience upon subsequent response systems over the brief period of seven years. Lessons learned from the Whittier experience clearly influenced operations during the Loma Prieta response, and lessons from both response systems influenced the more successful Northridge response operations. While specific communities involved in response to the Whittier Narrows Earthquake may have reverted to previous practices, some, such as the City and the County of Los Angeles, have significantly improved their response capacity, as shown in the Northridge operations. The transition appears permanent not so much at the specific local level as at the more generalized level of state (California) and federal (US) policy response.

The Maharashtra response system also has had a similar effect. While the event highlighted major gaps in scientific knowledge of the seismic region of the Deccan Plateau necessary for effective seismic policy in India, it also illustrated the importance of satellite and networked communications technology in disaster response in ways far more powerful than India's otherwise low technical structure would warrant. The Maharashtra case presents an extraordinary illustration of the critical function of communications in facilitating coordination across organizational and jurisdictional lines necessary for transition to a new state of sustainable performance. The critical issue for India is limited resources within country, which constrains action to reduce risk and may engender unacceptably high costs if sought from external sources.

The fourth subset of response systems — Northridge and Hanshin — hold the greatest potential for permanent transition to a balanced, creative 'edge of chaos' state in response to seismic risk. The difficulty with this state is its sustainability. It requires a continual exchange of information and resources with its immediate environment to maintain its currency and credibility. This can be done, but likely only with a fully-developed, advanced information infrastructure that facilitates ease of access, storage and analysis of large amounts of information. This information infrastructure may be used for other community purposes than seismic policy, but it appears essential to facilitate the information exchange required by this complex, interdisciplinary, interjurisdictional policy problem.

Conversely, the biggest obstacle to achieving a sustainable, creative capacity for managing seismic risk is likely to be resistance to making the substantial investment required to build and maintain a sociotechnical infrastructure capable of supporting inter-organizational and interjurisdictional learning in regard to the long-term policy problem of seismic risk. Such an infrastructure would require a continuous investment in organizational training and the maintenance of organizational knowledge bases, as well as investment in the hardware and software needed to operate the system.

The initial conditions reviewed for these four subsets of response systems demonstrate the striking role of information exchange in facilitating appropriate adaptation to the suddenly altered conditions caused by an earthquake. To the extent that the initial conditions support timely, accurate information exchange, the responsible organizations are more successful in achieving a coherent, sustainable response system in each of the eleven cases examined. Such a response system consists of parallel technical and organizational practices designed to enhance community capacity for solving problems related to seismic risk, with sufficient cultural openness to accept new techniques and methods.

NOTES

1. Interview, Agostin Barbabosa Kubli, Centro Interamericano de Estudios de Seguridad Social, Ministry of Health, Mexico City, October 9, 1985.
2. Physicians who worked in emergency medical response at Leninakan estimated close to 80,000 dead; other reports cite 45,000 dead. The actual figure may never be known.
3. *Excelsior*, Mexico City, 23 September, 1985, p. 1A.
4. *Excelsior*, Mexico City, 23 September, 1985, p. 23A.
5. *La Nación*, San Jose, Costa Rica, April 25, 1991: pp. 4A,6A,8A,11A.
6. Press conference conducted by Humberto Trejos, M.D., President, Comisión Nacional de Emergencia, April 25, 1991, 6:00 p.m. San Jose, Costa Rica.
7. Oktay Ergunay, Director, Earthquake Engineering Research Center. Interview, Ankara, Turkey, July 8, 1992.
8. Ermin Sahin, Minister of Interior, Interview, Ankara, Turkey, July 9, 1992.
9. "Socioeconomic Impacts and Emergency Response". Loma Prieta Earthquake Reconnaissance Report. *Earthquake Spectra*, Supplement to Vol. 6 (May 1990): 393–451.
10. Praveensingh Pardeshi, District Collector, Latur. Interview, Latur, India, December 22, 1993.
11. Dineshkumar Jain, District Collector, Solapur, Interview, Solapur, India, December 24, 1993.
12. *The Federal Response Plan*, for Public Law 93-288, as amended. Washington, DC: U.S. Government Printing Office, 1994-17-748/ 80726.
13. James Lee Witt, Director, Federal Emergency Management Agency,Interview, Pasadena, CA, January 31, 1994.
14. Testimony, Richard Andrews, Director, California Office of Emergency Services, Hearings held by the California Seismic Safety Commission, Van Nuys, CA, February 11, 1994.

Part II
Shared Risk in Practice:
The Evolution of Response Systems

CHAPTER FIVE

NONADAPTIVE SYSTEMS: SAN SALVADOR, ECUADOR AND ARMENIA

The Emergence of Disaster Response Systems

Disaster response systems evolved following each earthquake: San Salvador (1986), Ecuador (1987), and Armenia (1988). But while the response systems met immediate needs of relief and restoration in their respective communities, they failed to address the larger goal of seismic risk reduction. After the earthquakes, community practices were not noticeably changed in reference to the problem of recurring seismic risk. These response systems, termed "nonadaptive", left little lasting effect upon the capacity of their respective communities to mitigate or respond to seismic risk.

Nonadaptive systems, as described briefly in Chapter Four, are low on technical structure, low on flexibility, and, for the most part, low on cultural openness to new information and methods of response. Such systems have particular difficulty in accessing, storing, and communicating information to support interorganizational decision processes under the urgent, stressful conditions of disaster operations. This capacity is central to developing processes of information sharing and coordination of action that characterize effective disaster response. Without viable communication and information processes that enable the multiple components of a community response system to exchange information readily, the organizations are unable to comprehend the common threat to their environment. Without an increased awareness of a shared goal of risk reduction for areas of recurring seismic risk, the organizations participating in disaster operations tend to revert back to their previous states of performance prior to the earthquakes.

In this chapter, documentary data reporting actions taken by the three disaster response systems will be used to illustrate the extent to which the evolving systems made transitions through the successive stages of disaster response. All systems showed some evidence of performance in each of the four stages of information search, information exchange, organizational learning and adaptive performance. Yet, the initial information search processes used, and the techniques and infrastructure available to support that search, shaped the content and frequency of the exchange of information among the participating organizations. In turn, the content and frequency of information exchange affected the degree of organizational learning, which influenced the extent and timing of adaptive action mobilized in response to the disaster. These functions can be seen most clearly in brief profiles of the three cases.

San Salvador, October 10, 1986 (M = 5.4)

On October 10, 1986, a sharp earthquake struck the city of San Salvador at 11:49 a.m. The earthquake registered 5.4 on the Richter scale, a moderate shock but with unusually high vertical accelerations. The seismographic data indicate that the earthquake was caused by a vertical rupture of approximately six kilometers in length along a fault located directly beneath the city of San Salvador, ranging in depth from 3 to 11 kilometers (Harlow, Rymer, and White 1986, 2). The earthquake directly affected a 20-block area in the center of the city, damaging severely 33 Salvadorean government buildings, the United States Embassy, four of the six major hospitals in the city and a five-story downtown building with shops, offices and a cafeteria (United States Agency for International Development/El Salvador 1986).

Occurring just before noon on a busy Friday, the earthquake caused major destruction in an already fragile economic and social environment ravaged by civil war. The downtown area was filled with people shopping, working, meeting with friends and business associates, and otherwise engaged in the routine activities of an ordinary workday. In seconds, the earthquake caused a major urban disaster, leaving approximately 1,200 persons dead, 10,000 persons injured, more than 125,000 people homeless, and another 175,000 people with severe damage to their homes.[1] The earthquake affected more than one-fifth of the city's population, swollen to over 1 million by the vicissitudes of the civil war.

More than any other condition, the nation's involvement in the sixth year of a bitterly fought civil war limited its capacity for disaster response. With its resources, energy, and attention drained by the war, the national government had made virtually no investment in the reduction of known seismic risk. This fact colored the initial conditions in which the disaster response evolved, affecting seriously the information processes essential for self-organization and collective response. These limits on the sequence of information processes vital to effective disaster response are detailed briefly below.

Information Search

In October, 1986, communication within the city of San Salvador was conducted primarily by telephone. In the first hours immediately following the earthquake, the telephone lines were disabled, disrupting communication among government agencies responsible for protection of life and property, and between government and citizenry. The San Salvador *Cuerpos de Bomberos*, with legal responsibility for first response in emergencies, did not have radios. With headquarters located some distance from the downtown center, Fire personnel did not learn of the damage to the city center, or know about the collapse of the Ruben Dario Building with 300 persons trapped inside, until hours after the event. Unaware of the scale of the disaster, Salvadorean Fire personnel arrived at the Ruben Dario Building only after a representative of the National Government drove to their headquarters and summoned them to the scene of the most serious collapse.[2] The first persons on site were a search and rescue team from Guatemala, who had heard the news of the earthquake on short-wave radio, gathered their equipment, and drove immediately to the scene.

While lack of communications capacity hindered the information search capabilities of the local agencies, presence of communications capacity allowed other agencies to exercise influence on disaster operations disproportionate to their legal roles in the process. With staff trained and supplied to operate in the continuing disruption of the civil war, the U.S. Embassy had a stock of hand-held radios which were immediately put to use for communicating among U.S. staff. Communication with Salvadorean counterparts, however, was possible only in face-to-face meetings, difficult to arrange during the trauma of a disaster. The technical capacity for communication, de facto, gave U.S. professional staff the possibility for initiating action in response to disaster needs. The more difficult question involved the political constraints on the exercise of U.S. capacity in Salvadorean affairs vs. the ethical imperative to do whatever was necessary to save lives and assist the victims of a disaster.[3]

Other organizations sought to develop their own assessment of needs resulting from the disaster. The League of International Red Cross Societies, working in conjunction with the American Red Cross, sponsored a house-to-house survey of the most severely damaged *colonias*. Using a small army of interviewers, the Red Cross collected basic data on the number of families left homeless, the most urgent needs of these families, and the number, ages, and employment status of persons affected by the disaster.[4] This information was used by the Red Cross to inform their disaster relief operations, and was not available to other agencies. The information search process was slow, too slow, to inform Red Cross disaster relief operations in a timely manner.

The technical limitations on information search, compounded by the uneasy and temporary truce in the civil war, resulted in an incomplete and inconsistent account of disaster needs and actions in progress among the different actors involved in response operations.

Information Exchange

President José Napoleón Duarte, leading the response of the Salvadorean Government, took several measures to ensure appropriate information exchange. First, and most striking, Duarte negotiated a truce (albeit temporary) in the civil war, securing the consent of the *Frente Martí Farabundo para la Liberación Nacional* (FMLN), to reduce the level of hostility and propose collaboration to meet the humanitarian needs caused by the disaster.[5]

Second, Duarte, following the example of President Miguel de la Madrid of Mexico during the devastating 1985 Mexico City Earthquake (see Chapter Four), established a National Emergency Committee. President Duarte, seizing the opportunity to turn the disaster into an experience for common national effort, invited representatives of the major national organizations to participate in a National Emergency Committee (*Comité de Emergencia Nacional – COEN*) to share responsibility for the tasks of disaster operations.[6] Composed of diverse groups of business, labor, government and military organizations and divided by conflicting positions in the costly civil war, the organizations in COEN shared little common trust. The magnitude of the humanitarian need resulting from the devastating earthquake mobilized collective action during the immediate phase of response. But with no consistent means of information exchange and support, coordination based on shared information soon gave way to wary efforts by specific organizations to make separate contributions to disaster operations. In the general climate of distrust created by the civil war, information exchange among organizations participating in disaster response operations was guarded at best.

In particular, one major organization was not invited to participate in *COEN*. The Roman Catholic Church, a powerful force in Salvadorean society, was left to operate outside the coalition of nationally recognized organizations collaborating in the disaster response effort. The tensions between the Church and the Duarte Government erupted into open disagreement when a shipment of disaster relief supplies, sent by the Archbishop of San Francisco to the Archbishop of San Salvador, was delayed at the airport, presumably by order of the Duarte Government.[7] The apparent obstruction of relief supplies for political reasons reflected the degree of mistrust between the Church and the Government and the difficulty of overcoming this barrier to mutual cooperation even in the context of urgent needs created by the disaster.

Virtually all organizational participants recognized the need for information exchange to support coordinated response in disaster operations, and various efforts to initiate or improve information exchange were made. For example, the international search and rescue teams established an informal protocol for coordinating their work, relying upon common professional standards recognized by the international fire services. This informal protocol broke down, however, when the large Swiss team did not accept the informal standards developed locally, but operated only according to directives from its national headquarters in Bern.[8] Conversely, the international nongovernmental organizations, which played a major role in both collecting and distributing disaster relief goods, established weekly meetings to share information regarding their activities, the resources available, and the needs they discovered in the *colonias*.[9] Information exchange between government-directed disaster operations and the citizens was reported as seriously inadequate in a survey of citizens affected by the earthquake conducted shortly after the event.[10]

The cumulative burden of conflicting interests and wary perceptions among organizations participating in disaster response operations affected adversely the exchange of information vital to organizational learning. The effects of this sporadic pattern of information exchange affected the organizational learning processes both among organizations interacting within the response system and between the system and its wider environment.

Organizational Learning

An evolving disaster response system represents a process of interorganizational learning that produces a pattern of collective response to the disaster. The extent to which a specific response system achieves a sustained process of learning depends on the dynamic interaction among the participating organizations within the system, and between the system and its environment.

The N-K methodology, presented in Chapter 3, reveals the basic characteristics of the disaster response system that evolved following the San Salvador earthquake. Three of the measures are common to virtually all organizations participating in disaster operations. The goal (P) of the system is assumed to be the protection of life and property. The duration (D) of the period under study is three weeks immediately following the earthquake. The funding sources (S) are public, private or nonprofit. The remaining measures characterize the distinct features of the system. Findings are presented from a content analysis of news reports regarding disaster operations from *El Diario de Hoy*, San Salvador's principal newspaper, for the period, October 14 – November 3, 1986.[11] The content analysis identified a response system of 374 organizations (N = 374) that participated in disaster operations during this period. Table 9 summarizes that response system.

Table 9. Summary, Organizational Response System by Source of Funding and Jurisdiction, San Salvador, October 14 – November 3, 1986

	Public			Total Public	Nonprofit		Private		Total
	Int'l	Nat'l	Mun'l		Int'l	Nat'l	Int'l	Nat'l	
	N %	N %	N %	N %	N %	N %	N %	N %	N %
San Salvador	61 16.3	96 25.7	48 12.8	206 54.9	23 6.1	80 21.4	12 3.2	54 14.4	374 100.0

Table 10. Frequency Distribution: Types of Transactions in Disaster Response by Funding Source and

Type of Transaction	Public Organisations											
	International			National			Regional/State			Municipal		
	T	N	%	T	N	%	T	N	%	T	N	%
Emergency Response	23	28	3.4	18	21	2.6	0	0	0.0	1	1	0.1
Communication	2	2	0.3	5	6	0.7	0	0	0.0	1	1	0.1
Coordination of Response	2	2	0.3	12	15	1.8	0	0	0.0	1	1	0.1
Medical Care/Health	5	5	0.7	45	46	6.6	0	0	0.0	2	4	0.3
Damage/Needs Assessment	6	11	0.9	23	36	3.4	0	0	0.0	12	19	1.8
Certification of Deaths	0	0	0.0	3	3	0.4	0	0	0.0	3	3	0.4
Earthquake Assessment/Research	0	0	0.0	2	3	0.3	0	0	0.0	0	0	0.0
Security Prevention of Looting	0	0	0.0	11	21	1.6	0	0	0.0	0	0	0.0
Housing Issues	1	1	0.1	1	1	0.1	0	0	0.0	0	0	0.0
Disaster Relief (food, shelter, etc.)	8	10	1.2	31	47	4.5	0	0	0.0	8	12	1.2
Donations (money, goods, etc.)	57	68	8.3	9	10	1.3	0	0	0.0	3	3	0.4
Building Inspection	1	1	0.1	11	11	1.6	0	0	0.0	2	2	0.3
Building Codes Issues	0	0	0.0	1	1	0.1	0	0	0.0	0	0	0.0
Repair of Freeways, Bridges, Roads	0	0	0.0	2	2	0.3	0	0	0.0	0	0	0.0
Repair/Restore Utilities	0	0	0.0	11	16	1.6	0	0	0.0	0	0	0.0
Repair/Reconstruction/Recovery[a]	4	4	0.6	27	37	3.9	0	0	0.0	4	4	0.6
Transportation/Traffic Issues	0	0	0.0	12	13	1.8	0	0	0.0	2	2	0.3
Hazardous Materials Releases	0	0	0.0	0	0	0.0	0	0	0.0	0	0	0.0
Legal/Enforcement/Fraud	0	0	0.0	13	17	1.9	0	0	0.0	1	1	0.1
Political Dialogue/Legislation	0	0	0.0	8	8	1.2	0	0	0.0	0	0	0.0
Business Recovery	0	0	0.0	2	3	0.3	0	0	0.0	1	1	0.1
Economic/Business Issues	0	0	0.0	3	3	0.4	0	0	0.0	0	0	0.0
Visits by Officials	3	3	0.4	1	1	0.1	0	0	0.0	0	0	0.0
Education Issues	0	0	0.0	8	8	1.2	0	0	0.0	1	2	0.1
Government Assistance	0	0	0.0	0	0	0.0	0	0	0.0	0	0	0.0
Insurance Related Issues	0	0	0.0	1	1	0.1	0	0	0.0	0	0	0.0
Loans (Private and International)	1	1	0.1	1	5	0.1	0	0	0.0	0	0	0.0
Psychological/Counseling Services	0	0	0.0	1	1	0.1	0	0	0.0	0	0	0.0
Fundraising/Account Setup	2	2	0.3	5	5	0.7	0	0	0.0	4	4	0.6
Volunteers	0	0	0.0	1	1	0.1	0	0	0.0	0	0	0.0
Public Education	0	0	0.0	3	3	0.4	0	0	0.0	0	0	0.0
Relocation	0	0	0.0	18	52	2.6	0	0	0.0	0	0	0.0
Political/Special Interests	0	0	0.0	3	3	0.4	0	0	0.0	2	3	0.3
TOTAL:	115	138	16.8	292	400	42.6	0	0	0.0	48	63	7.0

T = Number of Transactions; N = Number of Actors; % = Percent of Total Transactions
[a] Not including freeways, bridges, roads or utilities
Note: El Salvador has no regional or state level
Source: *El Diario de Hoy*, San Salvador

Table 10 presents the matrix of types of transactions performed by type of jurisdiction and funding source. A total of 685 transactions were grouped into sixteen types of disaster response activities. Disaster response operations were conducted primarily by public agencies, with two-thirds (455, or 66.4%) of the reported transactions performed by public organizations. In turn, national level organizations carried out more than three-fifths, or 64.1%, of the public transactions.

Jurisdiction, San Salvador Earthquake, October 14 — November 3, 1986

Nonprofit Organizations						Private Organizations						TOTALS		
International			National			International			National					
T	N	%	T	N	%	T	N	%	T	N	%	T	N	%
3	3	0.4	10	12	1.5	0	0	0.0	2	2	0.3	57	67	8.3
0	0	0.0	4	5	0.6	0	0	0.0	1	1	0.1	13	15	1.9
1	1	0.1	5	8	0.7	0	0	0.0	0	0	0.0	21	27	3.1
1	1	0.1	7	7	1.0	0	0	0.0	1	2	0.1	61	65	8.9
1	1	0.1	17	31	2.5	0	0	0.0	1	3	0.1	60	101	8.8
5	5	0.7	4	4	0.6	0	0	0.0	0	0	0.0	15	15	2.2
0	0	0.0	1	1	0.1	0	0	0.0	0	0	0.0	3	4	0.4
0	0	0.0	0	0	0.0	0	0	0.0	0	0	0.0	11	21	1.6
0	0	0.0	0	0	0.0	0	0	0.0	0	0	0.0	2	2	0.3
0	0	0.0	34	43	5.0	0	0	0.0	2	4	0.3	83	116	12.1
7	7	1.0	8	8	1.2	2	3	0.3	7	13	1.0	93	112	13.6
2	3	0.3	8	8	1.2	0	0	0.0	0	0	0.0	24	25	3.5
0	0	0.0	1	1	0.1	0	0	0.0	0	0	0.0	2	2	0.3
0	0	0.0	0	0	0.0	0	0	0.0	0	0	0.0	2	2	0.3
0	0	0.0	0	0	0.0	0	0	0.0	0	0	0.0	11	16	1.6
0	0	0.0	14	17	2.0	0	0	0.0	3	3	0.4	52	65	7.6
0	0	0.0	2	2	0.3	0	0	0.0	0	0	0.0	16	17	2.3
0	0	0.0	0	0	0.0	0	0	0.0	0	0	0.0	0	0	0.0
1	1	0.1	0	0	0.0	0	0	0.0	0	0	0.0	15	19	2.2
0	0	0.0	10	10	1.5	0	0	0.0	0	0	0.0	18	18	2.6
0	0	0.0	9	9	1.3	0	0	0.0	5	19	0.7	17	32	2.5
0	0	0.0	4	4	0.6	0	0	0.0	2	2	0.3	9	9	1.3
0	0	0.0	0	0	0.0	0	0	0.0	0	0	0.0	4	4	0.6
0	0	0.0	3	3	0.4	0	0	0.0	4	10	0.6	16	23	2.3
0	0	0.0	0	0	0.0	0	0	0.0	0	0	0.0	0	0	0.0
0	0	0.0	0	0	0.0	0	0	0.0	0	0	0.0	1	1	0.1
0	0	0.0	0	0	0.0	0	0	0.0	0	0	0.0	2	6	0.3
0	0	0.0	0	0	0.0	0	0	0.0	1	1	0.1	2	2	0.3
2	2	0.3	18	43	2.6	6	7	0.9	3	3	0.4	40	66	5.8
0	0	0.0	1	1	0.1	0	0	0.0	0	0	0.0	2	2	0.3
0	0	0.0	1	1	0.1	0	0	0.0	0	0	0.0	4	4	0.6
0	0	0.0	3	3	0.4	0	0	0.0	0	0	0.0	21	55	3.1
0	0	0.0	3	3	0.4	0	0	0.0	0	0	0.0	8	9	1.2
23	24	3.4	167	224	24.4	8	10	1.2	32	63	4.7	685	922	100.0

Municipal organizations performed 10.5% of the public transactions, while international organizations performed 25.3% of the public actions involved in disaster response. These figures document the high level of international involvement in response to this disaster, as well as the surprisingly low level of actions reported for municipal and/or local organizations.

Table 10 also cites the relatively low number of transactions reported for "Communication" at 1.9% and "Coordination" at 3.1%. These figures document the limited effort invested in system functions that encompass information search and exchange. Such transactions are vital to organizational learning and to making the transition to self-organizing processes and collective action. The findings indicate that the highest percentage of transactions were devoted to giving aid to victims, 12.1%, which, if combined with two other related categories of donations and emergency services, constitutes 34% of the total transactions reported. This allocation of time and attention is consistent with the goal of the response system, protection of life and property.

The number of interactions between organizations operating within the response systems indicates the density of the system, and offers a gross measure of the degree of integration for the system. Table 11 presents the findings regarding the frequency and pattern of interactions among organizations participating in disaster response by jurisdictional level and funding source. Two characteristics emerge with striking clarity in these findings. First, the findings document the extent to which disaster operations were directed by the national government, with substantial international involvement. Of the total number of interactions (N = 335), more than three-fourths, 77.5%, involved national and international agencies. Of this total, 91 interactions, or 58.3%, involved national public agencies, 30 or 19.2% were with international agencies, while only a small proportion, 4.5%, were with municipal agencies.

Second, the findings show that nonprofit organizations also played a role in this response system, with 16.7% of the interactions, of which 13.5% involved national nonprofit organizations. International private organizations participated

Table 11. Frequency Distribution: Types of Interactions in Disaster Response, San Salvador Earthquake,

Type of Interaction	Public									Nonprofit		
	International			National			Municipal			International		
	K	N	%	K	N	%	K	N	%	K	N	%
Public												
International	9	18	5.8	12	21	7.7	0	0	0.0	2	4	1.3
National				38	53	24.4	12	37	7.7	6	12	3.8
Municipal							2	4	1.3	0	0	0.0
Nonprofit												
International										1	2	0.6
National												
Private												
International												
National												
Total Interactions	9	18	5.8	50	74	32.1	14	41	9.0	9	18	5.7

K = Number of Interactions; N = Number of Actors; % = Percent of Interactions column total
Source: *El Diario de Hoy*, San Salvador October 14 — November 3, 1986

in disaster response operations, but at 1.3%, represented a much smaller share of the activity. This pattern of interaction in disaster response can be partly explained by the hostile circumstances of the civil war, which led the national government to call for assistance from nonprofit organizations. The war also led the national government to tighten its control over relief goods entering the country and volunteers entering the disaster area. Subversive efforts were discovered in both, as medical supplies suitable for treating combat wounds were found among shipments of blankets. Volunteers with paramilitary training rather than social service experience were discovered among those offering to deliver disaster relief goods (*El Diario de Hoy*, October 21, 1986).

This pattern of findings indicates that the response system evolved with substantial contributions from international and nonprofit organizations, but that national governmental organizations retained the direction of disaster operations. Relatively little direct action was reported for local public organizations.

The nonlinear analysis of change in the frequencies of organizational actions identified from daily news reports confirms the irregular pattern of actions in disaster response and the dominant role of national public agencies in managing the response. The weak role of communication/coordination is shown both in the phase plane plots of this variable with the other five variables, and in the calculation of logistic equations to explain the variance in daily change for specific variables. Figure 1 shows the phase plane plot of the relationship between communication/coordination and emergency response. Figure 2 shows the marginal history chart, or daily record of change, for this relationship. In the first week, the two variables show quite divergent patterns, but settle into a usual pattern of

October 14 — November 3, 1986

			Private						Total		
National			International			National					
K	N	%	K	N	%	K	N	%	K	N	%
5	13	3.2	0	0	0.0	2	4	1.3	30	60	19.2
28	43	17.9	0	0	0.0	7	16	4.5	91	161	58.3
5	18	3.2	0	0	0.0	0	0	0.0	7	22	4.5
3	8	1.9	1	3	0.6	0	0	0.0	5	13	3.2
12	35	7.7	0	0	0.0	9	40	5.8	21	75	13.5
			1	2	0.6	1	2	0.6	2	4	1.3
						0	0	0.0	0	0	0.0
53	117	33.9	2	5	1.2	19	62	12.2	156	335	99.8

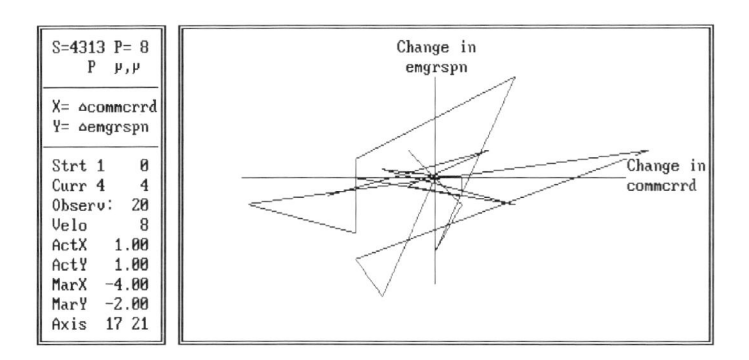

Figure 1. 1986 San Salvador Earthquake: Phase Plane Plot, Public Organizations, Emergency Response by Communication/Coordination

oscillation between change in communication/coordination and change in emergency response on the ninth day, relatively late in the response phase. Figure 3 shows the results of a logistic equation calculated for emergency response, with a value of $R^2 = 0.79$, k = 1.49, and F = 71.02. That is, nearly 80% of the variance in performance in emergency response can be explained by this equation. This is the highest value for R^2 that is found in the set of nonlinear analyses of organizational change for the San Salvador disaster response system. This function was complemented by a relatively strong finding for communication/coordination, with a value of $R^2 = 0.52$, F = 20.97, and $p < 0.001$. This finding likely reflects the early efforts made by the Duarte Government to establish a truce in the civil war and a unified approach, including the major societal organizations, to mobilize response to assist victims of the earthquake.

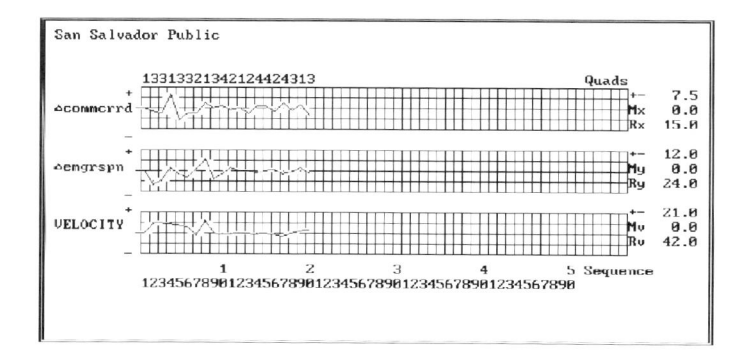

Figure 2. 1986 San Salvador Earthquake: Marginal History, Public Organizations, Emergency Response by Communication/Coordination

Chapter 5: Nonadaptive Systems 91

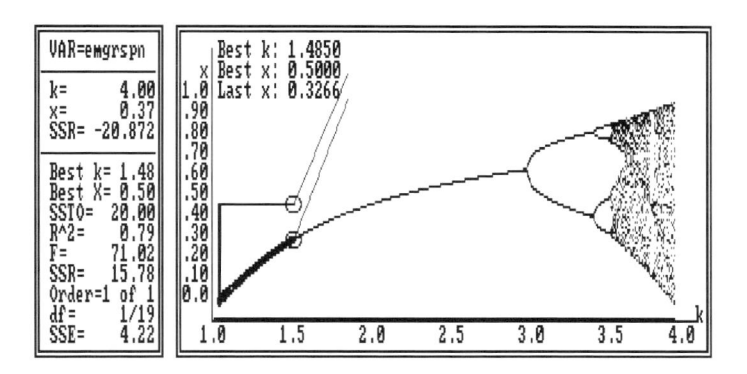

Figure 3. 1986 San Salvador Earthquake: Logistic Regression, Public Organization, Emergency Response

This function reinforced emergency response, and the two functions, working in concert, stabilized an otherwise weak response system.

Figure 4 shows the nonlinear logistic equation for nonprofit organizations engaged in emergency response, with $R^2 = 0.19$, $F = 4.42$, $k = 3.95$. This equation is statistically significant, but the high k value indicates that the function is in chaos. The function moves into chaos when the value of k exceeds 3.66. Figure 5 shows the nonlinear logistic regression equation for disaster relief, with $R^2 = 0.07$, $F = 1.43$, $k = 3.83$. This value is not significant, and the k value also registers in the chaotic range. These findings document the observed reports of tensions among nonprofit organizations participating in response operations during the first weeks of the response. Figure 6 shows a similar chaotic function for private organizations engaged in damage assessment. These findings reveal the instability that was present in the Salvadorean response system.

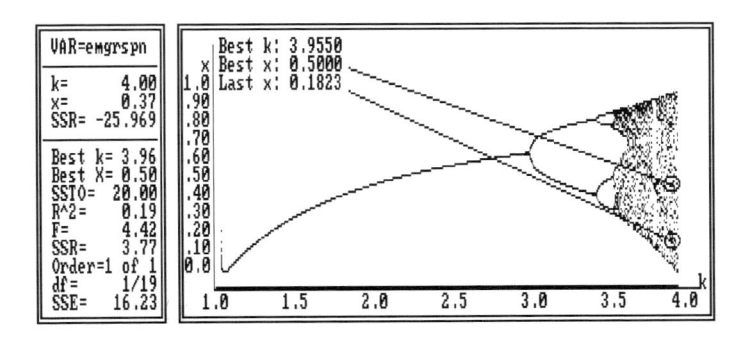

Figure 4. 1986 San Salvador Earthquake: Logistic Regression, Nonprofit Organizations, Emergency Response

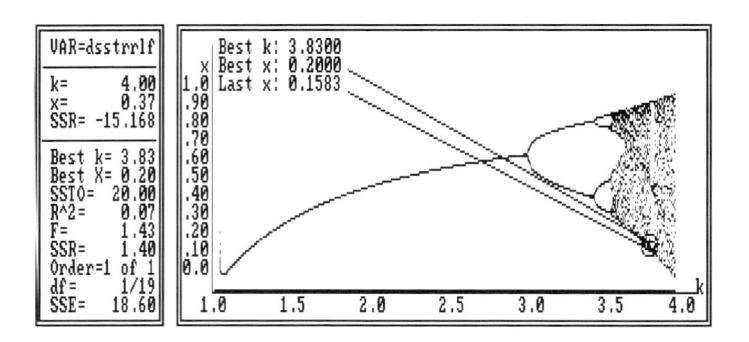

Figure 5. 1986 San Salvador Earthquake: Logistic Regression, Nonprofit Organizations, Disaster Relief

The initial conditions under which the system evolved, a nation at civil war, affected adversely the capacity of organizations at the local level to engage in effective response operations. Further, the kinds of information search and exchange activities essential to organizational learning and evolving processes of self-organization were constrained by hostilities and lack of both physical infrastructure and organizational development. Consequently, the organizations did not achieve a level of organizational learning sufficient to support informed, voluntary, collective action to reduce future seismic risk.

Ecuador, March 5, 1987 (M = 6.1, 6.9)

On Thursday evening, March 5, 1987, two earthquakes occurred in Napo Province, Ecuador. The first, $M_s = 6.1$, occurred at 8:54 p.m., rattling tea-cups in the capital city of Quito and causing momentary concern among its residents. The second, $M_s = 6.9$, struck little more than two hours later at 11:10 p.m., causing more serious damage to urban structures in Quito, especially the colonial

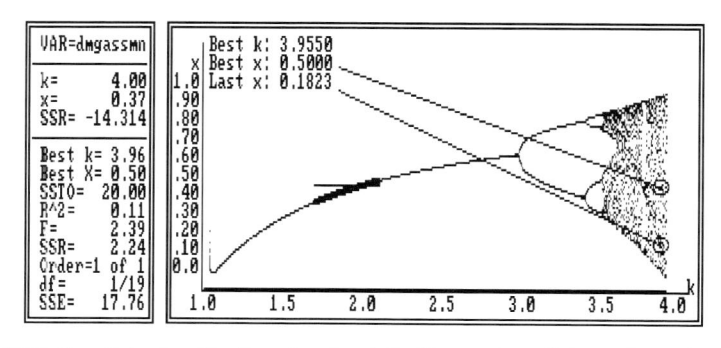

Figure 6. 1986 San Salvador Earthquake: Logistic Regression, Private Organizations, Damage Assessment

churches dating from the sixteenth century. Little information was available about the earthquakes other than that reported by scientific monitoring stations. The *Instituto Geofísico de la Escuela Politécnica Nacional* in Quito reported the magnitudes of the earthquakes and the location of the epicenter on the eastern slopes of the *Cordillera Real*, near the *Volcan Reventador*, about 75 km. northeast of Quito. Provincial cities close to the epicenter were strangely quiet.

Information Search

Concerned about possible damage to the Trans-Ecuadorian Pipeline that carried oil across the Andes, passing near the epicenter of the earthquake, President Leon Febres Cordero led a group of Cabinet ministers to investigate the status of the pipeline. On Friday, March 6, 1987, the group of national officials overflew the area of the epicenter and the course of the pipeline to assess possible damage. The group included the Ministers of Energy, Public Works, Finance, Health, Social Welfare and the Commander-in-Chief of the Armed Forces, as well as the Secretary of National Security, the National Director of Civil Defense, the Undersecretary for Environment, and the Director General of the Ecuadorian State Petroleum Corporation (*CEPE*).[12] When the group returned on the evening of March 6, 1987, they reported the extraordinary damage caused by the earthquakes.

The area near the Volcan Reventador had been inundated with heavy rains for the preceding month. Violent shaking from the earthquakes had denuded the mountainsides, with heavy vegetation sliding down the mountains into the rivers. The rivers, blocked by natural dams, literally changed their courses, and caused flash floods in the small towns located along the rivers. Whole towns were destroyed by the flash floods, resulting in more than 1,000 deaths. Other villagers fled to safer areas, but roads and bridges were destroyed. Transportation by land to the area was impossible; the rivers were filled with debris and were unnavigable. Telephone communications were destroyed, and local residents had no radios or means of communicating with outside sources of assistance. One evangelical mission had a radio and communicated to Quito news of the damage two days after the earthquake had struck.[13]

The March 5, 1987 earthquakes initiated a very complex set of interorganizational disaster response actions which is presented in more detail elsewhere (Comfort 1991a, 122–163). The earthquakes had three zones of impact, each requiring different forms of assistance, coordination and information. The urgent needs of these three zones interacted with one another to produce an extraordinary impact from this natural disaster upon the national economy and the wider Ecuadorian society. High financial costs for reconstruction of the Trans-Ecuadorian Pipeline and prolonged effects of the loss of 50% of its gross national income through the disruption of oil transport and subsequent suspension of oil production required substantial international efforts to assist this small Andean nation.[14] The response required for this disaster was at once broad geographically, complex organizationally, and expensive economically.

The earthquakes produced consequences of differing types and magnitudes in three geographic locations. The area of primary impact included Central Napo Province[15] and the epicenter located near the Reventador Volcano. This zone suffered the major loss of life due to the flash floods generated by massive landslides and debris flows. Official estimates placed the death toll for all three zones of the disaster at 1,000, with 5,000 left homeless or in need of resettlement.[16] Other estimates ranged from 300 to 1,000 dead.[17] While informed sources vary, all experts agree that it is not possible to determine precisely the number of dead, because no reliable census of persons living in the affected areas existed prior to the earthquakes.[18] This observation is particularly valid for central Napo Province, as it is largely undeveloped territory recently opened to colonists for settlement and cultivation.

Major damage occurred to the infrastructure, including destruction of approximately 30 kilometers of the Trans-Ecuadorian Pipeline, as well as destruction of approximately 40 kilometers of the main highway from Quito to Lago Agrio, secondary roads, the oil pumping station at El Salado, and seven bridges.[19] After the initial trauma of coping with loss of life and dislocation of families had lessened, the major problems driving organizational interaction in this area centered around the destruction of infrastructure. These problems included: (1) reconstruction of the oil pipeline in geologically unstable territory; (2) loss of oil revenues and its consequent impact upon the national economy; (3) loss of the highway, bridges, and secondary roads for safe travel and economic activity of the resident population; and (4) reorientation and resettlement of local residents, severely shaken emotionally and economically by the disaster, who were struggling to cope with questions regarding an uncertain future in a zone of high seismic risk.[20]

The zone of secondary impact from the earthquake was the Sierra, specifically the Andean highlands of Imbabura, Carchi and Pichincha Provinces. In this zone, the principal problem was housing. There were no reports of lives lost in the immediate occurrence of the earthquakes, but approximately 60,000 homes were damaged or rendered uninhabitable. The earthquakes produced differential effects for residents of differing economic status. It was an "earthquake for the poor,"[21] as the houses most severely damaged were those made of blocks of sun-dried mud, with no reinforcement or flexibility to withstand seismic movements.

The third zone of impact included the city of Lago Agrio and adjacent communities in eastern Napo Province. These communities suffered little structural damage and had no loss of life. The major problem generated by the earthquakes in this zone was isolation and economic deprivation, resulting from the destruction of the oil pipeline and the major route of land transportation, the highway from Quito to Lago Agrio.[22] These communities survived the initial event of the earthquakes without severe consequences, but the cumulative effects of long-term isolation, unemployment and lack of access to markets and supplies worsened over the prolonged period of time required for reconstruction of the infrastructure. The local economy, based upon oil production and agriculture, was devastated without means of transport.

The Indian communities along the Coca, Aguarico, Due, Salado and Papallacta rivers were especially vulnerable. Dependent upon the rivers for drinking water, nutrition — fish are an important staple in their diet — and transportation, these communities suffered serious deprivation in the loss of vital health and economic resources due to the pollution and obstruction of the rivers.[23] The disruption of existing economic, social and transportation systems caused by the earthquakes produced a cumulative economic and social disaster for the residents of this zone over the succeeding months. This interactive set of conditions steadily worsened over time and overwhelmed the local resources of the residents and communities of this zone. They could not cope with the isolating conditions generated by the earthquakes without external assistance.

The physical effects of the earthquakes generated different types of problems in three geographic zones which, in turn, escalated the impact of the disaster for the nation as a whole. The simultaneous needs of the populations in the three zones combined with the massive impact on the national economy from the loss of oil export revenues and the high costs of rebuilding the pipeline and transportation routes required resources beyond the capacity of Ecuador alone.

Information Exchange

President Leon Febres Cordero appealed to the international community for assistance to Ecuador in meeting the technical and economic needs generated by the disaster.[24] Some 22 nations responded to this call. Yet, the requirements for coordination and communication between participating nations and between the Ecuadorian levels of governmental jurisdiction in the simultaneous delivery of services to the three disaster zones escalated the complexity of organizational interaction still further.

The interaction of different needs between[25] the three zones of the disaster, each with its own degree of urgency, created a situation of extraordinary difficulty within this complex organizational environment. The process was dynamic, and in the early phase of disaster response, the parameters of needs, resources, personnel, and, therefore, action were uncertain. Under these conditions, any nationwide action in disaster management becomes an important measure of capacity. Problems reported in the coordination and delivery of disaster assistance are not surprising.[26] Rather, the complexity of this situation is so unusual that it merits particular attention in understanding the design and dynamics of inter-organizational coordination in disaster management.

The jurisdictional involvement in disaster response and recovery activities in the three zones of the disaster reflected this complexity. In Ecuador, the mission responsibility for disaster management on the national level lies with Civil Defense. This is a developing organization, first established in 1962.[27] Civil

Defense has undergone numerous changes of leadership and direction as it sought to meet the successive challenges of disaster management posed to the nation by the severe earthquake of 1976, the coastal floods of 1982–1983 and the Galapagos fire of 1984.[28] Although the legal responsibilities are now defined and a clear organizational structure exists, the capacity of this relatively new organization to take action is limited by the scarcity of resources and trained personnel throughout the intergovernmental system.

The demands upon the national Civil Defense organization in the March, 1987 disaster were greater than its capacity for delivery of services. The structure for a national Civil Defense organization exists at all jurisdictional levels — parochial, cantonal, provincial.[29] Yet, these organizations operate largely as an added responsibility for the existing parochial or cantonal councils. The presidents of the Municipal Councils are, for the most part, also the local directors of Civil Defense. At the cantonal level, the director's position may be held by an officer in the Ecuadorian Army, as in Lago Agrio where the Commander of the Battalon de Selva is also the director of canton's Civil Defense Council.[30] In both cases, the national Civil Defense organization sought links to existing agencies within the communities, for it has few resources of its own. Local governmental units, in particular, have little equipment and less training for disaster mitigation and preparedness. Although a nation of high seismic risk, disaster management in Ecuador has been limited by the scarcity of resources available for equipment and preparedness training.[31]

Communication is crucial in managing disaster operations over three geographic areas. Yet, the Civil Defense organizations at the parochial and cantonal levels in Quijos did not have radios to report the news of the disaster or the extent of the damage to the national organization in Quito. Telephone communications were disrupted by the disaster, and it took several days for the full extent of the damage to become known at the national level.[32] The only radio available for communication was a station operated by the Evangelical Mission in Quijos, which voluntarily served this important function by relaying messages to and from national offices in Quito.[33] Voluntary, religious and communal organizations that had resources and skills necessary for disaster assistance responded to the needs presented by the disaster through an informal network of personal and organizational contacts. These agencies sought to provide what assistance they could, but their capacity to do so was limited by lack of resources, equipment, and training for operation in disaster environments.

Recognizing the magnitude of the disaster and the extent of its impact upon the nation, President Leon Febres Cordero declared a national emergency in the provinces of Carchi, Imbabura, Napo and Pastaza under the provisions of Article 101 of the Law of National Security.[34] He called the ministers of Health, Finance, Public Works, Energy, Social Welfare, Environment, the director general of *CEPE*, the national director of Civil Defense, the commander-in-chief of the Armed Forces, and the director of the military Corps of Engineers to form a National Emergency Committee to assess the damage and plan the emergency response.[35] By this action, the president mobilized the highest officers of the Ecuadorian government in response to the disaster, giving it first priority in national affairs.

The international and voluntary organizations that participated in disaster operations, in principle, coordinated their activities with the National Emergency Committee. In practice, the complexity of the operations and lack of communications facilities between national offices and the provincial and cantonal fields of operation required that much of the work be done locally with limited contributions from the national level.

In disaster response, the primary organizational task is to weave a productive network of organizations, drawing resources and skills from all jurisdictional levels to meet the needs of the people in the disaster-affected communities. Accomplishing this task requires flexibility and resourcefulness at all levels of operation. In Ecuador, it was demonstrated most vividly by the use of *mingas*, or community work groups, to accomplish urgent, local needs.

Charitable organizations played an important role in this disaster, particularly at the community level. Ecuadorian organizations joined in a national campaign to offer voluntary contributions to disaster assistance in a remarkable demonstration of 'solidarity' with the disaster victims.[36] Links to the international community provided resources that were not immediately available in country to initiate the design and implementation of disaster assistance activities. The International Red Cross, Catholic Relief Services, World Vision, and other organizations were important sources of technical expertise, resources and personnel in the labor-intensive tasks of community assistance to disaster victims.

Information exchange occurred in response operations to this disaster, but most frequently within groups, rather than between sets of organizations or levels of government.

Organizational Learning

The National Emergency Committee established by the President of Ecuador drew upon the experience, expertise and facilities of the major ministries of the nation to direct organized action in disaster response and recovery. General Germán Ruiz, of the Ecuadorian Army, directed emergency operations in conjunction with General Antonio Moral Moral of the national Civil Defense Authority. Private and voluntary organizations supplemented the activities of these two entities by their work at the local, provincial, national, and international levels. Of the 241 organizations participating in disaster response, public organizations comprised nearly two-thirds, or 63.9%. Nonprofit organizations made up the next largest group, 29.5%, and private organizations comprised a small but significant proportion, 6.6%. Table 12 summarizes the response system for Ecuador by source of funding and jurisdiction.

Voluntary contributions from local, national and international sources joined limited public resources to meet needs for humanitarian assistance to the disaster

Table 12. Summary, Organizational Response System by Source of Funding and Jurisdiction, Ecuador, March 5 – 6, 1987

	Public								Total Public		Nonprofit				Private				Total	
	Int'l		Nat'l		State		Mun'l				Int'l		Nat'l		Int'l		Nat'l			
	N	%	N	%	N	%	N	%	N	%	N	%	N	%	N	%	N	%	N	%
Ecuador	54	22.4	79	32.8	6	2.5	15	6.2	154	63.9	12	5.0	59	24.5	6	2.5	10	4.1	241	100

victims. International sources provided financial resources for reconstructing the oil pipeline, bridges, and the highway, and relief from external debt obligations – all major costs exacerbated by the loss of revenue from oil exports. Obtaining international monetary credit required interorganizational coordination between nations, a problem that increased the complexity of response operations.

The striking characteristic of this disaster is the complexity of problems generated by the earthquakes and the interactive nature of those problems. The organizational action required to address these problems proved correspondingly complex. Consequently, action at national levels often did not reach the local levels, left largely to their own resources to rebuild damaged homes and lives. However, in the *parróquias* of the Sierra and the distant settlements of Napo Province, the disaster created policy-making situations at the local level, engaging most elements of the community in the innovative use of scarce resources to meet shared needs. For those residents living at marginal economic levels, the disaster threatened their basic existence and presented choices for either rebuilding in stronger ways or relocating to more stable territory. At separate levels of action, organizational learning did occur, but there was little interaction among the levels to characterize these activities as system-wide response.

Adaptive Behavior and/or Self-Organization

In this complex, dynamic set of conditions, a disaster response system of 241 organizations evolved in the three-week period following the earthquake. Composed of organizations that had other functions in routine operations, the response system distinguished its actions from others in the Ecuadorian context by the shared commitment of its members to meet the needs created by the earthquake's destructive effects.

Table 13 presents the matrix of transactions derived from the content analysis of news reports of disaster operations. This matrix provides striking evidence that the damage to the oil pipeline and the consequent economic crisis that it provoked displaced the humanitarian response to the immediate needs of the population in the damaged villages. Emergency response and medical care accounted for a small proportion, 5.4%, of the transactions reported, representing a combined total for local emergency response, medical care to the injured, and national emergency declarations. Disaster relief transactions constituted

7.1% of those reported in the newspapers, while international assistance represented 17.7%, the largest proportion of reported transactions for any single category. The set of transactions associated with damage assessment, inspection of buildings and building failures, represented 12.5% of the total transactions. Those involving economic status, economic assistance, rationing of gasoline and diesel fuel, repair of petroleum pipelines, and international assistance with petroleum-related infrastructure represented more than one-third, or 35.3%, of all transactions reported.

A surprisingly high proportion of the reported transactions, 7.1%, concerned strikes and civil unrest, indicating the severe stress placed upon the national economy and civil society by the consequences of the earthquake. Given the complexity of response operations in this disaster with three different areas of impact, the difficulty of communication and lack of coordination among them is reflected in the low proportion of reported transactions for communication, 1.6%, and coordination, 3.5%, for a combined total of 5.1%.

Response operations in this disaster were overwhelmingly carried out by public national organizations, 62.2%, during the first week, March 7—13, 1997. Next in frequency of mentions were public international organizations, including governments, at 16.7%, followed by nonprofit organizations at 12.1%. Local governments, provincial and municipal combined, accounted for 5.2% of the reported transactions. This pattern of dominance by national level organizations in disaster response operations is further documented in Table 14, which presents the matrix of interactions among the types of organizations. Nearly two-thirds, 64.2%, of the total number of interactions reported over the three-week period involved public national organizations, with 22.5% attributed to public international organizations. Striking is the small proportion of interactions, 2.7%, identified for public provincial and municipal governments. Nonprofit organizations were involved in 9.3% of all reported interactions in disaster operations.

The erratic relationship between communication and coordination and other emergency response functions is shown by the nonlinear analysis of change in the frequencies of disaster response functions reported by day for the three-week period following the disaster. Communication/coordination peaked late, on the fourth day following the earthquake. Aside from an initial effort to respond to information received, communication/coordination and emergency response activities occurred almost in a reverse pattern.

As the frequency of reported communication/coordination functions increased, the frequency of reported emergency response actions decreased. Figure 7, the marginal history chart for reported frequencies, shows this pattern. Emergency response, although low in frequency of reported actions taken by public organizations, did appear to be a stable function. Figure 8 presents the logistic equation for change in emergency response, with $R_2 = 0.85$, and $k = 1.2$. Change in damage assessment, reported for public organizations, also yielded a logistic equation, with $R_2 = 0.79$ and $k = 1.8$, shown in Figure 9. This function shows a bit more instability, but it is operational. Change in reported public actions for

Table 13. Frequency Distribution: Types of Transactions in Disaster Response by Funding Source and

| Type of Transaction | Public Organizations ||||||||||||
| | International ||| National ||| Regional/State ||| Municipal |||
	T	N	%	T	N	%	T	N	%	T	N	%
Emergency Response	0	0	0.0	6	5	0.9	1	1	0.2	0	0	0.0
Communication	0	0	0.0	6	5	0.9	0	0	0.0	0	0	0.0
Coordination of Response	3	3	0.5	15	14	2.4	0	0	0.0	1	1	0.2
Medical Care/Health	0	0	0.0	4	4	0.6	0	0	0.0	0	0	0.0
Damage/Needs Assessment	5	4	0.8	29	19	4.6	2	2	0.3	3	3	0.5
Certification of Deaths	0	0	0.0	4	3	0.6	1	1	0.2	0	0	0.0
State of Emergency Declarations	0	0	0.0	17	10	2.7	0	0	0.0	0	0	0.0
Evacuation	0	0	0.0	12	11	1.9	0	0	0.0	1	1	0.2
Earthquake Assessment/Research	0	0	0.0	8	4	1.3	0	0	0.0	0	0	0.0
Disaster Relief (food, shelter, etc.)	1	1	0.2	18	10	2.8	0	0	0.0	6	6	0.9
International Assistance/Donations	69	38	10.9	17	13	2.7	1	1	0.2	2	2	0.3
Building Inspection	16	15	2.5	0	0	0.0	1	1	0.2	0	0	0.0
Investigation of Building Failures	1	1	0.2	8	4	1.3	0	0	0.0	0	0	0.0
Repair/Restore of Gas Lines	13	6	2.1	29	9	4.6	0	0	0.0	0	0	0.0
Repair/Reconstruction/Recovery[a]	1	1	0.2	5	4	0.8	0	0	0.0	3	3	0.5
Transportation/Traffic Issues	3	3	0.5	14	9	2.2	1	1	0.2	0	0	0.0
Visit of Earthquake Site by Officials	2	2	0.3	8	5	1.3	0	0	0.0	0	0	0.0
Education Issues	0	0	0.0	5	3	0.8	0	0	0.0	0	0	0.0
Economic Assistance (loans, rates)	0	0	0.0	12	11	1.9	0	0	0.0	0	0	0.0
International Assistance with Gas	21	12	3.3	16	5	2.5	0	0	0.0	0	0	0.0
Rationing of Gas/Price of Gas	1	1	0.2	40	20	6.3	1	1	0.2	2	2	0.3
Econ. Status/National Econ. Policy	1	1	0.2	42	23	6.6	0	0	0.0	0	0	0.0
Strikes/Civil Unrest	1	1	0.2	13	7	2.1	0	0	0.0	0	0	0.0
TOTAL	138	89	21.8	328	198	51.8	8	8	1.3	18	18	2.8

T = Number of Transactions; N = Number of Actors; % = Percent of Total transactions
[a]Not including freeways, bridges, roads or utilities
Source: *Hoy*, Quito, Ecuador, March 7–31, 1987

Table 14. Frequency Distribution: Types of Interactions in Disaster Response by Funding Source and

| | Public ||||||||||||
| | International ||| National ||| Provincial ||| Municipal |||
	K	N	%	K	N	%	K	N	%	K	N	%
Public: International	2	5	1.3	31	46	20.5	0	0	0.0	0	0	0.0
Public: National				58	58	38.4	5	11	3.3	5	10	3.3
Public: Provincial							0	0	0.0	1	4	0.7
Public: Municipal										0	0	0.0
Nonprofit: Charity												
Nonprofit: Pol. Party/SIG												
Nonprofit: Professional												
Private												
Total	2	5	1.3	89	104	58.9	5	11	3.3	6	14	4.0

K = Number of Interactions; N = Number of Actors; % = Percentage of Total Number of Interactions
Pol. Party/SIG = Political Party/Special Interest Group

Chapter 5: Nonadaptive Systems

Jurisdiction, Ecuador Earthquake, March 7–31, 1987

Nonprofit Organizations									Private Organizations			TOTALS		
Charitable			Pol. Party/SIG			Professional									
T	N	%	T	N	%	T	N	%	T	N	%	T	N	%
4	3	0.6	0	0	0.0	1	1	0.2	0	0	0.0	12	10	1.9
0	0	0.0	0	0	0.0	0	0	0.0	4	4	0.6	10	9	1.6
1	1	0.2	2	2	0.3	0	0	0.0	0	0	0.0	22	21	3.5
0	0	0.0	1	1	0.2	0	0	0.0	0	0	0.0	5	5	0.8
2	2	0.3	1	1	0.2	1	1	0.2	0	0	0.0	43	32	6.8
2	2	0.3	0	0	0.0	0	0	0.0	0	0	0.0	7	6	1.1
0	0	0.0	0	0	0.0	0	0	0.0	0	0	0.0	17	10	2.7
1	1	0.2	0	0	0.0	0	0	0.0	0	0	0.0	14	13	2.2
0	0	0.0	0	0	0.0	0	0	0.0	0	0	0.0	8	4	1.3
15	13	2.4	3	3	0.5	1	1	0.2	1	1	0.2	45	35	7.1
11	8	1.7	4	4	0.6	6	5	0.9	2	2	0.3	112	73	17.7
0	0	0.0	1	1	0.2	2	2	0.3	4	4	0.6	24	23	3.8
0	0	0.0	0	0	0.0	3	2	0.5	0	0	0.0	12	7	1.9
0	0	0.0	0	0	0.0	0	0	0.0	10	3	1.6	52	18	8.2
1	1	0.2	8	7	1.3	2	2	0.3	0	0	0.0	20	18	3.2
0	0	0.0	0	0	0.0	0	0	0.0	0	0	0.0	18	13	2.8
0	0	0.0	0	0	0.0	0	0	0.0	0	0	0.0	10	7	1.6
0	0	0.0	0	0	0.0	0	0	0.0	0	0	0.0	5	3	0.8
0	0	0.0	0	0	0.0	0	0	0.0	2	2	0.3	14	13	2.2
0	0	0.0	0	0	0.0	0	0	0.0	0	0	0.0	37	17	5.8
0	0	0.0	3	1	0.5	1	1	0.2	1	1	0.2	49	27	7.7
0	0	0.0	5	5	0.8	2	2	0.3	2	2	0.3	52	33	8.2
0	0	0.0	27	15	4.3	3	3	0.5	1	1	0.2	45	27	7.1
37	31	5.8	55	40	8.7	22	20	3.5	27	20	4.3	633	424	100.0

Jurisdiction, Ecuador Earthquakes, March 7–31, 1987

Nonprofit									Private			Total		
Charity			Pol. Party/SIG			Professional									
K	N	%	K	N	%	K	N	%	K	N	%	K	N	%
0	0	0.0	1	6	0.7	0	0	0.0	0	0	0.0	34	57	22.5
8	21	5.3	5	12	3.3	7	19	4.6	9	17	6.0	97	148	64.2
0	0	0.0	0	0	0.0	0	0	0.0	0	0	0.0	1	4	0.7
0	0	0.0	1	3	0.7	2	6	1.3	0	0	0.0	3	9	2.0
2	5	1.3	1	2	0.7	4	8	2.6	1	2	0.7	8	17	5.3
			4	7	2.6	1	2	0.7	0	0	0.0	5	9	3.3
						1	2	0.7	0	0	0.0	1	2	0.7
									2	5	1.3	2	5	1.3
10	26	6.6	12	30	8.0	15	37	9.9	12	24	8.0	151	251	100.0

102   *Part II: Shared Risk in Practice*

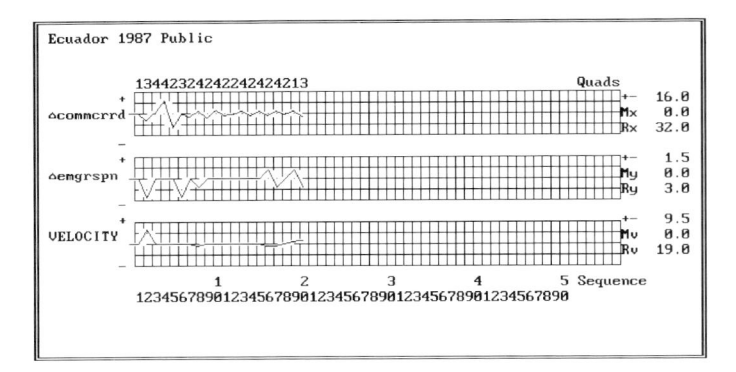

Figure 7. 1987 Ecuadorian Earthquakes: Marginal History, Public Organizations, Emergency Response by Communication/Coordination

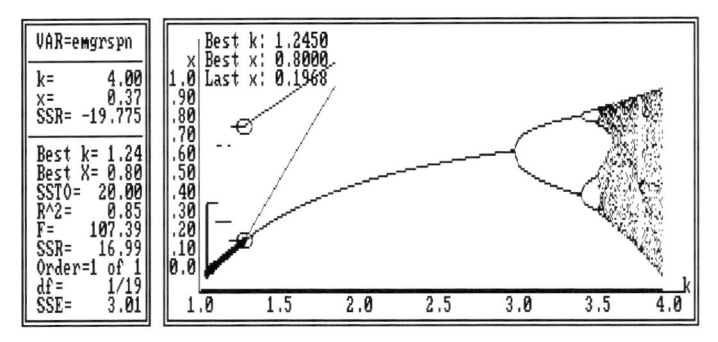

Figure 8. 1987 Ecuadorian Earthquakes: Logistic Regression, Public Organizations, Emergency Response

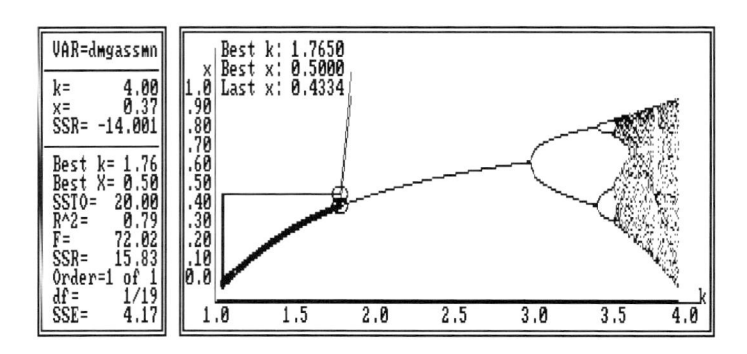

Figure 9. 1987 Ecuadorian Earthquakes: Logistic Regression, Public Organizations, Damage Assessment

Chapter 5: Nonadaptive Systems 103

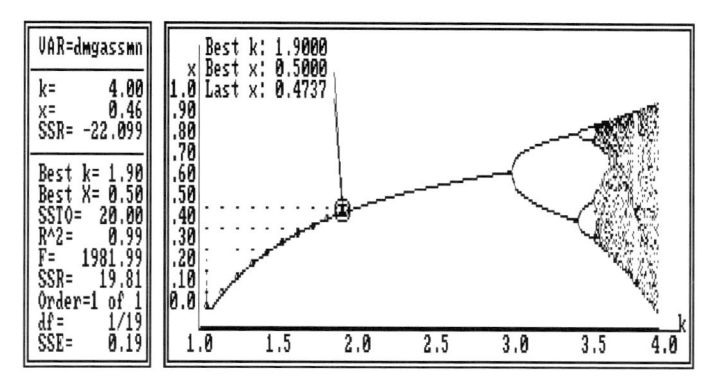

Figure 10. 1987 Ecuadorian Earthquakes: Logistic Regression, Private Organizations, Damage Assessment

disaster relief did not produce a logistic equation, nor did it show a coordinated pattern with communication/coordination.

Nonprofit organizations showed the strongest patterns of coordinated change in disaster relief, with little reported activity for the other functions. Private organizations showed some activity in disaster relief, and a volatile pattern of reported activity in recovery/reconstruction. Reported change in damage assessment actions involving private organizations produced a logistic equation with $R^2 = 0.99$, a remarkably strong fit, with k = 1.9, shown in Figure 10. In reported actions of recovery/reconstruction, the system slipped into chaos, as shown in Figure 11. The logistic equation calculated a value for $R^2 = 0.37$, but a k value of 3.995. The response system slides into chaos on this function when k reaches 3.66. These findings summarize the chaotic activity surrounding the destruction of the pipeline, the severe consequences of the disruption of oil production for the economy, and the secondary and tertiary effects of this disruption throughout

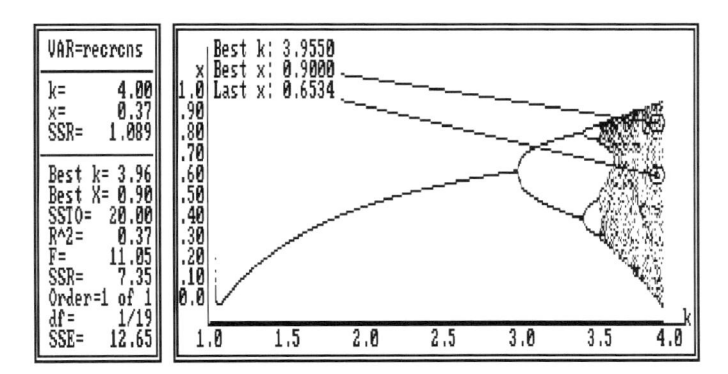

Figure 11. 1987 Ecuadorian Earthquakes: Logistic Regression, Private Organizations, Recovery/Reconstruction

the national economy. These findings demonstrate the increasing complexity of the Ecuadorian response system, as it evolved to meet the secondary and tertiary consequences from the earthquakes as they rippled throughout the society. The findings also document the lack of an adequate information infrastructure to support the interdependent interactions among organizations participating in response operations.

Only in Andean villages where communication occurred by word of mouth and coordination was supported by social tradition, did evidence of self-organization emerge through the formation of local work groups or *mingas*. Without an information infrastructure to support the timely, accurate exchange of information, organizational performance in the increasingly complex environment of this disaster clearly faltered.

Northern Armenia, December 7, 1988 (M = 6.9)

On a sunny Wednesday morning, December 7, 1988, a severe earthquake struck northern Armenia at 11:41 a.m., measuring 6.9 on the Richter scale. The earthquake had devastating effects on four cities in northern Armenia — Spitak, Leninakan, Kirovakan, Stepanavan — and 58 villages in the area, with the epicenter located near the town of Spitak. The cities varied in size. Spitak, closest to the epicenter, with a population of 20,000 was virtually destroyed. Leninakan, the major city in the region with a population of approximately 290,000, suffered 75% destruction. Kirovakan, the next largest city, had 170,000 residents, with destruction at approximately 25%. Stepanavan endured 67% destruction.[37]

Four minutes after the main shock, an aftershock of magnitude 5.8 Richter damaged buildings already weakened in the first temblor. In minutes, buildings collapsed; water, electricity and communication systems were knocked out. More than 25,000 persons were killed according to official reports; unofficial reports estimated over 55,000 dead.[38] at least 19,623 persons received pre-hospital medical care; approximately 12,500 persons received hospital care,[39] and over 514,000 persons were made homeless.[40] Thirty-three of the republic's 66 hospitals were severely damaged; 27 of this number were totally destroyed. Two hundred and fifty schools were damaged; 157 industrial plants were forced to suspend operations. Nearly one-third of Armenia's population of 3.5 million was affected to some degree by the earthquakes.[41] Governmental organizations, unprepared for such devastation, struggled to devise appropriate responses to the immensity of human needs that followed from the traumatic events.

Information Search

Severe damage to the communications infrastructure seriously hampered the information search processes following the Armenia Earthquake. Lack of communications proved the most critical problem in organizing disaster response. In 1988, three separate telephone systems were operating in Armenia. The civilian telephone system was shut down for weeks following the earthquake. The official governmental telephone system was restored within 7–10 days after the earthquake. The Civil Defense telephone system was restored very quickly, but it did not operate within the disaster zone. The military troops stationed in Armenia did not offer assistance to civilian groups in facilitating communications in disaster operations.[42]

In December, 1988, *perestroika* represented an emerging policy of openness in public communication in the Soviet Union. However, vestiges of centralized control remained firmly entrenched in practice. Civilian groups, such as ham radio operators who offered assistance with communications, were not allowed to transmit any information about the Armenian earthquake. The Republic's official leadership rejected the offer of the amateur radio operators' association to form a radio network among the damaged cities to support disaster operations.[43] As a result, Leninakan and other cities in the disaster zone were left essentially without communications to facilitate response operations for days following the earthquake. This critical lack of communication compounded difficulties of coordinating transport for the injured to hospitals, organizing search and rescue teams, identifying needs of the stricken populations accurately, and mobilizing the medical and emergency response to meet those needs.

The lack of physical infrastructure to support communication was compounded by the lack of civilian preparedness for natural disaster. In 1988, formal responsibility for protection of life and property in the event of disaster in Armenia, as in the rest of the Soviet Union, was assigned to Civil Defense, or the Soviet military. The plan, however, focused primarily on protection against nuclear attack, and steps for organizing a community for civilian response to an earthquake were largely undefined. Consequently, even if communications were available, which in most cases they were not, it was not clear who were the responsible parties to contact or how and where to contact them. This lack of organizational preparedness slowed the mobilization process and added confusion as civilian organizations rushed to help, but had no clear direction or points of contact.

The complexity of cultural differences in communication further compounded the lack of organizational preparedness for natural disaster in Armenia. The official language of the Soviet Union was Russian, but the citizens of Armenia spoke Armenian. While the large majority of Armenians spoke Russian as well, the most vulnerable of the population — the old, the very young — did not. The international teams spoke different languages and had to rely on interpreters or English as a common international language. Differences in language added to the complexity of the communications process, and inhibited the development of a common understanding of the scope of the problem, the most urgent needs, and a coherent strategy of action for the participating groups. Technical, organizational, and cultural constraints regrettably limited information search strategies in the first three weeks of disaster operations.

Information Exchange

Limited by inadequate information search strategies and the same technical, organizational and cultural constraints, the information exchange essential to mobilizing effective response to the damaged community proved seriously inadequate. This condition was acknowledged by all participants in response operations: physicians, Civil Defense, international rescue groups and observers, and lay witnesses.

Several unanticipated factors, however, contributed significantly to enhance information exchange, particularly in the international arena. Armenia, although a small republic of 3.5 million population, has a large and loyal diaspora, with significant groups of citizens of Armenian descent and heritage who have resettled in France, Germany, Britain, Italy, the United States, and other nations. These groups had maintained contact with friends and family in Armenia over the years, and, from their respective nations, acted to mobilize resources, aid and assistance for their "homeland" under the urgent conditions of disaster. These groups used their own resources to establish direct telephone contact with their counterparts in Armenia, to assess needs for assistance, and to organize resources externally to assist the stricken population.

Further affecting the information exchange processes, then-Premier Nikolai Gorbachev made an early decision to accept international aid for the Armenian disaster.[44] He sought to demonstrate to the world the new practices of *perestroika* in the Soviet Union, allowing unrestricted access for international rescue teams and aid groups to Armenia. For the first time, international groups responding to humanitarian needs in Armenia were allowed entry directly to Yerevan without visas. This unexpected openness greatly facilitated the exchange of information between Armenians and external groups. Some of the international teams, e.g. the Italian search and rescue team, brought their own satellite communications equipment with them. Such equipment enabled them to contact not only their home government in Rome, but also to provide supplementary communication for other teams and groups working in Armenia. Nonetheless, the exchange of information was still seriously limited among organizations engaged in response operations within Armenia.

A third factor seriously hampered information exchange for disaster response operations in Armenia. In December, 1988, the long-simmering dispute over Karabakh in the neighboring republic of Azerbaijan had erupted into an unrecognized state of guerrilla warfare. Karabakh is an enclave of residents of predominantly Armenian and Christian descent within the geographic boundaries of neighboring and Muslim Azerbaijan. Azerbaijan claimed the land and property of Karabakh; Armenians in Karabakh asserted their independence and wish to secede from Azerbaijan. The resulting conflict created hostile tensions between the two republics, and between the republics and the Soviet All-Union administration that presumably was responsible for maintaining order.

The tense conditions led to hostile relations between the citizens of the two republics and between the republics and Soviet military, whose presence was presumably intended to preserve order. This difficult situation added to the number of refugees from Karabakh who were seeking asylum in Armenia, and were located temporarily in the disaster zone, undocumented and likely illegal. These citizens of Karabakh could not be legally recognized by the formal administrative system, although they suffered injuries and death from the earthquake just as their Armenian relatives and friends. Such conditions hindered full information exchange out of distrust among the hostile parties.

Organizational Learning

Despite the limiting technical and organizational conditions, organizational learning did occur among separate organizations involved in disaster response in Armenia. A disaster response system of 198 organizations evolved following the Armenian earthquake. In Armenia SSR, nearly two-thirds of the organizations participating in response to the disaster were public, 127, or 64.1%. The response system also included a sizeable group of non-governmental organizations, 58, or 29.3%, with a substantial representation of international non-governmental organizations. A very small number of private organizations, 13, or 6.6%, all international, made up the remainder. Table 15 summarizes the organizational response system for Armenia by jurisdiction and funding source.

Many of the nonmilitary organizations directly involved in disaster response reported instances of learning from their experiences in the shattered disaster zone and used these insights to reorganize their activities and reallocate resources accordingly. The system of triage adopted by paramedics at the Leninakan Airport offers a vivid, if painful, example of such learning from experience. Since seventeen of the eighteen hospitals in Leninakan were severely damaged or destroyed, with their medical staff among the victims, most of the severely injured were transported to Yerevan for treatment. With approximately 18,000 injured patients (Nechaev and Reznik 1990, 7), the number waiting for treatment and transportation overwhelmed the limited capacity of the airport and the available ground transportation. Although there was no shortage of physicians — nearly 9,000 physicians volunteered from all regions of the Soviet Union to assist in

Table 15. Summary, Organizational Response System by Funding Source and Jurisdiction, Armenia, December 8–29, 1988

	Public				Total Public	Nonprofit		Private		Total
	Int'l	Union	Republic	Local		Int'l	Nat'l	Int'l	Nat'l	
	N %	N %	N %	N %	N %	N %	N %	N %	N %	N %
Armenia	33 16.6	39 19.7	40 20.2	15 7.6	127 64.1	34 17.1	24 12.1	13 6.6	0 0.0	198 100

medical care for the victims — they could do little for the victims without clean water, electricity, or adequate medical equipment and facilities.

Facing the painful task of how to allocate extremely scarce resources in life-threatening conditions, the medical staff devised a form of classifying the patients according to the severity of their injuries and the order of treatment required. Red tags meant urgent care; yellow tags meant condition stable, but serious; green tags meant minor injuries that could be treated locally; and black tags meant no hope. This color-coded triage system, easily identifiable, enabled a range of workers to coordinate their activities accordingly. This included the physicians, helicopter crews, ambulance drivers, nurses and paramedics, many of whom were from different regions and spoke different languages. It provided a means of organizing the work flow to maximize time and resources under urgent conditions.

Other critical tasks in disaster operations, such as search and rescue operations to extricate live victims from collapsed buildings, were severely hampered by the lack of communications, the lack of heavy equipment, such as cranes and bulldozers, the lack of a prepared plan for community response, and the lack of basic information regarding local resources and lifeline systems.

The All-Union government accepted responsibility for disaster response and recovery in the Armenian earthquake disaster for the first time in Soviet experience. Previous natural disasters had been either excluded from international news reports or treated as the responsibility of the local republic. Despite very generous response from all other republics of the Soviet Union in terms of volunteers and contributions of aid, the lack of an information and organizational infrastructure to coordinate a timely response exacerbated the magnitude of the disaster. The findings are striking in the documentation of information content and exchange as the most critical missing component in the effort to improve capacity for organizational learning and interorganizational coordination in disaster conditions. Greater awareness of the need for more developed and current preparedness plans in communities exposed to seismic risk proved the greatest lesson drawn by the organizations involved in this painful experience.

Adaptive Behavior and/or Self-Organization

The Armenia response system documented an interesting point regarding the evolution of complex, adaptive systems. Breaking out the set of public organizations, the largest subset, 40, or 20.2%, were from the republic of Armenia, with national or All-Union organizations a very close second at 39, or 19.7%. These findings reflect dramatically the shift in Soviet administrative policy, which prior to Armenia had declared natural disasters a responsibility of the republics, with little if any response from the Union, or central government. In this disaster, not only did the Central Government assume primary responsibility for mobilizing the response operations, but Premier Gorbachev also welcomed international assistance for the first time to a natural disaster within the Soviet Union. This decision elicited a warm response from the international community, with public, private and nonprofit organizations from 30 nations participating in response operations. Of the total number of organizations identified for the response system, 80, or 40.4%, were international. This fact also reflects the outpouring of assistance from the loyal Armenian diaspora.

The smallest subset, 15, or 7.6%, was municipal organizations. This finding confirms the basic tenet of adaptive, self-organizing systems, in which actions taken at the local level govern the evolution of the system. If few organizations are capable of action at the local level, the system is not likely to adapt effectively to a sudden, urgent change in the environment. Inadequate action at the local level hinders the further evolution of the response system. This condition is shown first by the official reports released by Civil Defense and later documented by findings from the content analysis of local news reports.

Interview data established a sketchy timeline of disaster response actions, with details missing. Civil Defense Headquarters in Armenia reported the following actions and response times, shown in Table 16.

In this report, there is no mention of local capacity in response actions, nor any expectation that local actions would be taken. Although each city had local organizations for fire suppression, traffic control, and medical care, there was no formal effort to engage them in response operations. All-Union Civil Defense assumed responsibility for operations in any disaster, overlooking the important factor of local knowledge in facilitating response operations.

Table 16. Log: Response Operations, 1988 Armenia Earthquake

December 7, 1988:
11:41 a.m.:	Earthquake occurs; 6.9 Richter scale; epicenter at Nalban, near Spitak
11:42 a.m.:	Civil Defense, Kirovakan reported severe damage from earthquake to Civil Defense Headquarters, Armenia SSR in Yerevan via Civil Defense telephone network
11:44 a.m.:	Civil Defense, Leninakan reported severe damage from earthquake to Civil Defense Headquarters, Armenia SSR in Yerevan via Civil Defense telephone network
11:45 a.m.:	Aftershock occurs; 5.8 Richter scale; causing buildings damaged in main shock to collapse
11:46 a.m.:	Civil Defense Headquarters, Yerevan informed Armenian Government of earthquake; initiated intelligence procedures
12:36 a.m.:	Report from helicopter overflight of Leninakan; took pictures, but could not distinguish accurately the extent of the damage
2:00 p.m.:	No communication existed between the three major cities of Leninakan, Kirovakan and Yerevan
3:40 p.m.:	Military units tasked to organize rescue operations in Leninakan, Spitak, Kirovakan and Stepanavan; local units entered stricken area
4:30 p.m.:	Medical team from Yerevan Surgical Institute arrives at Hospital #1, Leninakan
6:30 p.m.:	Main military units entered disaster area to begin rescue operations

December 8, 1988:
Ground troops report detailed assessment of damage in Leninakan: communications totally disrupted; 154 km. of roads destroyed; 70 km. of railroads destroyed; water, electricity facilities destroyed; waste disposal disrupted.[a]

[a]Civil Defense briefing, Civil Defense Headquarters, Yerevan, Armenia SSR, March 21, 1989.

Table 17. Frequency Distribution: Types of Transactions in Disaster Response by Funding Source and

Type of Transaction	Public Organizations											
	International			National			Regional/State			Municipal		
	T	N	%	T	N	%	T	N	%	T	N	%
Emergency Response	1	1	0.4	12	25	4.6	1	1	0.4	1	3	0.4
Communication	5	6	1.9	17	23	6.5	5	9	1.9	0	0	0.0
Coordination of Response	0	0	0.0	17	29	6.5	3	6	1.1	0	0	0.0
Medical Care/Health	1	1	0.4	3	5	1.1	8	9	3.1	4	5	1.5
Damage Needs Assessment	0	0	0.0	3	6	1.1	3	5	1.1	0	0	0.0
Certification of Deaths	0	0	0.0	1	1	0.4	0	0	0.0	0	0	0.0
Earthquake Assessment/Research	0	0	0.0	1	2	0.4	2	2	0.8	0	0	0.0
Security/Prevention of Looting	0	0	0.0	0	0	0.0	3	3	1.1	0	0	0.0
Housing Issues	1	2	0.4	1	1	0.4	0	0	0.0	0	0	0.0
Disaster Relief (food, shelter, etc.)	1	1	0.4	15	22	5.7	10	15	3.8	6	7	2.3
Donations (money, goods, etc.)	26	37	9.9	1	1	0.4	1	1	0.4	0	0	0.0
Building Inspection	0	0	0.0	0	0	0.0	0	0	0.0	0	0	0.0
Building Code Issues	0	0	0.0	0	0	0.0	1	1	0.4	0	0	0.0
Repair of Freeways, Bridges, Roads	0	0	0.0	0	0	0.0	0	0	0.0	0	0	0.0
Repair/Restore Utilities	0	0	0.0	0	0	0.0	1	1	0.4	0	0	0.0
Repair/Reconstruction/Recovery[a]	0	0	0.0	3	6	1.1	0	0	0.0	1	1	0.4
Transportation/Traffic Issues	0	0	0.0	2	2	0.8	0	0	0.0	0	0	0.0
Hazardous Materials Releases	0	0	0.0	0	0	0.0	0	0	0.0	0	0	0.0
Legal/Enforcement/Fraud	0	0	0.0	1	1	0.4	0	0	0.0	0	0	0.0
Political Dialogue/Legislation	0	0	0.0	0	0	0.0	0	0	0.0	0	0	0.0
Business Recovery	0	0	0.0	0	0	0.0	0	0	0.0	0	0	0.0
Economic/Business Issues	0	0	0.0	0	0	0.0	0	0	0.0	0	0	0.0
Visits by Officials	2	2	0.8	3	3	1.1	2	3	0.8	0	0	0.0
Education Issues	0	0	0.0	0	0	0.0	1	1	0.4	0	0	0.0
Government Assistance	0	0	0.0	6	6	2.3	4	4	1.5	3	4	1.1
Insurance Related Issues	0	0	0.0	0	0	0.0	0	0	0.0	0	0	0.0
Loans (private and international)	0	0	0.0	0	0	0.0	0	0	0.0	0	0	0.0
Psychological/Counseling Services	0	0	0.0	0	0	0.0	0	0	0.0	0	0	0.0
Fundraising/Account Setup	1	2	0.4	3	5	1.1	0	0	0.0	1	2	0.4
Volunteers	1	1	0.4	0	0	0.0	0	0	0.0	0	0	0.0
TOTAL:	39	53	14.9	89	138	34.0	45	61	17.2	16	22	6.11

T = Number of Transactions; N = Number of Actors; % = Percent of Total Transactions
[a] Not including freeways, bridges, roads or utilities
Source: *Sovetakan Hayastan*, December 8–29, 1988. Yerevan, Armenia

Table 17 presents the frequency distribution of types of transactions performed by the subsets of organizations engaged in disaster response. The data are drawn from a content analysis of news reports from *Sovetakan Hayastan*, the governmental newspaper published in Yerevan, Armenia, from December 8–27, 1988.[45] The content analysis identified twenty-one types of transaction in a total of 262 reported transactions that were performed by the organizational response system. The largest proportion, 25.2%, involved donations of money and goods, while the second largest subset, 16.0%, concerned disaster relief. Medical/health care to the injured constituted the third largest subset, 10.3%. Reported transactions

Chapter 5: Nonadaptive Systems

Jurisdiction, Armenia Earthquake, December 8–29, 1988

Nonprofit Organizations							Private Organisations							Totals		
International			National			International			National					
T	N	%	T	N	%	T	N	%	T	N	%	T	N	%
2	2	0.8	5	8	1.9	0	0	0.0	0	0	0.0	22	40	8.4
0	0	0.0	1	1	0.4	0	0	0.0	0	0	0.0	28	39	10.7
0	0	0.0	0	0	0.0	0	0	0.0	0	0	0.0	20	35	7.6
3	4	1.1	0	0	0.0	0	0	0.0	0	0	0.0	19	24	7.3
0	0	0.0	0	0	0.0	0	0	0.0	0	0	0.0	6	11	2.3
0	0	0.0	0	0	0.0	0	0	0.0	0	0	0.0	1	1	0.4
1	1	0.4	0	0	0.0	0	0	0.0	0	0	0.0	4	5	1.5
0	0	0.0	0	0	0.0	0	0	0.0	0	0	0.0	3	3	1.1
0	0	0.0	0	0	0.0	0	0	0.0	0	0	0.0	2	3	0.8
1	1	0.4	9	12	3.4	0	0	0.0	0	0	0.0	42	58	16.0
20	24	7.6	11	14	4.2	7	7	2.7	0	0	0.0	66	84	25.2
0	0	0.0	0	0	0.0	0	0	0.0	0	0	0.0	0	0	0.0
0	0	0.0	2	2	0.8	0	0	0.0	0	0	0.0	3	3	1.1
0	0	0.0	0	0	0.0	0	0	0.0	0	0	0.0	0	0	0.0
0	0	0.0	0	0	0.0	0	0	0.0	0	0	0.0	1	1	0.4
0	0	0.0	0	0	0.0	0	0	0.0	0	0	0.0	4	7	1.5
0	0	0.0	0	0	0.0	0	0	0.0	0	0	0.0	2	2	0.8
0	0	0.0	0	0	0.0	0	0	0.0	0	0	0.0	0	0	0.0
0	0	0.0	0	0	0.0	0	0	0.0	0	0	0.0	1	1	0.4
0	0	0.0	0	0	0.0	0	0	0.0	0	0	0.0	0	0	0.0
0	0	0.0	0	0	0.0	0	0	0.0	0	0	0.0	0	0	0.0
0	0	0.0	0	0	0.0	0	0	0.0	0	0	0.0	0	0	0.0
0	0	0.0	0	0	0.0	0	0	0.0	0	0	0.0	7	8	2.7
0	0	0.0	0	0	0.0	0	0	0.0	0	0	0.0	1	1	0.4
0	0	0.0	0	0	0.0	0	0	0.0	0	0	0.0	13	14	5.0
0	0	0.0	0	0	0.0	0	0	0.0	0	0	0.0	0	0	0.0
0	0	0.0	0	0	0.0	0	0	0.0	0	0	0.0	0	0	0.0
0	0	0.0	0	0	0.0	0	0	0.0	0	0	0.0	0	0	0.0
5	5	1.9	1	1	0.4	4	6	1.5	0	0	0.0	15	21	5.7
1	1	0.4	0	0	0.0	0	0	0.0	0	0	0.0	2	2	0.8
33	38	12.6	29	38	11.1	11	13	4.2	0	0	0.0	262	363	100.0

involving communications constituted 28, or 10.7% of the total number, and transactions involving coordination made up 20, or 7.6% of the total. Although the actual number of transactions is small, these findings document the relatively new importance placed upon official briefings to a global public in the Gorbachev era. They also facilitated the evolution of a disaster response system.

Table 18 presents data on the types of interactions reported among the subsets of organizations involved in disaster response. The findings show that the largest number of interactions, 37 out of a total of 118, or 46.8%, involved Union organizations, while only 2 reported interactions involved municipal level organi-

Table 18. Frequency Distribution: Types of Interactions in Disaster Response by Funding Source and Jurisdiction, Armenia Earthquake, December 8–29, 1988

Type of Interaction	Public International K N %	Public Union K N %	Public Republic K N %	Public Municipal K N %	Nonprofit International K N %	Nonprofit National K N %	Private International K N %	Private National K N %	Total K N %
Public									
International	5 9 6.3	1 2 1.3	0 0 0.0	0 0 0.0	7 14 8.9	1 2 1.3	1 2 1.3	0 0 0.0	15 29 19.0
Union		16 18 20.3	16 15 20.3	0 0 0.0	2 3 2.5	3 6 3.8	0 0 0.0	0 0 0.0	37 42 46.8
Republic			9 9 11.4	3 6 3.8	0 0 0.0	3 8 3.8	0 0 0.0	0 0 0.0	15 23 19.0
Municipal				2 4 2.5	0 0 0.0	0 0 0.0	0 0 0.0	0 0 0.0	2 4 2.53
Nonprofit									
International					1 2 1.3	3 6 3.8	1 2 1.3	0 0 0.0	5 10 6.3
National						4 7 5.1	0 0 0.0	0 0 0.0	4 7 5.1
Private									
International							1 3 1.3	0 0 0.0	3 1.3
National								0 0 0.0	0 0 0.0
Total Interactions	5 9 6.3	17 20 20.3	25 24 31.6	5 10 6.3	19 29 12.7	14 29 17.7	3 7 3.8	0 0 0.0	79 118 100.0

K = Number of Interactions; % = Percent of Total Interactions
Source: *Sovetakan Hayastan*, Yerevan, Armenia, December 8–19, 1988

Chapter 5: Nonadaptive Systems 113

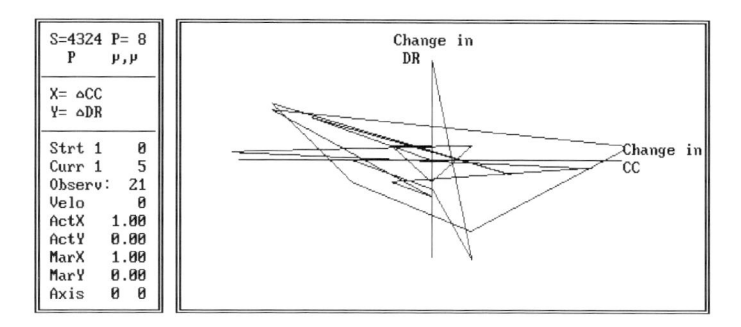

Figure 12. 1988 Armenia Earthquake: Phase Plane Plot, Public Organizations, Emergency Response by Communication/Coordination

zations. Republic organizations were identified in 15, or 19.0%, of the interactions, with international organizations identified in 15, or 19.5%, of the interactions. These findings provide further documentation of the dominance of the organizational response by All-Union organizations.

Examining the frequency data from the content analysis using nonlinear techniques, the findings corroborate the earlier evidence of a response system struggling to cope with an overwhelmingly difficult situation without the necessary infrastructure for communication and coordination. Public, governmental organizations carried the burden of emergency response in the disaster, but the phase plane plots show a lack of communication and coordination on all other functions: damage assessment, disaster relief, recovery/reconstruction, and financial assistance. Figure 12 shows the phase plane plot for emergency response by communication/coordination. The plot shows erratic swings in performance, with emergency response decreasing as communication/coordination increases. Figure 13 shows the marginal history of the changes in emergency response in relation to the changes in communication/coordination. The chart reveals opposing shifts between the two functions in the first four days, with

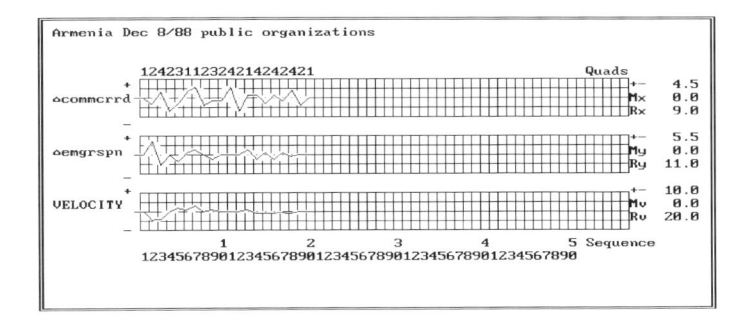

Figure 13. 1988 Armenia Earthquake: Marginal History, Public Organizations, Emergency Response by Communication/Coordination

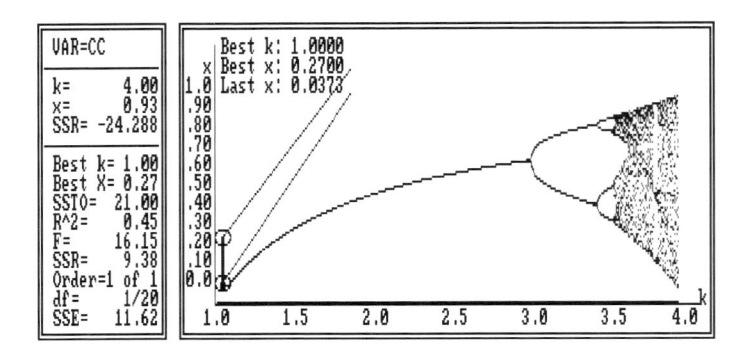

Figure 14. 1988 Armenia Earthquake: Logistic Regression, Public Organizations, Communication/Coordination

emergency response lessening over the 21 day period. The logistic regression analysis did find a logistic equation for emergency response, but the R^2 value is very low at 0.03, with an F value at 0.53, which is not statistically significant. Figure 14 reports a logistic regression equation for the combined function of communication/coordination, with $R^2 = 0.45$ and $F = 16.15$. This finding must be considered cautiously, however, given the low number of observations on which it is based, 25 out of a total of 262 reported transactions.

For the non-governmental organizations, the functions of emergency response and disaster relief showed the most activity. A logistic equation was calculated for emergency response, with $R^2 = 0.43$, and $F = 14.98$. Again, these are moderate values, but interesting in that logistic equations could not be calculated for any of the other five disaster response functions. The group of 13 international private organizations was too small to register in the analysis.

The content analysis identified reports of activity only in disaster relief by international organizations that sent contributions late in the response process, first on the seventh day, and again on the 17th day. The data for Armenia show a relatively small disaster response system evolving under the dominant direction of Soviet All-Union organizations, with substantial involvement from Republic of Armenia organizations and relatively little from municipal organizations of the affected cities. These findings are consistent with reported initial conditions that documented the lack of an information infrastructure to support adaptation to seismic risk at the local level, and to create a basis for self-organizing activity in the event of a destructive earthquake.

Conclusions

Findings from this subset of three disaster response systems show consistencies among them. Table 19 summarizes these findings for this subset of nonadaptive response systems.

Table 19. Summary, Frequency Distributions, Nonadaptive Disaster Response Systems by Funding Source and Jurisdiction

	Public								Total Public		Nonprofit				Private				Total	
	Int'l		Nat'l		State		Mun'l				Int'l		Nat'l		Int'l		Nat'l			
	N	%	N	%	N	%	N	%	N	%	N	%	N	%	N	%	N	%	N	%
San Salvador	61	16.3	96	25.7	0	0.0	48	12.8	206	55.0	23	6.1	80	21.4	12	3.2	54	14.4	374	100.0
Ecuador	54	22.4	79	32.8	6	2.5	15	6.2	154	63.9	12	5.0	59	24.5	6	2.5	10	4.1	241	100.0
Armenia	33	16.6	39	19.7	40	20.2	15	7.6	127	64.1	34	17.1	24	12.1	13	6.7	0	0.0	198	100.0

All three systems were relatively small. In all three, national organizations took the primary role in directing and organizing disaster response operations. All three response systems further showed a strong international component, and a very small municipal or local component, where first response is vital in the time-dependent actions of emergency response. These characteristics indicate a relatively weak set of initial conditions of preparedness, information infrastructure, and capacity for action at the local level. All three response systems were characterized as being low on technical structure, low on flexibility, and low, for the most part, on cultural openness to new information and methods of response.

NOTES

1. United States Agency for International Development (USAID)/El Salvador. 1986. "Assessment of Damages Resulting from the San Salvador Earthquake of October 10, 1986". San Salvador: USAID Mission Reports, Salvador Disaster, November: 5–11.
2. Antonio Godoy, Deputy Chief, Cuerpo de Bomberos, City of San Salvador. Interview. San Salvador, October 21, 1986.
3. Julia Taft, Director, Office of Foreign Disaster Assistance, USAID, statement at Natural Hazards Conference, Boulder, Colorado, July, 1986.
4. Direct observation of Red Cross field operations and interviews with Red Cross staff. San Salvador, October 19–20, 1986.
5. *El Diario de Hoy*, (San Salvador) 17 October, 1987.
6. US Office of Foreign Disaster Assistance, Situation Report No. 6, Friday, October 17, 1986, 2:00 p.m.; interview, Paul Bell, Coordinator, OFDA Operations Team, San Salvador, October 17, 1986.
7. US Embassy, 1986. "The Question of Private Relief to the Archdiocese of San Salvador". Bulletins One and Two, San Salvador, El Salvador, October 16.
8. For a fuller account of this incident, please see Comfort, L. K. "The San Salvador Earthquake" in Rosenthal, Uriel, Michael T. Charles, and Paul t'Hart, eds. 1989. *Coping with Crises: The Management of Disasters, Riots and Terrorism*. Springfield, IL: Charles C. Thomas, Publisher:330–331.
9. Direct observation of information exchange by representatives of international and national nongovernmental organizations regarding disaster assistance, San Salvador, October 20–22, 1986. These meetings were skillfully supported by Paul Bell, Coordinator of the US Office of Foreign Disaster Assistance Office located in Costa Rica. An experienced disaster manager fluent in Spanish and knowledgeable about the Salvadorean conditions, Mr. Bell facilitated the exchange of information between the international and national nonprofit organizations.

10. This survey was carried out in collaboration with Senor Hector Armando Maldonado, San Salvador, El Salvador, in November, 1986.
11. The content analysis included news reports from *El Diario de Hoy*, San Salvador's principal newspaper, for three weeks immediately following the earthquake of October 10, 1986. The first day of the analysis begins on October 14, 1986, as the earthquake disrupted the electricity and operation of the presses at *El Diario de Hoy*, located in the heavily damaged section of the city. No papers were published for the three-day period immediately following the earthquake, October 11–13, 1986. However, the coverage on October 14, 1986 was devoted almost entirely to news reports on the earthquake. Consequently, news reports from October 14, 1986 represent the cumulative accounts for the first three days of response operations.
12. *Hoy*, (Quito, Ecuador) 8 March, 1987: p. 1.
13. The physical characteristics of the earthquake are documented in detail in Robert L. Schuster, Technical Editor. 1991. *The March 5, 1987 Ecuador Earthquakes: Mass Wasting and Socio-economic Effects*. Washington, DC: National Research Council, Committee on International Disasters. Vol. 5.
14. *Hoy*, (Quito, Ecuador). Accounts of ecological, technical, economic, social, political, cultural and international impacts of the earthquakes were reported daily in the two major Quito newspapers, *Hoy* and *El Comercio*, during the month of March, 1987 and continuing during the succeeding months. The author read both papers daily during the period of her field study, June 14 – July 15, 1987, and sought to obtain back issues of both papers for the month of March, 1987. Regrettably, she was unable to obtain a complete set of back issues of *El Comercio* for this period. Consequently, the newspaper references in this analysis are drawn primarily from *Hoy* during March, 1987, but refer to both newspapers during June and July, 1987. To counter any possible bias from a single source, the author sought to find at least two references for critical points in the analysis.
15. See map of disaster zones presented in Appendix B.
16. United Nations Economic Commission for Latin America and the Caribbean – ECLAC. "The Natural Disaster of March, 1987 in Ecuador and its Impact on Social and Economic Development," Report #87-4–406, 6 May, 1987, p.1.
17. *Hoy*, (Quito, Ecuador) 9 March, 1987. Lower figures were also reported in the house-by-house censuses conducted by the municipalities. In central Napo Province, the zone of primary impact from the disaster, the assessment team of CATEC/Catholic Relief Services, also conducted a house-by-house census of need. CATEC (Corporación de Apoyo a la Tecnología y a la Comunicación) joined with Catholic Relief Services in designing and conducting an emergency assistance project in Napo Province. The two organizations are voluntary relief organizations financed by contributions from Catholic parishioners and operating with an international mission of social service. However, there were no complete records of residents living in the area prior to the earthquake, leaving in doubt the actual number of persons killed in the disaster. "Summary of Relief Program," Catholic Relief Services, Quito, Ecuador, June 15, 1987. Program Director, Catholic Relief Services. Interview, Quito, Ecuador, July 12, 1987.
18. General Antonio Moral Moral, National Director of the Civil Defense, cited in *Hoy*, Quito, Ecuador, 9 March, 1987, p.1. See also United Nations ECLAC Report #87-04-406, op.cit.: p.1.
19. *Hoy*, (Quito, Ecuador) 10 March, 1987, p. 3A.
20. *Hoy*, (Quito, Ecuador). Also, professional interviews with local government officials in Baeza, Borja, and El Chaco, July 9, 1987 and with the Project Director, "Proyecto Emergencia in Napo Province," Catholic Relief Services/CATEC in Quito, July 12, 1987.
21. *Hoy*, (Quito, Ecuador) 10 March, 1987, p. 9. Interview, Field Representative, USAID/OFDA, Quito, Ecuador, June 28, 1987.
22. Interview, Padre, Misión Carmelita, Lago Agrio, Ecuador, June 29, 1987; interview, Director of the CEPE-Texaco Consortium, Quito, Ecuador, July 3, 1987. The CEPE-Texaco Consortium was established between the Government of Ecuador and Texaco Oil Company to manage oil production and shipment from a given location in eastern Napo Province.

23. Interview, Padre, Misión Carmelita, Lago Agrio, June 29, 1987; interview, Field Director for Indian Services, Catholic Relief Services, Quito, Ecuador, July 7, 1987.
24. *Hoy*, (Quito, Ecuador) 11 March, 1987, p. 1. Interview, United States Ambassador, Quito, Ecuador, July 6, 1987.
25. The term 'between' is used in a statistical sense to connote the type of variance that exists between member organizations of a given set, in contrast to the type of variance that exists 'within' each member organization. In this analysis, the set of organizations includes all organizations that participated in the Ecuadorian disaster operations. Variance 'between' organizations may be explained by distinctive characteristics or attributes of individual organizations. Variance 'within' organizations is assumed to be distributed randomly. The total variance for the set of organizations is the sum of the 'between' or explained variance and the 'within' or random variance. This analysis is seeking to identify the types of characteristics that contribute to variance between organizations that participated in the Ecuadorian disaster operations.
26. *Hoy*, (Quito, Ecuador) 8 March, 1987, p.1; 9 March, 1987, p.1; 11 March, 1987, p.9A. These reports were confirmed in interviews during June – July, 1987 with informed observers from both Ecuador and the United States, who had participated in disaster assistance operations in March, 1987. A survey of residents of the disaster zones also confirmed difficulties and delays in the distribution of disaster assistance. It is important to identify where the difficulties exist in the process, without making judgments as to cause, before the process can be redesigned for improved performance.
27. Coordinator, *COEN* (National Center of Emergency Operations). Interview, Quito, Ecuador, July 7, 1987. See also *Secretaría General del Consejo de Seguridad Nacional, Dirección Nacional de Defensa Civil,"Ley de Seguridad Nacional"*, 1987, p.1.
28. Coordinator, National Emergency Operations Center and Director, National Council of the Civil Defense Authority. Interview, Quito, Ecuador, July 7, 1987.
29. The Ecuadorian political sub-divisions, from largest to smallest, are: province, *canton, parróquia*. The rural *parróquia* normally contains a dozen or so villages.
30. Comandante de Battalon de Selva, Lago Agrio, and Director of Civil Defense, Canton of Lago Agrio. Interview, Lago Agrio, Ecuador, June 30, 1987.
31. The linkage between disaster management and development deserves further study. The example of Ecuador is an especially interesting case.
32. Comandadura de Defensa Civil, Province of Pichincha. Interview, Quito, Ecuador, June 17, 1987.
33. Pastor, Iglesia del Pacto Evangélico del Ecuador, Fundación Adelanto Comunitario Ecuatoriano, Quito, Ecuador, July 13, 1987.
34. *Hoy*, (Quito, Ecuador) 7 March, 1987, p. 6A.
35. *Hoy*, (Quito, Ecuador) 7 March, 1987, p.6A.
36. *Hoy*, (Quito, Ecuador) 22 March, 1987, p.1. Interview, professor of sociology, Universidad Católica, Quito, Ecuador, June 22, 1987.
37. These figures were presented by General Leo Melkomov, Director, Civil Defense, Armenia SSR, in a briefing for members of the Disaster Reanimatology Study Group in Yerevan, Armenia SSR, March 21, 1989.
38. Reports of the number of dead in this earthquake varied widely. Armenian-American observers reported an estimated 55,000 dead. The British Embassy reported an estimated 50,000 dead. Russian physicians who organized the medical response for Leninakan estimated 80,000 dead. Interview, Pittsburgh, PA, September 23, 1989.The actual number may never be known.
39. Briefing, Prof. Gabrielyan, Ministry of Health, Armenia SSR, Yerevan, Armenia, March 21, 1989. Two researchers, Nechaev and Resnick, report slightly different figures, but in the same range.
40. Briefing, Director, Armenia Civil Defense, Civil Defense Headquarters, Yerevan, Armenia SSR, March 21, 1989.
41. Civil Defense briefing, Civil Defense Headquarters, Yerevan, Armenia SSR, March 21, 1989.

42. Political Affairs Officer, U.S. Embassy, Moscow, USSR. Interview, March 15, 1989. The officer was stationed in Yerevan to assist with disaster response from December 13, 1988 — January 10, 1989. He reported his direct observations of the disaster response operations. This information was corroborated by General Leo Melkomov. Civil Defense, Armenia SSR, in his briefing in Yerevan, on March 21, 1989.
43. RADIO Magazine published reports of the efforts made by amateur radio operators to offer their services to assist in disaster operations. Published in Moscow, April, 1990. In Russian.
44. Political Affairs Officer, U.S. Embassy. Interview, Moscow, USSR, March 15, 1989.
45. I gratefully acknowledge Dr. Arthur Melkonian, Armenian Ministry of Health, who did the content analysis of newspapers from Yerevan, Armenia.

CHAPTER SIX

EMERGENT ADAPTIVE SYSTEMS: MEXICO CITY, COSTA RICA, ERZINCAN

Beginning Adaptation

Three disaster response systems — Mexico City (1985), Costa Rica (1991), and Erzincan, Turkey (1992) — fall in the category of emergent adaptive systems, which reveals the beginning of adaptation. Adaptation represents organizational learning in response to environmental change that leads to a change in performance for the entire system of organizations engaged in disaster operations. It is an important initial phase in a community's capacity to mobilize resources and reallocate energies in response to sudden, urgent needs. In each of the three cases, the dynamic processes initiated by the community to meet immediate needs were not sustained much past the first traumatic weeks, but these processes nonetheless created a memory of shared experience in response to common trauma that could be mobilized to support action to reduce risk from future events.

Emergent adaptive systems, described briefly in Chapter Four, are characterized by low technical structure, medium organizational flexibility, and emerging openness to new cultural meanings of seismic risk in their respective communities. The built infrastructure in each of these three disaster-stricken communities revealed serious inadequacies in design and construction. The organizations responding to the events reflected different cultures, economies and political systems, and showed varying degrees of flexibility in being able to adapt to the damaging conditions of a major earthquake. But openness to a new interpretation of roles and redefinition of meaningful action for participating organizations and groups in light of the destructive events proved to be the critical difference between the response systems in this category and those in the category of "nonadaptive systems". In the emergent adaptive category, each response system demonstrated capacity to build a sense of shared understanding with, and to elicit a commitment to action from, the wider society, depending upon the available means of communication and access to information.

Information processes for these three disaster response systems revealed the basic set of information functions: 1) information search; 2) information exchange; 3) organizational learning; and 4) adaptive behavior and/or self organization. While each system differed in the particular mix and strength of these functions, each also demonstrated evidence of the set of information functions and the interdependence among them. The initial information search function influenced the exchange of information among the participating organizations, shaping the subsequent degree of organizational learning and adaptive behavior demonstrated by each system. This interaction among the set of functions will be shown in brief profiles of the three disaster response systems.

Mexico City: September 19, 1985 (M = 8.1)

At 7:19 a.m. on Friday, September 19, 1985, a powerful earthquake measuring 8.1 on the Richter scale struck Mexico City, as the densely populated urban area was awakening to morning activity. The epicenter of the earthquake, located some 200 miles southwest of the capital off the Pacific Coast of the state of Michoacan, sent strong ground motion waves through the weak soil structure of the ancient lake bed on which much of modern Mexico City is built. Amplified by the soft soils, the shock waves toppled high-rise buildings, collapsed hospitals, caved hotels and schools inward, and severely damaged governmental structures. In the city of 18 million people, a total of 954 buildings collapsed, with tens of thousands of people trapped inside.[1] The official death toll listed an estimated 10,000 dead. Unofficial estimates cited more than 20,000 dead.[2] Collapsed urban infrastructure created extraordinarily difficult demands for the rescue of live persons trapped inside, with fallen concrete structures creating almost impossible barriers to extrication. International rescue teams with special training and equipment flew to Mexico City to participate in the time-critical effort to save lives.[3]

Two characteristics of this earthquake had profound effects upon the emergence of the evolving response system. First, the severe shock damaged nine governmental ministries.[4] With the loss of offices, telephones, and records from these ministries, the government's capacity to function in response to the disaster was seriously reduced. Second, the communications towers of Telefonos de Mexico (TELMEX) collapsed in the earthquake, silencing telephone communication with the rest of the world, and severely restricting it within the city. These two conditions affected in important ways the evolution of information processes to support disaster operations.

Information Search

The damage to telephone communications and the destruction of major ministries disabled the national government's ability to gather timely, accurate information about the state of the city and the needs of the residents in the affected colonias. The same conditions, conversely, created an opportunity for the emergence of new sources of energy and organization that contributed to the developing organizational response system. Without visible presence of the normally strong central government, volunteers emerged to perform essential tasks of search and rescue, ministration of first aid care to injured victims, and distribution of water due to damaged water mains. Students spontaneously took responsibility for directing traffic, assessing buildings for damage, and organizing food and shelter for homeless victims. Information gathering was local, and families, friends and neighbors formed informal networks of communication and assistance within their own colonias.[5] Yet, the magnitude of the damage and the large numbers of people affected by the disaster created an enormous trauma for the nation. Out of this trauma emerged a response system that engaged ordinary people in responsible roles in disaster operations to assist their families, neighbors and community. But without continuing processes of communication and ready access to timely information, this system could not be sustained over time.

On the national level, the international news organizations played an unanticipated, but important role in communicating news of the event to the external world. With international communications disrupted, an enterprising CBS news correspondent based in Mexico City filmed extensive footage of the September 19, 1985 disaster on videotape, hired a Lear Jet and flew to Laredo, Texas to present the first photos of the damaged city on the 6:00 p.m. evening news.[6] Satellite communications then beamed the photos to television stations around the world, and international journalists, as well as plane-loads of international aid, converged on the stricken city. In the first ten days following the earthquake, at least 350 international news organizations had representatives covering the unfolding drama of disaster operations in Mexico City.[7]

This sudden influx of professional journalists and representatives of voluntary charitable organizations, often working with the recognition and support of their respective governments, had a substantial effect upon the search for information in the initial stages of disaster operations. Facilitated in communicating their work by advanced information technology, the large number of international journalists had an extraordinary effect in opening the news coverage of the disaster to the Mexican people as well as to the rest of the world. A normally cautious Mexican national administration fell suddenly under the spotlight of the world press, and sought to manage the response operations to the disaster as effectively as possible. Then-President Miguel de la Madrid, presiding over a nation reeling from economic dislocation and heavy external debt prior to the earthquake, recognized the event as an opportunity to gain good will and cooperation from the international community. He needed to restructure Mexico's external debt in order to rebuild the capital city's damaged structures.[8] In response to scrutiny from the international press as well as demands from his own people, President de la Madrid took several innovative steps to demonstrate national capacity to organize disaster operations. These steps included, first, allowing an unusual range of freedom in investigative reporting to both international and Mexican journalists to document the efforts made by the Mexican Government in organizing immediate response to the disaster. Second, as both international and Mexican journalists reported a remarkable level of spontaneous, voluntary action by students and other informal, community groups in organizing the distribution of relief goods, provision of medical care, and temporary shelter to families made homeless by the event, Government officials, for the most part, acknowledged these activities as legitimate contributions to disaster response and allowed them to go forward.[9]

Third, and central to forging a national base of collaboration and support for disaster operations, President de la Madrid formed the National and Metropolitan Emergency Commission that included representatives of major governmental ministries, business corporations, labor unions, and the Church in organizing a national effort to meet immediate needs and rebuild the damaged infrastructure. By engaging the major public, private, and nonprofit organizations of the nation in a common effort at response and recovery from the destructive event, the President widened the information search to include the perspectives and assessment of each of the participating organizations in determining the needs and allocating available resources to response and recovery operations. The President also charged this national-level committee with the responsibility of documenting the significant amounts of international aid that were donated to Mexico, acknowledging each nation for its contribution, and seeking to ensure that the contributions were directed promptly to the intended recipients. Mindful of the painful historical experience following the Managua, Nicaragua Earthquake of 1976 when the Nicaraguan Government was accused of mismanaging international aid, President de la Madrid sought to demonstrate the performance of a competent, professional, national administration in managing international assistance. His strategy of recovery depended upon gaining the respect and willing support of the international community. These three interacting conditions produced a much broader information search than conventional governmental policy-making.[10]

Information Exchange

The actual exchange of information resulting from these information search processes, however, showed a substantial discrepancy between policy and practice. Although information regarding governmental policy and intended operations was widely distributed, access to the information and actual mechanisms for obtaining assistance proved much more limited. This tension, reported relatively widely in reference to many aspects of disaster operations — from obtaining flashlights for search and rescue to distributing water to the neighborhoods to providing temporary shelter for those left homeless — is vividly illustrated in reference to the distribution of international disaster assistance.

Both donor nations and the Mexican people perceived the prompt and equitable distribution of international aid as a primary measure of governmental

Table 20. Perceived Benefits of International Aid

"Mexico has received in these last two months (November 1985) much gratuitous international aid to alleviate the problems generated by the earthquakes. How beneficial, in your opinion, has this international assistance been?"

	N	%
Very beneficial	341	46.8
Beneficial	222	30.5
So-So	62	8.5
Not so beneficial	47	6.5
No benefit	42	5.8
Don't know	14	1.9
	728	100.0

Valid N = 714

competence and responsible action in reference to urgent human needs. Results from a survey of 728 residents of damaged areas of Mexico City provide a measure of the degree of information exchange that residents of these damaged areas of the city actually perceived.[11] Tables 20–23 reveal a rapidly narrowing funnel of information that leads to action.

As cited in Table 20, the large majority of respondents, 563 out of a sample of respondents, or 77.3%, perceived international aid to be beneficial to Mexico in coping with disaster-related problems, yet only 35.4% reported access to information about how to get international aid. This number dropped still further, to 22.9%, when respondents were asked whether information about how to obtain aid was sufficient. The number of respondents who reported actually receiving aid — either for themselves or their families — dropped still further to 76 cases, or 10.4% of all respondents. Finally, only a very small proportion, 6.5%, or 47, of the respondents reported actual contact with an international organization engaged in disaster assistance.

Table 21. Perceived Availability of Information Regarding International Aid in District

"Was there information in your district about how to get access to this (international) aid, if necessary?"

	N	%
Yes	258	35.4
No	433	59.5
Don't know	34	4.7
Not applicable	3	0.4
	728	100.0

Valid N = 691

Table 22. Perceived Sufficiency of Information Regarding International Aid in District

"In your opinion, was the information given in your district about how to ask for, and receive aid sufficient?"

	N	%
Sufficient	167	22.9
Insufficient	253	34.8
Don't know	49	6.7
Inapplicable	259	35.6
	728	100.0

Valid N = 420

Table 23. Reception of International Aid

"In fact, did you or your family or any of your neighbors receive any aid of this type?"

	N	%
Yes	76	10.4
No	641	88.1
No response	11	1.5
	728	100.0

Valid N = 717

Table 24. The Relationship Between Availability and Sufficiency of Information Regarding International Aid

Sufficiency of Information

Availability of information in district	Sufficient	Insufficient	Total N	%
Yes	159	93	252	
	63.1[a]	36.9		60.7
	95.8[b]	37.3		
	34.3[c]	22.4		
No	7	156	163	
	4.3	45.7		
	4.2	62.7		
	1.7	37.8		39.3
Total Cases	166	249	415	
Total Percent	40.0	60.0		100.0

Chi Square = 142.592; Sig. = 0.000; DF = 1
[a] = row percent; [b] = column percent; [c] = percent of total cases

Table 24 reports that 60% of the respondents considered information regarding international aid to be insufficient in their districts. Table 25 shows that nearly two-thirds of the respondents, 65.8%, reported that they did not receive international aid, and that information about international aid was not available in their districts. Finally, Table 26 shows that only 31 of the 54 respondents who actually received international aid perceived the information to be sufficient, while the

Table 25 The Relationship Between Availability of Information and Reception of International Aid

Availability of Information in District

Reception of international aid	Yes	No	Total N	%
Yes	47 62.7[a] 18.5[b] 6.9[c]	28 37.3 6.6 4.1	75	11.0
No	207 34.2 81.5 30.4	399 65.8 93.4 58.6	606	89.0
Total Cases	254	427	681	
Total Percent	37.3	62.7		100.0

Chi Square = 23.193; Sig. = 0.000; DF = 1
[a] = row percent; [b] = column percent; [c] = percent of total cases

Table 26. The Relationship Between Reception of International Aid and Sufficiency of Information

Reception of International Aid

Sufficiency of aid	Yes	No	Total N	%
Sufficient	31 18.8[a] 57.4[b] 7.5[c]	134 81.2 37.3 32.4	165	40.0
No	23 9.3 42.6 5.6	225 90.7 62.7 54.5	248	60.0
Total Cases	54	359	413	
Total Percent	13.1	86.9		100.0

Chi Square = 7.89; Sig. = 0.005; DF = 1
[a] = row percent; [b] = column percent; [c] = percent of total cases

large majority of respondents reported that the information about aid was inadequate.[12]

These findings show major discrepancies between stated policy and actual practice in the distribution of international disaster assistance by the evolving response system. Such discrepancies suggest gaps in the process of information exchange between the organizations receiving and cataloguing disaster assistance as it entered the country and the actual families who needed food, clothing, and household goods in the city's damaged neighborhoods. The process of information exchange, organized at the national level, appeared to have inadequate links for actual delivery at the colonia level.

Organizational Learning

The number and types of organizations engaged in the Mexico City disaster operations shaped the process of organizational learning among them. Returning to the N-K methodology, the number of organizations participating in disaster operations, the types of transactions they conducted, and the frequency of interactions among the participating organizations characterized the process of organizational learning. The patterns that evolved among them also provide insight into the incentives and constraints on the process. Based on reports of disaster operations from *Excelsior*, a major Mexico City newspaper, from September 20 – October 11, 1985, the content analysis identified a disaster response system of 500 organizations (N = 500) that evolved following the September 19, 1985 earthquake.

Table 27 shows that the response was largely public, with 264, or 52.8% of the organizations coming from this sector. Within this public subset, national organizations represented the largest group, with 117, or 44.3% of the total of 264 organizations. International organizations represented more than one-third, or 34.1%, of the public subset, while state and municipal organizations together made up 21.6%. Nonprofit organizations, including those funded from both national and international sources, made up nearly one third, or 31.8%, of the organizations participating in disaster operations. Private organizations represented the smallest group with 15.4% of the total.

Table 27. Frequency Distribution: Disaster Response System by Funding Source and Jurisdiction, 1985 Mexico City Earthquake

Public				Total Public	Nonprofit		Private		Total
Int'l	Nat'l	State	Mun'l		Int'l	Nat'l	Int'l	Nat'l	
N %	N %	N %	N %	N %	N %	N %	N %	N %	N %
90 18.0	117 23.4	38 7.6	19 3.8	264 52.8	59 11.8	100 20.0	28 5.6	49 9.8	500 100

Table 28 presents the matrix of types of transactions performed by level of jurisdiction and funding source. The full response system performed a total of 641 transactions, as reported in *Excelsior*. Of those, the overwhelming proportion, 78.8%, were performed by public organizations, with nonprofit organizations engaged in 17.5%, and private organizations in 3.7.% of the reported transactions. Examining the type of transactions performed, the largest proportion, 29.8%, was reported for donations of money and goods to be distributed to the people directly affected by the earthquake.

The second largest proportion, or 7.8%, of reported transactions involved provision of food, shelter, medical care to people displaced or harmed as a consequence of the earthquake. Interestingly, the third largest proportion, or 7.6%, of the transactions involved coordination of activities in mobilizing response operations. Communication followed coordination closely, claiming 6.7% of the transactions, while emergency response and repair/ reconstruction activities each constituted 5.8% of the response activities reported in *Excelsior*. These findings show a marked increase in reported transactions for communication and coordination from the category of "nonadaptive" response systems, reflecting not only the size and complexity of the Mexico City response system, but also the active engagement of organizations operating in collective response to the city's need.

The measure, K, represents the number of interactions among organizations that collaborate to accomplish a shared task. Table 29 presents the matrix of interactions among types of organizations participating in disaster operations. The pattern of interactions revealed primarily a public response, with public organizations involved in 88.9% of the total number of 219 reported interactions. Of these, the largest proportion was between national public agencies and all other types of organizations, 67.4%, with the next largest group of interactions between international public agencies and all other types of organizations, 11.4%. Nearly one-half, 31.1%, of the interactions involving national public organizations were with other national public organizations, with the next largest proportion, 16.9%, occurring between national public organizations and provincial public organizations, treating Mexico City as a province. Nonprofit organizations were involved in 10.5% of the total interactions, and private organizations were involved in 0.9% of the total number of interactions among organizations.

These findings document the significant international involvement in disaster response, and also the low level of local governmental participation, with only 1.4% of the total number of interactions attributed to the municipal level. In part, this finding reflects the severity of the earthquake and the massive damage to the capital city, but it also indicates the limited role of local government, which likely had critical information about local conditions, in the overall response process.

Table 28. Frequency Distribution: Types of transactions in Disaster Responses by Funding Source and

Type of Transaction	Public Organizations												
	International			National			State			Municipal		
	T	N	%	T	N	%	T	N	%	T	N	%
Emergency Response	8	10	1.2	16	25	2.5	3	7	0.5	2	2	0.3
Communication	3	8	0.5	18	38	2.8	7	15	1.1	2	3	0.3
Coordination of Response	6	9	0.9	27	81	4.2	1	3	0.2	2	3	0.3
Medical Care/Health	2	6	0.3	10	17	1.6	0	0	0	0	0	0
Damage/Needs Assessment	0	0	0.0	17	20	2.7	1	1	0.2	0	0	0
Certification of Deaths	0	0	0.0	11	30	1.7	2	5	0.3	3	5	0.5
Earthquake Assessment/Research	0	0	0.0	3	8	0.5	0	0	0	0	0	0
Security/Prevention of Looting	0	0	0.0	4	7	0.6	0	0	0	0	0	0
Housing Issues	0	0	0.0	4	9	0.6	1	2	0.2	0	0	0
Disaster Relief (food, shelter, etc.)	8	18	1.2	25	66	3.9	0	0	0	0	0	0
Donations (money, goods, etc.)	133	138	20.7	23	36	3.6	2	3	0.3	2	2	0.3
Fundraising	1	1	0.2	0	0	0.0	0	0	0.0	0	0	0.0
Building Inspection	1	4	0.2	5	12	0.8	1	3	0.2	0	0	0
Building Code Issues	0	0	0.0	1	2	0.2	0	0	0	0	0	0
Repair of Freeways, Bridges, Roads	0	0	0.0	1	4	0.2	0	0	0	1	2	0.2
Repair/Restore Utilities	0	0	0.0	16	34	2.5	1	1	0.2	2	6	0.3
Repair/Reconstruction/Recovery[a]	1	3	0.2	27	59	4.2	0	0	0	2	4	0.3
Transportation/Traffic Issues	0	0	0.0	2	4	0.3	0	0	0	0	0	0
Legal/Enforcement Issues	0	0	0.0	15	34	2.3	1	2	0.2	1	2	0.2
Legislation/Legislative Processes	0	0	0.0	10	31	1.6	1	3	0.2	0	0	0
Business Recovery/Agriculture	0	0	0.0	4	9	0.6	0	0	0	0	0	0
Economic/Business Issues	0	0	0.0	4	6	0.6	0	0	0	0	0	0
International Loans	12	26	1.9	1	1	0.2	0	0	0	0	0	0
Economic Policy Issues	10	29	1.6	5	13	0.8	0	0	0	0	0	0
Decentralization Issues	0	0	0.0	2	5	0.3	0	0	0	0	0	0
Visits by Officials	0	0	0.0	1	1	0.2	0	0	0	0	0	0
Education Issues	0	0	0.0	6	17	0.9	0	0	0	1	3	0.2
Emergency Education	0	0	0.0	6	14	0.9	0	0	0	1	1	0.2
Government Assistance	0	0	0.0	10	26	1.6	1	1	0.2	0	0	0
Psychological/Counselling Services	0	0	0.0	2	3	0.3	0	0	0	0	0	0
Volunteer Activities	0	0	0.0	3	8	0.5	0	0	0	0	0	0
TOTAL	185	252	28.9	279	620	43.5	22	46	3.4	19	33	3.0

T = Number of Transactions; N = Number of Actors; % = Percent of Transactions column total
[a] Not including freeways, bridges, roads or utilities
Source: *Excelsior*, Mexico City, DF

Adaptive Behavior and/or Self-Organization

The nonlinear analysis of change in the frequencies of organizational actions reported in daily news reports supports the dominant role of the national public organizations in disaster operations and the evolving disaster response system. While some functions were clearly more important than others, actions by public national organizations were reported for each of the six major functions of disaster operations — emergency response, damage assessment, communication/coor-

Jurisdiction, Mexico City Earthquake, September 20 – October 11, 1985

Nonprofit Organizations						Private Organizations						Total		
International			National			International			National						
K	N	%	K	N	%	K	N	%	K	N	%	K	N	%
2	4	0.3	5	6	0.8	0	0	0	1	1	0.2	37	55	5.8
0	0	0	11	15	1.7	1	1	0.2	1	1	0.2	43	81	6.7
1	9	0.2	12	30	1.9	0	0	0	0	0	0.0	49	135	7.6
0	0	0	0	0	0.0	0	0	0	0	0	0.0	12	23	1.9
0	0	0	5	8	0.8	0	0	0	0	0	0.0	23	29	3.6
0	0	0	1	2	0.2	0	0	0	0	0	0.0	17	42	2.7
0	0	0	0	0	0.0	0	0	0	0	0	0.0	3	8	0.5
0	0	0	0	0	0.0	0	0	0	0	0	0.0	4	7	0.6
0	0	0	4	9	0.6	0	0	0	0	0	0.0	9	20	1.4
0	0	0	16	33	2.5	0	0	0	1	3	0.2	50	120	7.8
4	4	0.6	15	19	2.3	6	6	0.9	6	6	0.9	191	214	29.8
1	1	0.2	0	0	0.0	3	7	0.5	2	4	0.3	7	13	1.1
0	0	0	0	0	0.0	0	0	0	0	0	0.0	7	19	1.1
0	0	0	0	0	0.0	0	0	0	0	0	0.0	1	2	0.2
0	0	0	0	0	0.0	0	0	0	0	0	0.0	2	6	0.3
0	0	0	3	6	0.5	0	0	0	1	1	0.2	23	48	3.6
0	0	0	7	11	1.1	0	0	0	0	0	0.0	37	77	5.8
0	0	0	0	0	0.0	0	0	0	0	0	0.0	2	4	0.3
0	0	0	2	6	0.3	0	0	0	0	0	0.0	19	44	3.0
0	0	0	2	5	0.3	0	0	0	0	0	0.0	13	39	2.0
0	0	0	4	9	0.6	0	0	0	0	0	0.0	8	18	1.2
0	0	0	4	9	0.6	0	0	0	1	1	0.2	9	16	1.4
0	0	0	0	0	0.0	0	0	0	0	0	0.0	13	27	2.0
0	0	0	4	11	0.6	1	1	0.2	0	0	0.0	20	54	3.1
0	0	0	1	2	0.2	0	0	0	0	0	0.0	3	7	0.5
0	0	0	0	0	0.0	0	0	0	0	0	0.0	1	1	0.2
0	0	0	2	5	0.3	0	0	0	0	0	0.0	9	25	1.4
0	0	0	0	0	0.0	0	0	0	0	0	0.0	7	15	1.1
0	0	0	0	0	0.0	0	0	0	0	0	0.0	11	27	1.7
0	0	0	2	3	0.3	0	0	0	0	0	0.0	4	6	0.6
0	0	0	4	5	0.6	0	0	0	0	0	0.0	7	13	1.1
8	18	1.3	104	194	16.2	11	15	1.7	13	17	2.0	641	1195	100.0

dination, disaster relief, recovery/reconstruction, and financial assistance. This finding indicated that a disaster response system evolved, if not fully developed.

Led by public organizations, the response system also involved nonprofit and private organizations. Nonprofit organizations were primarily engaged in disaster relief and to some extent, emergency response. Private organizations played a smaller role, active primarily in the functions of damage assessment and recovery/reconstruction.

130   *Part II: Shared Risk in Practice*

Table 29. Frequency Distribution: Types of Interactions in Disaster Response by Funding Source

Type of Interaction	Public											
	International			National			Provincial			Municipal		
	K	N	%	K	N	%	K	N	%	K	N	%
Public: International	4	7	1.8	13	26	5.9	2	4	0.9	0	0	0
Public: National				68	69	31.1	37	42	16.9	5	9	2.3
Public: State							7	12	3.2	3	5	1.4
Public: Municipal										1	2	0.5
Nonprofit: International												
Nonprofit: National												
Private: National												
Private: National												
TOTAL	4	7	1.8	81	95	37.0	46	58	21.0	9	16	4.1

K = Number of Interactions; N = Number of Actors; % = Percent of Total Interactions
Source: Excelsior, Mexico City, Mexico D.F.

Figure 15 shows the phase plane plot of change in reports of emergency response actions by change in reports of communication/coordination actions for public organizations. The pattern reveals early chaotic behavior between the variables, settling into a more regular pattern in the later period of response.

Figure 16 shows the marginal history of this interaction, with reports of communication/coordination activities declining on Day 2 following the earthquake, but rising again sharply on Day 3. Reported activities drop again on Day 4, but rise on Day 6. Change in reported emergency response activities is roughly similar, and especially interesting is the sharp rise on Day 6. This surge in emergency response actions is late in terms of life-saving activities, but likely reflective of the strong international response to the difficult heavy rescue requirements reported in the damaged city. The interaction between the two variables then set-

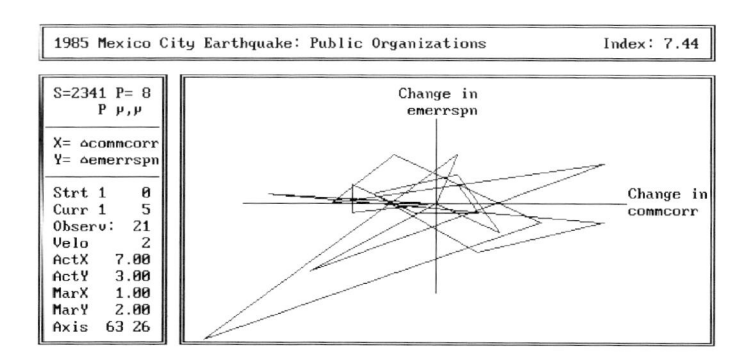

Figure 15. 1985 Mexico City Earthquake: Phase Plane Plot, Public Organizations, Emergency Response by Communication/Coordination

and Jurisdiction, Mexico City Earthquake, September 20 – October 11, 1985

Nonprofit						Private						Total		
International			National			International			National					
K	N	%	K	N	%	K	N	%	K	N	%	K	N	%
4	9	1.8	2	4	0.9	0	0	0.0	0	0	0	25	50	11.4
4	15	1.8	22	40	10.0	4	14	1.8	7	20	3.2	147	209	67.4
0	0	0.0	6	10	2.7	0	0	0.0	3	5	1.4	19	32	8.7
0	0	0.0	2	5	0.9	0	0	0.0	0	0	0	3	7	1.4
3	10	1.4	5	6	2.3	5	9	2.3	1	2	0.5	14	18	6.4
			3	4	1.4	3	4	1.4	3	4	1.4	9	21	4.1
						0	0	0	2	4	0.9	2	4	0.9
												0	0	0.0
11	34	5.0	40	69	18.3	12	27	5.5	16	35	7.3	219	341	100.0

tles into a more regular pattern, with reports of communication/coordination activities leading shifts in emergency response. Interaction between the functions of communication/coordination and disaster relief show a slightly different pattern in the marginal history record presented in Figure 17. Initial reports of disaster relief activities are followed by a sharp increase in reported communication/coordination activities. This shift may be explained by the return of technical capacity as telephone lines were repaired and communications restored. The pattern shifts after Day 3 to reveal that when communication/coordination activities decline, disaster relief activities also decline. In terms of enlisting broad support for disaster relief activities in disaster operations, communication/coordination activities were clearly influential.

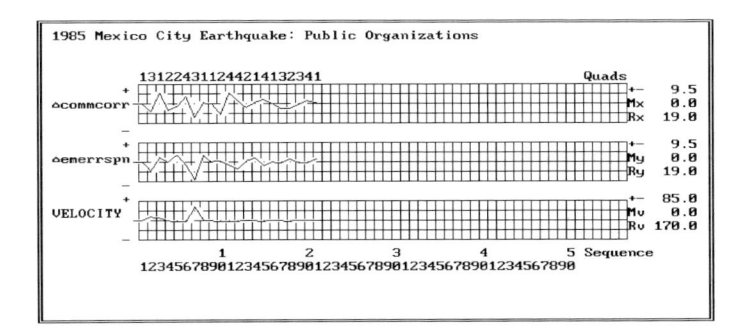

Figure 16. 1985 Mexico City Earthquake: Marginal History, Public Organizations, Emergency Response by Communication/Coordination

132   Part II: Shared Risk in Practice

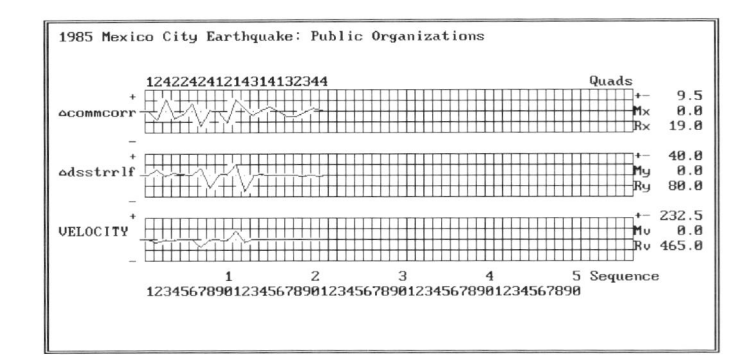

Figure 17. 1985 Mexico City Earthquake: Marginal History, Public Organizations, Disaster Relief by Communication/Coordination

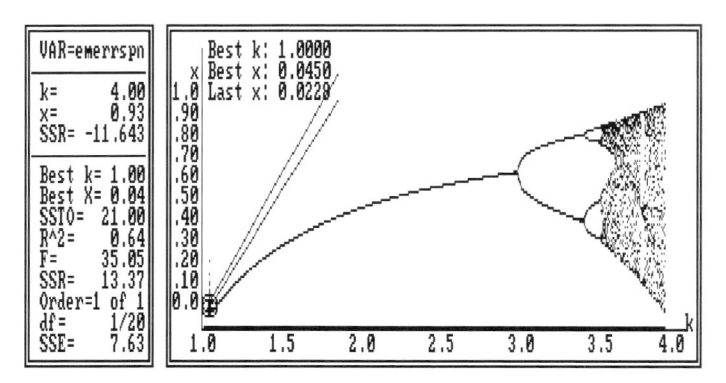

Figure 18. 1985 Mexico City Earthquake: Logistic Regression, Public Organizations, Emergency Response

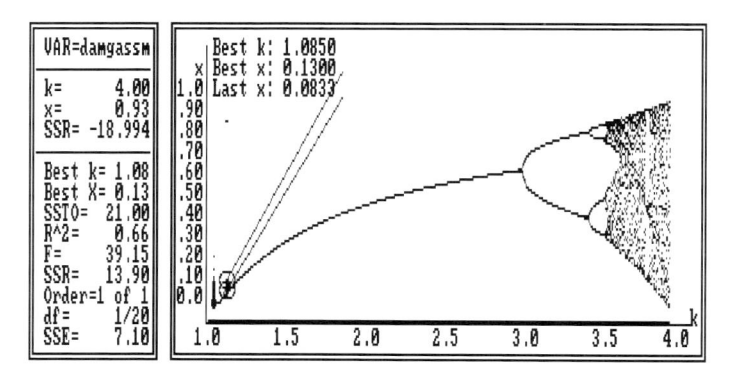

Figure 19. 1985 Mexico City Earthquake: Logistic Regression, Public Organizations, Damage Assessment

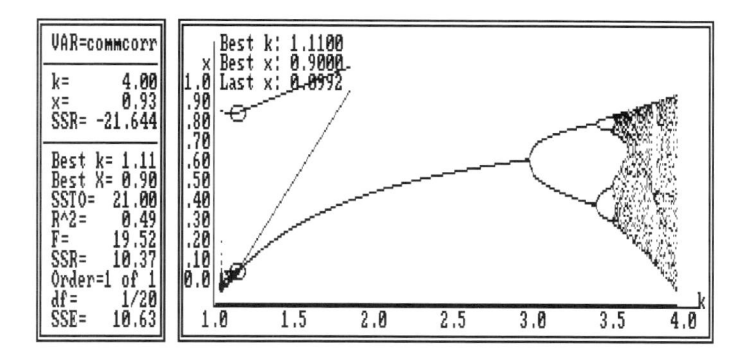

Figure 20. 1985 Mexico City Earthquake: Logistic Regression, Public Organizations, Communication/Coordination

Figures 18–20 report the findings from a logistic regression analysis of reported activities by public organizations for the three functions of emergency response, damage assessment, and recovery/reconstruction. The logistic equation calculated for reported emergency response activities produced an $R^2 = 0.64$, $F = 35.1$, $p < 0.001$ and $k = 1.0$. That is, 64% of the variance in reported emergency response activities is explained by this equation. Damage assessment received a slightly higher R^2 value at 0.66, $F = 39.5$, $p < 0.001$, $k = 1.0$. Recovery/reconstruction showed a comparable R^2 value of 0.61, $F = 31.3$, $p < 0.001$, $k = 1.0$.

Communication/coordination, a function which grouped together two distinct but related activities in disaster operations, shows a lower R^2 value of 0.49, but it is statistically significant with $F = 19.5$, and $k = 1.11$. Logistic equations were found for disaster relief and financial assistance with R^2 values of 0.28 and 0.09 respectively. The relatively low R^2 value for disaster relief is consistent with the survey findings of perceived discrepancies in the distribution of goods donated by the international community. The equation for financial assistance was not statistically significant. Table 30 summarizes findings from the logistic regression analysis.

The logistic regression analysis for nonprofit organizations produced an equation for the response function of disaster relief ($R^2 = 0.45$; $F = 16.2$; $p < 0.001$; $k = 1.0$) and Damage Assessment, ($R^2 = 0.24$; $F = 6.42$; $p < 0.05$; $k = 2.84$). It also produced a logistic equation for Financial Assistance ($R^2 = 0.13$; $F = 2.99$), which was not statistically significant and had a k value of 3.995 indicating that the function had slipped into chaos. No equations were found for emergency response, communication/coordination or recovery/ reconstruction activities for nonprofit organizations, indicating that these functions were not widely observed in actions by nonprofit organizations.

For private organizations, logistic equations were found for two variables. Emergency response reported very high values, showing the immediate response

Table 30. Summary, Logistic Regression Analysis, Public Organizations, Mexico City Earthquake, 1985

Response Function	R^2	F	p	k
Emergency Response	0.64	35.1	0.001	1.0
Damage Assessment	0.66	39.5	0.001	1.1
Communication/Coordination	0.49	19.5	0.001	1.11
Disaster Relief	0.28	7.96	0.05	1.0
Recovery/Reconstruction	0.61	31.3	0.001	1.0
Financial Assistance	0.09	1.74	NS	1.05
Subgroup Total	0.79	74.4	0.001	1.05
Grand Total[a]	0.82	92.1	0.001	1.0

[a]Includes total mentions for Public, Private and Nonprofit Organizations

of private hospitals to the emergency situation ($R^2 = 0.91$; $F = 203.2$; $k = 1.77$); while disaster relief reported an equation that was not statistically significant ($R^2 = 0.11$; $F = 2.59$; $k = 1.0$). No equations were found for the remaining variables of damage assessment, communication/coordination; recovery/reconstruction or financial assistance, indicating that few activities by private organizations in these functions were reported in the newspapers.

The initial conditions of rapid dissemination of information regarding the earthquake and its destructive consequences to the international press influenced the information search and exchange activities, particularly at the national and international level. The content analysis of newspapers reported little activity by local governmental organizations, and the survey results cited a substantial discrepancy between policy and practice in the distribution of international assistance. The mechanisms for translating policy into action at the local level were not yet in place, despite numerous instances of spontaneous action by informal, neighborhood groups. A large and active organizational system evolved in response to the Mexico City Earthquake, but without fully-developed information processes, little change in behavior occurred among organizations that would prevent losses in future earthquakes. The activities undertaken in disaster operations represented, for the most part, innovative responses to meet immediate needs, rather than a long-term institutional commitment to reduce seismic risk for the region.

Costa Rica, April 22, 1991 (M = 7.4)

On Monday, April 22, 1991, at 3:57 p.m., an earthquake struck the Valle de Estrella, Costa Rica on the Caribbean slope of the Cordillera de Talamanca, close to the southeastern border with Panama. The earthquake registered 7.4 on the Richter scale of surface wave magnitude,[13] the most powerful earthquake recorded in this century of Costa Rica's significant seismic history. In terms of threat to life, the earthquake's major impact fell outside of the heavily populated area of the Meseta Central. Although different figures were cited for the number of dead and the number of injured, the most consistent figures reported were 47 dead and 198 persons seriously injured in Costa Rica (EQE International 1991, Aguirre 1991, Lavell 1993, Sarkis 1993). In Limon, a city of approximately 75,000 residents that suffered the heaviest impact, buildings were largely one and two story wood-frame, concrete block, or concrete frame structures. Only one structure in Limon, the three-story International Hotel, completely collapsed, killing one man who was trapped inside. Eight other deaths were reported in Limon. More deaths occurred in the small towns of Talamanca (18) and Matina (20), where the structures were not as well built.

Search and rescue operations in Limon and the surrounding towns were largely carried out at the local level by family, friends, local police and fire departments in the first few hours after the earthquake. Trained urban search and rescue teams arrived from Switzerland and Great Britain with search dogs and special equipment, but by the time they arrived on Friday, April 26, 1991, there was no longer need for their services. Fires did break out, the most damaging at the RECOPE refinery near Moin, but local emergency response organizations effectively brought them under control. The overall cost of damage caused to infrastructure, losses in export, commercial wood, commercial soils, housing and social infrastructure were estimated at US $965 million (Bermudez 1993,3-5). This sum represents approximately 7% of Costa Rica's Gross National Product, a substantial loss for a nation of 2.6 million people.

By observable criteria, in April, 1991, Costa Rica had one of the most advanced emergency planning organizations with high potential for emergency response in Latin America. The Comision Nacional Emergencia (CNE) was operating from modern, well-designed offices in San Jose, with a professional staff that included experts in geology, engineering, medicine, and computer science affiliated with the University of Costa Rica, major national industries, such as the Refineria Costarricense de Petroleo (RECOPE), and major hospitals in San Jose. The National Emergency Plan assigned the primary responsibility for managing and coordinating all activities relating to disaster to the CNE, which reported directly to the President of the Republic.[14] The CNE had recently invested in a $2 million computerized emergency information system, and was engaged in developing a hazards vulnerability analysis for the entire country.[15]

The CNE had established working relationships with international agencies located in San Jose that were also seeking to improve disaster preparedness and response: US Regional Office of Foreign Disaster Assistance, Regional Office of the Pan American Health Organization, and Regional Office of the League of Red Cross Societies. It had organized planning exercises for its staff and affiliated institutions at the national level. In short, the CNE had a well-trained staff seeking to carry out the mission of their agency as they understood it. The staff had developed an ambitious agenda for a small nation that was vulnerable to a range of serious hazards. It had prudently allocated scarce resources and early efforts in preparedness and training to the area of highest perceived risk and heaviest concentration of population, the Meseta Central which included the capital city of San Jose.

When the earthquake occurred on April 22, 1991, the CNE assumed its legal obligations to coordinate response to the disaster. But by April 25, 1991, the third day after the severe earthquake, it was clear that there had been inadequate information gathered to support effective response to the city of Limon, the isolated towns of Limon province, and the canton of Turrialba, areas that suffered the heaviest damage.[16] President Rafael Calderon assumed direct control of disaster operations and placed two of his Cabinet ministers, the Minister of Agricultura y Ganaderia and the Minister of Vivienda y Asentamientos Humanos, in charge of disaster operations in the province of Limon. The CNE's role was redefined to support the government ministries in the conduct of disaster response and recovery operations.[17]

This set of events, which effectively reversed the role of the CNE according to the National Emergency Plan and the expectations of its president and executive director, illustrated vividly the dynamics involved in disaster response. Clearly, existing information processes had failed to provide the CNE with timely, accurate information to support response action, but why and how had this occurred, and what conditions were specific to this earthquake in contrast to the procedures outlined in the formal National Emergency Plan?

Information Search

The processes for gathering and analyzing information to support decision making at the national level in response to local needs were not fully in place at the CNE. Its major investment in a computerized information system was relatively new, and most of the data for the area affected by the earthquake — the city and towns in the province of Limon and the Valle de Estrella — was not yet entered into the computerized knowledge base for the system.

The CNE relied largely on the national telephone system for communication with outlying cities and towns, which went down immediately in some areas and was overloaded in others. Its radio system did not have the capacity to communicate across the mountains to the Atlantic coastal city of Limon and the smaller towns of Bataan, Matina, Sixaola and others in the affected area, nor did it have the transport capability to send helicopters immediately on reconnaissance flights to assess the damage. Nor did the local units of the Guardia Civil, Costa Rica's civilian response organization, have advanced communications capability. Local and provincial committees of the CNE were not yet developed and could not provide the two-way exchange of information regarding assessment of damage and communication of needs essential to mobilize national response action at the local level. The CNE had inadequate means for direct exchange of information between the stricken areas and its central office in San Jose and had little capacity for organizing local action in these outlying areas.

Ironically, the news media had both better equipment and better means of transportation for information search and damage assessment than the CNE, and early seized the lead in reporting the consequences of the earthquake to the wider population.[18] These reports, while timely, were made from a journalist's perspective and did not provide the kind of systematic, professional assessment of damage to the infrastructure and needs of the earthquake-affected populations essential for effective disaster operations. Yet, the news media played a significant role in creating both national and international awareness of the disaster and in mobilizing support for disaster relief and recovery. This role was similar to that played by the international media in the 1985 Mexico City Earthquake.

Without adequate transportation and communication facilities to support an initial damage assessment, the CNE's information search regarding the impact of the earthquake in these outlying cities and towns yielded delayed, vague, and incomplete reports that provided little basis for informed action. The President's actions in assuming lead responsibility for disaster response and operations reflected his ability to bring wider resources to the task and to obtain a more timely, accurate, and detailed assessment of needs in the provincial regions.

Information Exchange

When President Calderon assumed direct control of disaster operations, the formal organization of the CNE became a participating member of the disaster response system, rather than the active coordinator and manager of the evolving system. Working under urgent demands for action, separate groups of organizations formed around common tasks and carried out their functions, often crossing jurisdictional boundaries within groups, but with relatively little interaction among the groups. Each group instead reported directly to the President. At least seven distinct subsets of organizations were identified that performed their own assessment of needs in the disaster-affected areas, and organized their actions accordingly. These subsets included organizations representing the basic disciplines involved in disaster management: public policy and management, medical response, engineering, public health, and information processes, as well as two additional perspectives that proved especially important in this disaster: transportation and agriculture/ commerce/industry.

The degree of information exchange among organizations participating in response operations was stronger within each subgroup than among the set of subgroups. For example, transportation proved a crucial element of both emergency response and medical response, and the airport itself became an important locus of operations management and information exchange in this disaster. Local organizations established an operations headquarters at the airport, with radio communications, a fax machine, and telephones. This communications capability enabled direct communications with national ministries and organizations located in San Jose, as well as communications via radio to those villages that could receive and send messages. Personnel at this airport headquarters office recorded incoming supplies and voluntary assistance from disaster relief organizations — public, private, and nonprofit — as well as reports of needs from outlying areas. In a spontaneous effort to match the flow of incoming supplies to reported needs from the disaster-affected towns and villages in the coastal region, this hastily established operations office organized a de facto communications exchange and record-keeping system that provided an important basis for informed decision. While this operations office provided a temporary means of information exchange among participating organizations that facilitated immediate disaster operations, the infrastructure necessary to produce a lasting change in adaptive behavior to reduce risk to the community was not in place.

Professional guidance from the US Office of Foreign Disaster Assistance, which had established trusted relationships of long standing with the CNE, Red Cross, PAHO, ONUCA, and the Costa Rican ministries, served an important function in supporting the organization and operation of this office.[19] Accordingly, disaster relief supplies were received, stored, and dispensed to outlying communities from the airport in an increasingly ordered manner, as disaster operations progressed. As stated above, medical services to isolated towns in the disaster-affected area were coordinated from the airport through available air transport.

In contrast to the clear and rapid identification, collection, and analysis of information for bridge reconstruction, an engineering function that affected transportation, quick assessment of damage and the development of an action strategy took a different form in reference to housing, a second engineering function. Approximately 850–1,000 homes were destroyed, leaving an estimated 3,500 people homeless. Shelters were established in parks and other public places, but the dominant response was to consider housing a matter for private or nonprofit action. There appeared to be little coordination of housing services, except for the distribution of supplies of plastic and other materials extended through the Red Cross and other non-governmental organizations.[20] Most persons who suffered damaged or destroyed housing did not have insurance, and struggled to cope with their losses with help from family and friends. The housing damage was more diffuse than the infrastructure losses, affecting individual families who had no real means of articulating their needs collectively, and no strong leadership emerged at the local level to press for assistance in meeting these needs. Information remained scattered, and policy makers moved to more urgent, that is, more sharply articulated, demands.

Organizational Learning

Returning to the N-K methodology, content analysis of reported actions taken in disaster operations documented an actual disaster response system of 245 organizations that contrasted in practice with the formal structure outlined in the National Emergency Plan. From the news stories reported in *La Nacion*, a national newspaper published in San Jose, I identified the organizations engaged in disaster response by jurisdiction, source of support, and type of transaction for the period, April 23, 1991 – May 15, 1991.[21] The number of mentions for each organization over the total period provides a rough approximation of intensity of engagement of public, private, and nonprofit organizations in the disaster operations process, as reported in the newspapers.[22]

The response system, as documented by actions reported in *La Nacion* and summarized in Table 31, was predominantly public, with 142, or 58.0%, of the organizations involved representing different jurisdictional levels. Of that number, national organizations represented the largest share, 74 out of 142, or 52.1%. The second largest number of organizations, 52, or 36.6%, was international, and provincial and municipal organizations together made up the remaining portion of 16, or 11.3%. Nonprofit organizations represented one-fifth, or 20.3%, of the total response system, and private organizations composed 21.7%.

Table 32 presents the set of 378 transactions reported for the 1991 Costa Rica response system. The largest proportion, 57, or 15.1%, involved disaster relief activities, nearly half of which were initiated by the national government, but conducted with strong support from international and nonprofit organizations. The second largest category of reported transactions was donations of money and goods to the disaster-stricken communities. International organizations – public, private and nonprofit – played a substantive role in this category, which registered 47, or 12.4%, of the total transactions.

These categories were followed by communication, 11.1%, and coordination, 10.6%, in reported transactions to mobilize disaster operations. Combined, these categories constitute 20.9% of the total transactions reported for disaster operations, and reflect two characteristics of disaster response. First, the news reports document the substantial effort made by the Calderon Government to mobilize national response quickly to meet the needs of the stricken communities and to contain the disruption to the nation's economy, caused by the damage to the transportation infrastructure. Secondly, these reports verify the concerted effort made to secure international assistance in meeting the urgent needs of the small nation, both in terms of immediate assistance for afflicted families, and the larger, longer-term costs of rebuilding the damaged infrastructure to mitigate risk from future seismic events. For a nation of its size, 2.6 million people, and severe

Table 31. Frequency Distribution, Disaster Response System by Funding Source and Jurisdiction, 1991 Costa Rica Earthquake

Public								Total Public		Nonprofit				Private				Total	
Int'l		Nat'l		Prov.		Mun'l				Int'l		Dom.		Int'l		Dom.			
N	%	N	%	N	%	N	%	N	%	N	%	N	%	N	%	N	%	N	%
52	21.2	74	30.2	6	2.4	10	4.1	142	58.0	15	6.1	35	14.2	16	6.6	37	15.1	245	100

Part II: Shared Risk in Practice

Table 32. Frequency Distribution: Types of Transactions in Disaster Response by Funding Source and

Type of Transaction	Public Organizations											
	International			National			Provincial			Municipal		
	T	N	%	T	N	%	T	N	%	T	N	%
Emergency Response	2	3	0.5	4	9	1.1	1	3	0.3	0	0	0.0
Communication	2	8	0.5	31	54	8.2	0	0	0.0	1	4	0.3
Coordination of Response	7	18	1.9	23	43	6.1	0	0	0.0	0	0	0.0
Medical Care/Health	0	0	0.0	9	15	2.4	0	0	0.0	0	0	0.0
Damage/Needs Assessment	4	12	1.1	25	51	6.6	2	2	0.5	1	2	0.3
Certification of Deaths	1	2	0.3	8	15	2.1	1	1	0.3	0	0	0.0
Earthquake Assessment/Research	3	6	0.8	19	41	5.0	0	0	0.0	0	0	0.0
Security/Prevention of Looting	0	0	0.0	3	6	0.8	1	1	0.3	0	0	0.0
Housing Issues	1	4	0.3	1	1	0.3	0	0	0.0	0	0	0.0
Disaster Relief (food, shelter, etc.)	16	48	4.2	23	45	6.1	0	0	0.0	2	2	0.5
Donations (money, goods, etc.)	14	75	3.7	9	68	2.4	0	0	0.0	1	11	0.3
Building Inspection	0	0	0.0	0	0	0.0	0	0	0.0	0	0	0.0
Building Code Issues	0	0	0.0	0	0	0.0	0	0	0.0	0	0	0.0
Repair of Freeways, Bridges, Roads	2	3	0.5	7	7	1.9	0	0	0.0	0	0	0.0
Repair/Restore Utilities	0	0	0.0	15	21	4.0	0	0	0.0	0	0	0.0
Repair/Reconstruction/Recovery[a]	2	4	0.5	8	17	2.1	1	4	0.3	0	0	0.0
Transportation/Traffic Issues	1	1	0.3	4	6	1.1	0	0	0.0	0	0	0.0
Hazardous Materials Releases	0	0	0.0	0	0	0.0	0	0	0.0	0	0	0.0
Fraud	0	0	0.0	0	0	0.0	0	0	0.0	0	0	0.0
Legal/Enforcement Issues	0	0	0.0	3	5	0.8	0	0	0.0	0	0	0.0
Legislation/Legislative Process	0	0	0.0	1	1	0.3	0	0	0.0	0	0	0.0
Business Recovery	0	0	0.0	0	0	0.0	0	0	0.0	0	0	0.0
Economic/Business Issues	0	0	0.0	3	6	0.8	0	0	0.0	0	0	0.0
Visit by Officials	0	0	0.0	0	0	0.0	0	0	0.0	0	0	0.0
Education Issues	0	0	0.0	5	10	1.3	0	0	0.0	0	0	0.0
Government Assistance	0	0	0.0	4	8	1.1	0	0	0.0	0	0	0.0
Insurance related Issues	0	0	0.0	1	2	0.3	0	0	0.0	0	0	0.0
Loans by Private Banks	22	59	5.8	7	19	1.9	0	0	0.0	0	0	0.0
Psychological/Counseling Services	0	0	0.0	1	1	0.3	0	0	0.0	0	0	0.0
Volunteer Activities	0	0	0.0	0	0	0.0	0	0	0.0	0	0	0.0
TOTAL:	77	243	20.4	214	451	56.6	6	11	1.6	5	19	1.3

T = Number of Transactions; N = Number of Actors; % = Percent of Transactions Column Total
[a] Not including freeways, bridges, roads or utilities
Source: *La Nacion*, San Jose, Costa Rica, April 23 — May 15, 1991

seismic risk, Costa Rican leaders were aware that the risk could only to be managed effectively if the costs were shared with a wider community.

Table 33 presents the interactions reported among the participating organizations, a total of 166. The interactions involved organizations cooperating with one another to achieve a shared objective in the response operations. The number of interactions and the number of organizations engaged in collaborative tasks provide a measure of the density or interconnectedness of the response system. Of the total number of 166 reported interactions, the largest proportion, 97, or

Chapter 6: Emergent Adaptive Systems 141

Jurisdiction, Costa Rica Earthquake, April 23 – May 14, 1991

Nonprofit Organizations						Private Organizations						TOTALS		
International			National			International			National					
T	N	%	T	N	%	T	N	%	T	N	%	T	N	%
1	2	0.3	3	11	0.8	0	0	0.0	1	2	0.3	12	30	3.2
3	3	0.8	5	5	1.3	0	0	0.0	0	0	0.0	42	74	11.1
5	8	1.3	5	12	1.3	0	0	0.0	0	0	0.0	40	81	10.6
0	0	0.0	2	4	0.5	0	0	0.0	0	0	0.0	11	19	2.9
1	1	0.3	1	3	0.3	0	0	0.0	0	0	0.0	34	71	9.0
0	0	0.0	1	1	0.3	0	0	0.0	0	0	0.0	11	19	2.9
1	2	0.3	2	2	0.5	0	0	0.0	0	0	0.0	25	51	6.6
0	0	0.0	1	1	0.3	0	0	0.0	0	0	0.0	5	8	1.3
0	0	0.0	0	0	0.0	0	0	0.0	0	0	0.0	2	5	0.5
4	6	1.1	12	20	3.2	0	0	0.0	0	0	0.0	57	21	15.1
8	28	2.1	5	14	1.3	4	13	1.1	6	11	1.6	47	220	12.4
0	0	0.0	0	0	0.0	0	0	0.0	0	0	0.0	0	0	0.0
0	0	0.0	0	0	0.0	0	0	0.0	0	0	0.0	0	0	0.0
1	3	0.3	0	0	0.0	0	0	0.0	0	0	0.0	10	13	2.6
0	0	0.0	1	9	0.3	0	0	0.0	0	0	0.0	16	30	4.2
1	1	0.3	1	1	0.3	0	0	0.0	0	0	0.0	13	27	3.4
0	0	0.0	0	0	0.0	0	0	0.0	0	0	0.0	5	7	1.3
0	0	0.0	0	0	0.0	0	0	0.0	0	0	0.0	0	0	0.0
0	0	0.0	0	0	0.0	0	0	0.0	0	0	0.0	0	0	0.0
0	0	0.0	0	0	0.0	0	0	0.0	0	0	0.0	3	5	0.8
0	0	0.0	0	0	0.0	0	0	0.0	0	0	0.0	1	1	0.3
0	0	0.0	0	0	0.0	0	0	0.0	0	0	0.0	0	0	0.0
0	0	0.0	0	0	0.0	0	0	0.0	0	0	0.0	3	6	0.8
0	0	0.0	0	0	0.0	0	0	0.0	0	0	0.0	0	0	0.0
0	0	0.0	0	0	0.0	0	0	0.0	0	0	0.0	5	10	1.3
0	0	0.0	0	0	0.0	0	0	0.0	0	0	0.0	4	8	1.1
0	0	0.0	0	0	0.0	0	0	0.0	0	0	0.0	1	2	0.3
0	0	0.0	0	0	0.0	0	0	0.0	0	0	0.0	29	78	7.7
0	0	0.0	0	0	0.0	0	0	0.0	0	0	0.0	1	1	0.3
1	4	0.3	0	0	0.0	0	0	0.0	0	0	0.0	1	4	0.3
26	58	6.9	39	83	10.3	4	13	1.1	7	13	1.9	378	891	100.0

58.4%, involved national agencies, and within that proportion, the largest component involved national organizations (62 of a total 97). The next largest proportion, 52, or 31.1%, involved the set of international organizations, with public organizations making up the largest component, 46, or 88.5%, of that set. Public provincial and municipal organizations combined were involved in 2, or 1.2%, of the interactions reported for disaster response operations.

Comparing the number of interactions by funding source, private organizations were engaged in the smallest number of interactions, 6, or 3.6%, of the total.

Table 33. Frequency Distribution: Interactions by Funding Source and Jurisdiction, Costa Rica Earth-

Type of Interaction	Public										Private									
	Internat'l			National			Local			Internat'l			National			Local		
	K	N	%	K	N	%	K	N	%	K	N	%	K	N	%	K	N	%
Public: International	7	9	4.2	38	35	22.9	1	2	0.6	1	2	0.6	0	0	0.0	0	0	0.0
Public: National				41	33	24.7	11	15	6.6	2	4	1.2	3	6	1.8	3	6	1.8
Public: Local							1	2	0.6	0	0	0.0	1	2	0.6	0	0	0.0
Private: International										0	0	0.0	1	2	0.6	2	4	1.2
Private National													0	0	0.0	0	0	0.0
Private: Local																1	2	0.6
Nonprofit: International																		
Nonprofit: National																		
Nonprofit: Local																		
Total Interactions	7	9	4.2	79	68	47.6	13	19	7.8	3	6	1.8	5	10	3.0	6	12	3.6

K = Number of Interactions; N = Number of Actors (only counted once); % = Percent of Total Interactions
Source: *La Nacion*, San Jose, Coata Rica, April 23 – May 15, 1991.

Nonprofit organizations were involved in slightly more, 5.4%, while public organizations were clearly the dominant group, with 151, or 91%, of the total. Within the public group, two types of organizations played distinctive roles in the Costa Rican response system. These were financial and research organizations, both national and international. Although small in number, these organizations are identified distinctly for their example in bridging national and international objectives in managing the reconstruction and recovery phase of the disaster and the mitigation of future seismic risk.

Despite the number of reported transactions involving communication, 11.1%, existing information processes did not function well to serve the CNE's formal role of coordination of disaster operations following this disaster. The shift in coordination from interactive cooperation among participating agencies to direct supervision by the President affected the patterns of organizational learning within the response system. Although the CNE remained actively involved in disaster response and recovery activities, the information processes in this disaster appeared fragmented and operated largely within functional groups that reported directly to the President. Within each group, substantial experience and expertise were marshaled to address specific types of problems and to devise practical courses of action. Yet, among the separate groups there appeared to be little exchange of information or coordination of action. This lack of coordination between functional groups appeared to constrain the effective performance of participating organizations and to generate an unusual degree of distrust and animosity, especially among organizations with interdependent responsibilities. This pervasive distrust and barely concealed hostility among different sets of organizations participating in the common task of disaster operations inhibited frank, candid communication among them and diminished the willingness and capacity quake, April 23 – May 14, 1991

Nonprofit									Total									Grand Total		
International			National			Local			International			National			Local			All Categories		
K	N	%	K	N	%	K	N	%	K	N	%	K	N	%	K	N	%	K	N	%
5	9	3.0	0	0	0.0	0	0	0.0	13	20	7.8	38	35	22.9	1	2	0.6	52	57	31.1
12	14	7.2	18	23	10.8	7	14	4.2	14	18	8.4	62	62	37.3	21	35	12.6	97	115	58.1
0	0	0.0	0	0	0.0	0	0	0.0	0	0	0.0	1	2	0.6	1	2	0.6	2	4	1.2
0	0	0.0	0	0	0.0	0	0	0.0	0	0	0.0	1	2	0.6	2	4	1.2	3	6	1.8
1	2	0.6	1	2	0.6	0	0	0.0	1	2	0.6	1	2	0.6	0	0	0.0	2	4	1.2
0	0	0.0	0	0	0.0	0	0	0.0	0	0	0.0	0	0	0.0	1	2	0.6	1	2	0.6
4	7	2.4	0	0	0.0	3	4	1.8	4	7	2.4	0	0	0.0	3	4	1.8	7	11	4.2
			1	2	0.6	1	3	0.6	0	0	0.0	1	2	0.6	1	3	0.6	2	5	1.2
						0	0	0.0	0	0	0.0	0	0	0.0	0	0	0.0	0	0	0.0
22	32	13.2	20	27	12.0	11	21	6.6	32	47	19.3	104	105	62.7	30	52	18.0	166	204	100.0

of the participating organizations to explore and execute the most appropriate, feasible, and efficient alternatives for action in response and recovery operations. The tensions among the organizations were documented in the news accounts, and likely contributed to the relative frequency of news reports about communication and coordination.

The results show a remarkable discrepancy between the formal inter-jurisdictional structure of national, provincial, and local Committees of Emergency Preparedness described in the National Emergency Plan and the actual organizational activity reported in the news stories. Attention appeared to focus on a small number of organizations within the two categories of national and international organizations. For example, five national organizations received over half (51.4%) of the reported mentions in the content analysis of newspapers, and four international actors received over half (50.1%) of the mentions in their respective groups. Municipal response ranked a distant third with only 5.2% of the mentions, and provincial response received the least coverage with 2.4% of the total mentions. The local structures for emergency response were clearly underdeveloped in this region.

These findings must be interpreted in the context of continuing economic and social development for Costa Rica (Lavell 1991; Maskrey and Lavell 1993). Although the design for emergency management coordinating committees at the municipal and provincial levels of jurisdiction exists formally in the National Emergency Plan, these committees were not sufficiently developed in practice to play an active role in emergency response. Significant differences in organizational development, training, equipment, and investment of resources between the central government in San Jose and the provincial and municipal governments resulted in an underdeveloped organizational structure in the Atlantic

Region with little capacity to mitigate or respond to disaster (Maskrey and Lavell, 1993). Local needs, under these conditions, could only be met by national and international action.

Adaptive Behavior

Frequencies of reported organizational activity cited in *La Nacion* provide the basis for a nonlinear analysis of change in organizational behavior over a three-week period following the 22 April, 1991 earthquake. While these findings are not drawn from official records, they provide the most consistent public data available on disaster operations. These data show that the disaster response system emerged relatively slowly, exhibiting a peak of activity on Day 3. Logistic regression equations were calculated for each of the six basic disaster response functions performed by public organizations, displaying the initial capacity for a full disaster response organization. The relatively simple design of the buildings — one and two-story wood-frame structures — and the time of the earthquake — 3:57 p.m. — allowed most inhabitants to escape quickly.

Communication/coordination proved to be the strongest emergency function performed by public organizations in the Costa Rican disaster operations, producing an R^2 value of 0.52, F = 21.78, and k = 1.0, as shown in Figure 21. Equations were also calculated for Disaster Relief, Recovery/Reconstruction, Damage Assessment, Emergency Response, and Financial Assistance, revealing a functioning emergency response system, although the findings were relatively weak and one equation, Financial Assistance, was not statistically significant. These findings, summarized in Table 34, document beginning efforts in five of the six basic emergency response functions for the public organizations during the three-week period immediately following the disaster.

Nonprofit organizations also participated in the response system, particularly in the function of disaster relief. The nonlinear analysis of reported organizational activity for Disaster Relief produced an R^2 value of 0.57, F = 26.53, k =

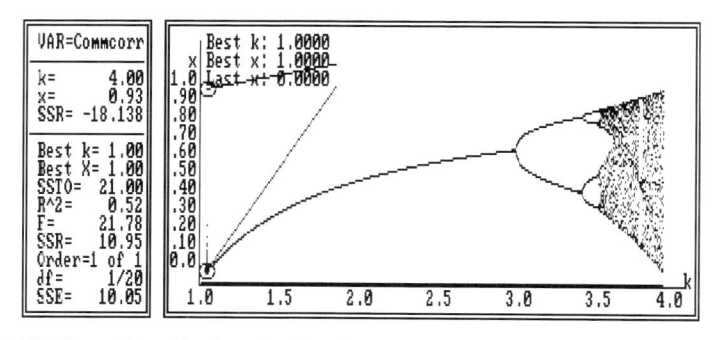

Figure 21. 1991 Costa Rica Earthquake: Logistic Regression, Public Organizations, Communication/Coordination

Chapter 6: Emergent Adaptive Systems

Table 34. Logistic Regression Analysis, Public Organizations, 1991 Costa Rica Earthquake

Function	R^2	F	p	k
Emergency Response	0.23	6.04	0.05	1.11
Damage Assessment	0.22	5.62	0.05	1.00
Communictaion/Coordination	0.52	21.78	0.001	1.00
Disaster Relief	0.25	6.61	0.05	1.00
Recovery/Reconstruction	0.42	14.54	0.01	1.00
Financial Assistance	0.07	1.48	NS	1.00
Sub Group, Public Organisations	0.55	24.82	0.001	1.00
Grand Total, Public, Private, Nonprofit	0.57	26.48	0.001	1.00

1.0, shown in Figure 22. Private organizations reported most activity in the functions of Disaster Relief and Recovery/Reconstruction, but these functions did not produce logistic equations. A logistic equation was found for Communication/Coordination, $R^2 = 0.17$, $F = 4.09$, $k = 3.99$, but the F value is not statistically significant, and the k value shows the function to be in chaos. Consequently, this analysis shows only limited participation by the private sector in response operations.

The response and recovery system that evolved in the Costa Rican disaster operations was clearly nonlinear,[23] marked by discontinuities in communication, coordination, and organization in contrast to its predesigned, centralized, linear National Emergency Plan. That is, the response system was sensitive to the initial conditions of the disaster affected area — the city of Limon and the towns, villages, ports, and banana plantations in the area — and continued to evolve in unpredictable ways. Within this nonlinear system were sets of subsystems operating separately with reasonable stability and purpose. There was little coordination of action or shared information among them, but they did represent significant actions taken by separate groups in a self organizing approach to disaster

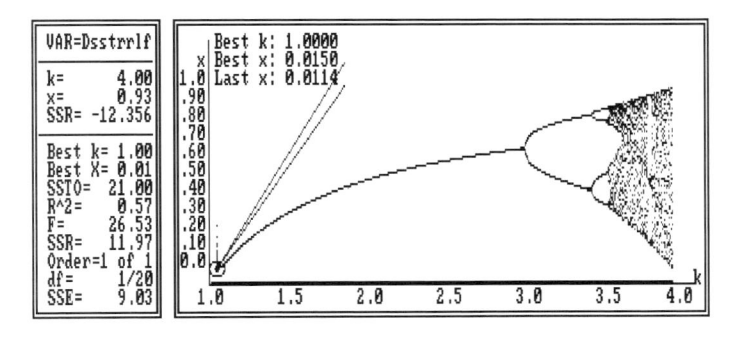

Figure 22. 1991 Costa Rica Earthquake: Logistic Regression, Nonprofit Organizations, Disaster Relief

response. These actions constituted the beginning steps toward an emergent, adaptive, disaster response system.

Erzincan, Turkey, March 13, 1992 (M = 6.8)

At 7:19 p.m. on Friday evening, March 13, 1992, Muslim families in Erzincan were engaged in the normal activities of their daily lives in this dusty city of approximately 100,000 residents in the North Anatolian region of Turkey. Mothers were preparing the evening meal to end a day of Ramadan fasting. Men were playing bridge and discussing the city's affairs in local social halls. Others were gathering in the mosques for the evening call to prayer. Without warning, a powerful earthquake rocked the buildings and shattered the calm evening of Ramadan feasting, conversation and prayer. Registering 6.8 on the Richter scale of magnitude, the earthquake resulted in a loss of 683 lives by official estimates. Unofficial reports estimated the loss in lives at approximately 2,000. Approximately 1500 persons were reported injured. More than 200 buildings, primarily of masonry and reinforced concrete construction, totally collapsed or sank through their first floors. Additional hundreds of buildings were damaged, leaving an estimated 50,000 to 60,000 residents homeless. The estimated loss in damage to property and public infrastructure was calculated at over $3.4 billion. Disaster claimed the city for the second time in little more than fifty years.

In seconds, the earthquake altered the lives, security and comfort of two-thirds of the population of the city. No resident was left untouched, if not harmed physically, then economically, socially or emotionally. Occurring in a region of widely recognized seismic risk, the devastating effects of the earthquake were sobering from the perspective of public policy and planning.

Two conditions in this disaster seriously affected the capacity of the community to mobilize an effective response system following the disaster. First, the disaster had a powerful impact upon the population of this small city in a region long known for seismic risk. The psychological trauma of observing death or severe injury to close family members or friends can significantly alter an individual's capacity for response, especially for those with no prior training for, or experience of, trauma. When that trauma is amplified across the network of social, emotional, and economic ties that bind together a community, the effect may seriously disable that community's ability to respond decisively to the event. The ability to respond effectively depends upon a rapid process of organizational and interorganizational communication and learning that activates a complex disaster response system. The performance of this system depends upon the ability of people to take responsible, informed action under urgently stressful conditions, an extraordinarily difficult task. To make the problem more complex, individual people respond differently under stressful conditions, and are likely to generate different patterns of action and organizational learning in the process, creating a turbulent, dynamic effect within the system. Given these initial conditions, information processes vital to the rapid mobilization of response to the event were affected in critical ways.

Relatively little research has examined the effects of initial conditions, including widespread direct observation of death and injury, upon a community's capacity to organize disaster response.[24]

The disabling effects of post traumatic stress are well known, but we have not yet considered the impact upon entire communities subjected to trauma. A devastating natural disaster receives the highest rating of 6, or catastrophic, on the "Severity of Psychological Stressors Scale"[25]. Given the clinical definition of post-traumatic stress, characteristic symptoms include a numbing of general responsiveness and increased arousal that would affect one's ability for interorganizational coordination. Common sources of trauma include:

"a serious threat to one's life or physical integrity, a serious threat or harm to one's children, spouse, or other close relatives and friends; sudden destruction of one's home or community; or seeing another person who has recently been, or is being seriously injured or killed..."[26]

The community of Erzincan offers a unique setting in which to examine the effect of trauma upon interorganizational coordination, since it has endured over 1,000 earthquakes in its history with three major earthquakes in this century alone. Responsible officials and residents of the city are well aware of the seismic risk to the community, and Turkey has an established national earthquake plan. Yet, as reports from the 1992 earthquake document, the community of approximately 100,000 people suffered an extraordinarily heavy physical, economic, and psychological loss.

Second, the community's medical facilities were heavily damaged and medical staff were themselves victims in this disaster. Disaster medical services require interactive coordination and support between internal and external actors in an integrated emergency response system. Heavy losses in medical staff and almost total destruction of a major hospital created an numbing sense of loss and vulnerability among those seeking medical care. This perceived vulnerability accentuated further the sense of trauma that fell over the community, and affected in important ways the information processes central to the evolution of response.

Information Search

Information search processes in this disaster operated on two levels — national and local — using markedly different mechanisms that had profoundly different effects upon the mobilization of an effective response system. Sophisticated investments in seismic monitoring and modeling of seismic effects at the national level informed policy making in important ways. Within minutes after the earthquake struck in the early evening of March 13, 1992, national officials in Ankara learned of the earthquake through reports from a sensitive seismic monitoring system operated by the National Earthquake Research Center in Ankara. Within 30 minutes, scientists at the Research Center estimated the losses in lives and property in Erzincan, using a sophisticated damage assessment model. National officials immediately activated the national emergency plan, and within hours of the earthquake, the Prime Minister and other national ministers flew to Erzincan with resources and supplies to aid the devastated city.

But in Erzincan, city officials and residents were seriously disabled by the event. Local Civil Defense and response teams with specific knowledge of the buildings and streets were slow to mobilize, as their members were absorbed in ensuring the security of their families. Outside rescue teams, from neighboring cities and neighboring countries, converged on the small city. With little knowledge of the built infrastructure and characteristic patterns of the population, the external teams were less than efficient. Paradoxically, many of the community residents capable of giving local guidance to external teams were themselves suffering from the disaster. Not until twenty-four hours after the event did information begin to accumulate with the degree of specificity required to inform action. In the time-critical task of search and rescue for live survivors, this was an extraordinarily costly delay.

A survey of 185 respondents interviewed in Erzincan and Ankara four months after the earthquake[27] shows the gap between information and action in mobilizing disaster response. The characteristics of the sample — age, sex, and role in the disaster response process — are presented in the Appendix. Briefly, the majority of respondents were men, 71.2%, and almost two-thirds of the sample were lay people or citizens, 65.8%. Medical personnel comprised 17.1% of the sample, and local coordinators made up the third largest group, 11.4%. The findings document the severe impact of the earthquake upon the community and the capacity of its members to respond.

Table 35 reports the startling exposure to trauma for the majority of respondents. Over 60%, 101 respondents, had observed closely someone who was seriously injured or dead. Two-thirds of those responding, or 76, reported finding someone who was dead. Nearly one-third of those responding stated they had found someone alive who later died. Over 65 respondents reported caring for at least one seriously injured person. Given the representative design of the sample, these results show an overwhelming exposure to psychological trauma for the community of Erzincan. This exposure was doubtless amplified because of the relatively small size of the city, with many overlapping professional, business, friendship, and familial relationships.

Table 36 reports the types of information sought by a relatively small proportion of the sample, 40 out of 185 respondents, as a basis for action. This finding indicates that a majority of respondents may have been so numbed by the experience that it was difficult for them to think of immediate action.

Table 37 reports a significant association between the physical ability to help others and observation of seriously injured or dead persons, and a more severe trauma, actually finding a dead person. Of 65 respondents reporting first aid

Table 35. Frequency Distribution: Exposure to Traumatic Stress

"Did you observe anyone closely who was seriously injured or dead?"

	n	%
Yes	101	62.3
No	61	37.7
Total	162	100.0

(missing cases = 23)

"Did you find anyone who was dead?"

	n	%
Yes	76	66.1
No	39	33.9
Total	115	100.0

(missing cases = 70)

"Did you find anyone alive who later died?"

	n	%
Yes	38	32.8
No	78	67.2
Total	116	100.0

(missing cases = 69)

"How many seriously injured victims did you look after?"

	n	%
31 or more	12	15
21–30	5	6.3
11–20	9	11.3
6–10	11	13.7
2–5	15	18.7
One	13	16.3
None	15	18.7
Total	80	100.0

(missing cases = 105)

training, 55, or 84.6%, had observed persons who were seriously injured or dead. This high proportion indicates that virtually all medical personnel were exposed to the highest level of catastrophic trauma.

These findings reveal the disabling effect of trauma upon a community stricken by disaster, a condition not usually acknowledged in a rational emergency plan. More important, it underlies the critical nature of the local response to the timely evolution of a disaster response system.

Table 36. Frequency Distribution: Information Sought as a Basis for Action

"What information did you seek as a basis for action?"

	n	%
Access to telephone, radio	2	4.0
Risk of fire	22	44.0
Safety of structure	1	2.0
Injury to others	7	14.0
Injury to self	8	16.0
Other	10	20.0
Total	50	100.0

(missing = 135)

Table 37. Cross-tabulation: Capacity for Action by Exposure to Traumatic Stress

	Yes n	Yes %	No n	No %	Total n	Total %
Observed seriously injured or dead						
Able to help others						
Yes	69	79.3	16	26.7	85	57.8
No	18	20.7	44	73.3	62	42.2
Total	87	100.0	60	100.0	147	100.0
Found dead persons						
Able to help others						
Yes	60	86.9	17	48.6	77	74.0
Total	69	100.0	35	100.0	104	100.0

Chi Square = 38.223; $p < 0.0000$

Information Exchange

Without adequate information available at the local level, efforts to facilitate the exchange of information – technical, organizational, economic or cultural – between jurisdictions and among organizations are limited. The exchange worked well among ministries at the national level, who were trained in the National Emergency Plan and followed it carefully. But the transfer of information diminished across jurisdictional boundaries – e.g. from the national to the provincial levels, and especially across three levels to the municipal level. Information exchange functioned more easily within levels of jurisdiction than between jurisdictions. The exchange worked least well between the international organizations that arrived to offer assistance to the damaged community. Many of them came with no local support and no translators. In an already chaotic disaster environment, efforts to seek and/or give information often proved frustrating and unproductive.[28]

Organizational Learning

Despite breaks in communication and gaps in exchange of information, a distinct response system emerged to meet the needs of the residents of Erzincan. The content analysis of news reports[29] identified a response system of 171 organizations, of which the largest number, 134, or 78.4%, was public organizations. The next largest number, 29, or 16.9%, was nonprofit organizations, and the smallest number, 8, or 4.7%, represented private organizations. In the breakdown of public organizations by jurisdictions, international organizations represented the largest proportion, 53, or nearly 40% of the total number of public organizations. National organizations were second, 40, or 29.9%, and municipal organizations third, with 28, or 20.9%. Provincial organizations comprised the smallest proportion, 13, or 9.7% of the total number of 134 public organizations. These findings are summarized in Table 38.

The transactions performed by the set of organizations participating in disaster operations are presented in Table 39. Both newspapers and expert reports document the substantial international contribution to the Erzincan response system. International organizations represented not only the highest proportion of organizations in the response system, but also international assistance represented the highest proportion of transactions reported in disaster operations. The next highest category was emergency response, with 15.2% of the reported transactions, and third was medical/health care to the injured, with 11.8% of the transactions. Coordination received 6.6% of the reported transactions, and communication, reflecting other observations, received the low proportion of 1.7%. Table 40 cites the interactions reported among the participating organizations. These figures portray a relatively small response system, largely directed by the national government, with strong contributions of international assistance and relatively weak participation at the local level. Despite the strong initial response from the national level, the system revealed an inability to function without the basic knowledge from the local level. The gap in communication between the national and local levels of governmental agencies adversely affected the performance of the whole response system.

Table 38. Frequency Distribution: Disaster Response System by Funding Source and Jurisdiction, 1992 Erzincan, Turkey Earthquake

Public								Total Public		Nonprofit				Private				Total	
Int'l		Nat'l		State		Mun'l				Int'l		Nat'l		Int'l		Nat'l			
N	%	N	%	N	%	N	%	N	%	N	%	N	%	N	%	N	%	N	%
53	31.0	40	23.4	13	7.6	28	16.4	134	78.4	25	14.6	4	2.3	1	0.5	7	4.0	171	100

Table 39. Frequency Distribution: Types of Transactions in Disaster Response, Erzincan, Turkey Earthquake, March 13 – June 2, 1992

| Type of Transaction | Public Organizations ||||||| Municipal || Nonprofit Organizations || Private Organizations || Total ||
| | International || National || Provincial || | | | | | | | |
	T	N	%	T	N	%	T	N	%	T	N	%	T	N	%	T	N	%	T	N	%
Emergency Response	28	10	8.0	14	9	4.0	5	4	1.4	1	1	0.3	3	3	0.9	2	2	0.6	53	29	15.2
Communication	1	1	0.3	2	1	0.6	1	1	0.3	1	1	0.3	0	0	0.0	1	1	0.3	6	5	1.7
Coordination of Response	8	3	2.3	10	8	2.9	3	2	0.9	2	2	0.6	0	0	0.0	0	0	0.0	23	15	6.6
Medical Care /Health	0	0	0.0	4	2	1.1	7	6	2.0	29	21	8.3	1	1	0.3	0	0	0.0	41	30	11.8
Damage/Needs Assessment	4	4	1.1	5	4	1.4	7	7	2.0	0	0	0.0	3	2	0.9	0	0	0.0	19	17	5.5
Certification of deaths	0	0	0.0	2	2	0.6	3	2	0.9	0	0	0.0	0	0	0.0	0	0	0.0	5	4	1.4
State of Emergency Declarations	0	0	0.0	1	1	0.3	0	0	0.0	0	0	0.0	0	0	0.0	0	0	0.0	1	1	0.3
Earthquake Assessment/Research	3	1	0.9	1	1	0.3	0	0	0.0	0	0	0.0	0	0	0.0	0	0	0.0	4	2	1.1
Disaster Relief (Food, Water, etc)	2	2	0.6	5	5	1.4	2	2	0.6	3	3	0.9	5	5	1.4	1	1	0.3	18	18	5.2
Building Code Issues	0	0	0.0	4	4	1.1	0	0	0.0	1	1	0.3	0	0	0.0	0	0	0.0	5	5	1.4
Repair/Restore of Utilities	0	0	0.0	9	4	2.6	5	4	1.4	4	3	1.1	0	0	0.0	0	0	0.0	18	11	5.2
Repair/Reconstruction/Recovery	1	1	0.3	1	1	0.3	0	0	0.0	0	0	0.0	0	0	0.0	0	0	0.0	2	2	0.6
Transportation/Traffic Issues	0	0	0.0	5	5	1.4	0	0	0.0	1	1	0.3	0	0	0.0	1	1	0.3	7	7	2.0
Visits by officials	2	2	0.6	8	4	2.3	0	0	0.0	0	0	0.0	0	0	0.0	0	0	0.0	10	6	2.9
International Assistance	67	47	19.3	1	1	0.3	0	0	0.0	0	0	0.0	28	25	8.0	2	2	0.6	98	75	28.2
Economic Assistance (loans, rates)	0	0	0.0	9	6	2.6	0	0	0.0	0	0	0.0	0	0	0.0	2	2	0.6	11	8	3.2
Report on governmental response	0	0	0.0	13	9	3.7	1	1	0.3	0	0	0.0	0	0	0.0	0	0	0.0	14	10	4.0
Protest against govern.response	0	0	0.0	6	5	1.7	4	1	1.1	3	1	0.9	0	0	0.0	0	0	0.0	13	7	3.7
TOTAL	116	71	33.3	100	72	28.7	38	30	10.9	45	34	12.9	40	36	11.5	9	9	2.6	348	252	100.0

T = Number of Transactions; N = Number of Actors; % = Percent of Total Transactions

Sources: *Cumhuriyet*, and professional reports: Mitchell, William. "Social Impacts and Emergency Response; Turkish Earthquake Renaissance Report."Preliminary Draft, Eartquake Engineering Research Institute, Oakland, CA, 1992. Gurkan, Polat. "Preliminary Field Reconnaissance Report on the Erzincan Earthquake of 13 March 1993." Gurkan, Polat and Oktay, Ergunay."The Turkish Disaster Management Seminar of the UNDRO under the UNDP, 2 – 5 June 1992."

Table 40. Frequency Distribution: Types of Interactions in Disaster Response, Erzincan, Turkey Earthquake, March 13 – June 2, 1992

	Public Organizations									Nonprofit Organizations		Private Organizations		Total	
	International		National		Provincial		Municipal						
	K	%	K	N	%	K	N	%	K	N	%	K	N	%	K	N	%	K	N	%
Public: International	5	100.0	9	10	13.6	0	0	0.0	0	0	0.0	8	14	12.1	0	0	0.0	22	39	33.3
Public: National			16	18	24.2	15	17	22.7	3	6	4.5	2	3	3.0	0	0	0.0	36	44	54.5
Public: Provincial						0	0	0.0	6	10	9.1	0	0	0.0	0	0	0.0	6	10	9.1
Public: Municipal									0	0	0.0	0	0	0.0	0	0	0.0	0	0	0.0
Nonprofit												1	2	1.5	3	1	1.5	2	5	3.0
Private															0	0	0.0	0	0	0.0
Total	5	15	7.6	25	28	37.9	15	17	22.7	9	16	13.6	11	19	16.7	1	3	1.5	66	98	100.0

K = Number of Interactions; N = Number of Actors; % = Percentage of Total Number of Interactions

Source: *Cumhuriyet* and professional reports: Mitchell, William. "Social Impacts and Emergency Response; Turkish Earthquake Renaissance Report." Preliminary Draft, Earthquake Engineering Research Institute, Oakland, CA, 1992. Gurkan, Polat. "Preliminary Field Reconnaissance Report on the Erzincan Earthquake of 13 March 1992." Gurkan, Polat and Oktay, Ergunay. "The Turkish Disaster Management Seminar of the UNDRO under the UNDP, 2 – 5 June 1992."

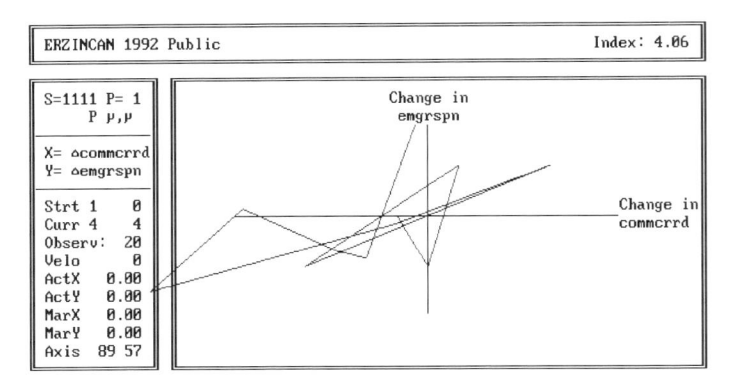

Figure 23. 1992 Erzincan, Turkey Earthquake: Phase Plane Plot, Public Organizations, Emergency Response by Communication/Coordination

Adaptive Behavior

The nonlinear analysis of frequencies of organizational action following the earthquake confirms the relatively weak response system identified from the sources available. Figure 23 shows a phase plane plot of the relationship between emergency response and communication and coordination. An initial spike in emergency response, likely representing the response from Ankara, is followed by a decline in communication/coordination that increases sharply only on the 10th day following the earthquake. Figure 24 shows the marginal history of the relationship between the two variables. Emergency response activities appear to sometimes follow, sometimes lead communication/coordination. The measure of velocity indicates that only on Days 6 and 7 were the two emergency functions operating with reasonable coordination. Figure 25 shows the phase plane plot for disaster relief and communication/coordination, confirming the slow initial

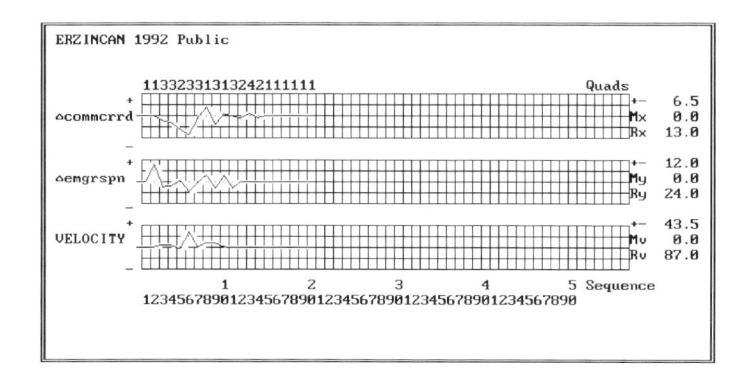

Figure 24. 1992 Erzincan, Turkey Earthquake: Marginal History, Public Organizations, Emergency Response by Communication/Coordination

Chapter 6: Emergent Adaptive Systems 155

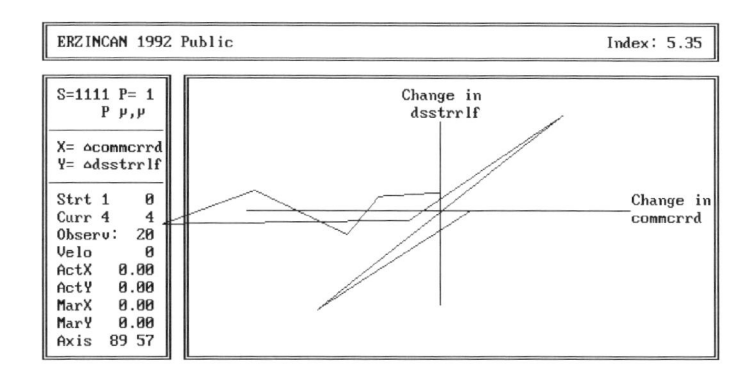

Figure 25. 1992 Erzincan, Turkey Earthquake: Phase Plane Plot: Public Organizations, Emergency Response by Communications/Coordination

response. Figure 26 documents the daily activity in disaster relief as peaking on Day 9 in the second week following disaster, late into the response period.

Indicating a relatively weak response system, logistic regression equations were found for two emergency functions — Communication/Coordination (R^2 = 0.88, F = 145.64, k = 1.2) and Recovery/Reconstruction (R^2 = 0.24, F = 5.97, k = 1.2). An equation was also found for Financial Assistance, but the F value was not significant. Figure 27 presents the map of the logistic equation for Communication/Coordination, which combines the stronger performance of Coordination with the weaker performance of Communication to produce a high value for R^2. Nonlinear analysis of emergency functions for nonprofit organizations revealed one logistic regression equation for Disaster Relief (R^2 = 0.14; F = 3.05; k = 1.0), but it was not statistically significant.

Even though private organizations comprised only 4.7% of the total response system, the frequencies reported for actions by these eight organizations revealed

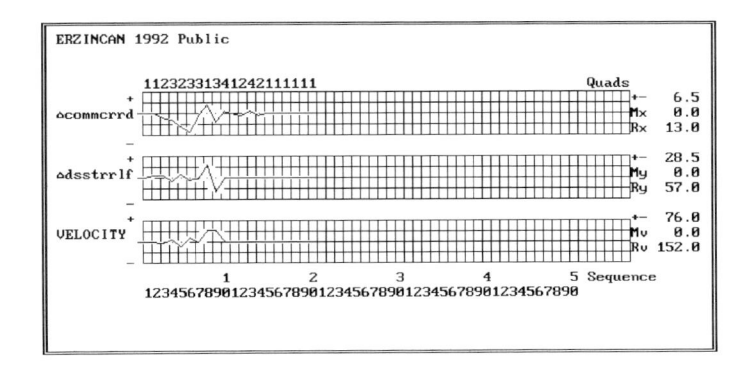

Figure 26. 1992 Erzincan, Turkey Earthquake: Marginal History, Public Organizations, Disaster Relief by Communication/Coordination

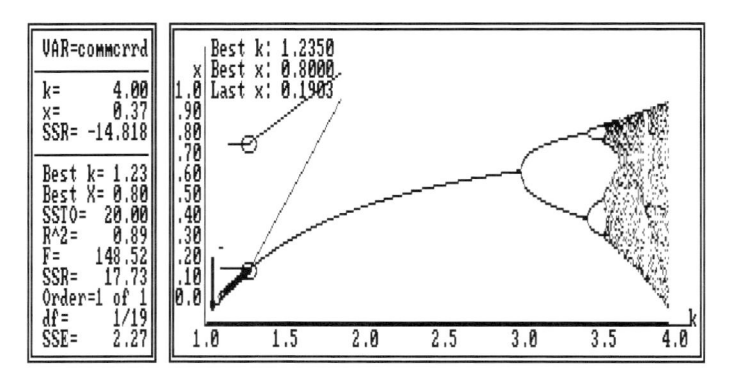

Figure 27. 1992 Erzincan, Turkey Earthquake: Nonlinear Logistic Regression, Public Organizations, Communication/Coordination

one logistic equation, Communication/Coordination ($R^2 = 0.94$; $F = 277.6$; $p < 0.000$, $k = 1.1$). Those private organizations that were involved clearly sought to coordinate their actions with other organizations participating in response operations.

The distinctive characteristics of this response system include the significant use of sophisticated information technology by the national administration in Ankara, but also the startling lapse in effectiveness of this technology when it was not matched with organizational capacity at the local level. The reasons for this lapse at the local level appear exacerbated by the disabling effect of the trauma upon the entire community, a condition that had not been anticipated in prior emergency planning.

Conclusion

The three response systems in this category each have distinctive strengths that are countered by significant weaknesses. While the elusive balance needed for continuing organizational learning and growth is not sustained in any of the three systems, the experience of shared collaboration, even if brief, underscores the importance of coherent interjurisdictional planning and development of seismic policy. Table 41 summarizes the distribution of organizations by funding source and jurisdictional levels for the three response systems profiled in this chapter.

Reviewing the distributions among the three disaster response systems, three similarities are apparent which are characteristic of this category. First, the response systems are strongly national in their composition, with national governments taking the lead in organizing the response and mobilizing action. Second, this strong national participation is complemented by a significant contribution from international organizations, stronger in each case than the subnational levels of state/provincial and municipal governments, and in the case of Erzincan, more numerous than the national agencies. This condition

Table 41. Summary, Frequency Distributions, Emergent Adaptive Systems, by Funding Source and Jurisdiction

	Public								Total Public		Nonprofit		Private		Total	
	Int'l		Nat'l		State		Mun'l									
	N	%	N	%	N	%	N	%	N	%	N	%	N	%	N	%
Mexico City	90	23.4	117	23.4	38	7.6	19	3.8	264	52.8	159	31.8	77	15.4	500	100
Costa Rica	52	21.2	74	30.2	6	2.4	10	4.1	142	57.9	50	20.4	53	21.7	245	100
Erzincan	53	31.0	40	23.4	13	7.6	28	20.9	134	78.4	29	16.9	8	4.7	171	100

runs counter to the requirements for self organization that acknowledge the importance of local organizations and conditions in mobilizing effective community response. Third, in each of the response systems, communication and coordination are weak and/or fragmented, a condition that creates gaps in the flow of information and feedback on organizational actions necessary to support the rapid evolution of an effective system. These gaps are shown consistently by the nonlinear analysis of phase plane plots and logistic regression equations. They also explain the temporary nature of the spontaneous actions taken in disaster operations by the three response systems, and the inability of the emergent systems to adapt their performance to reduce seismic risk to their communities in more permanent ways.

NOTES

1. *Excelsior*, Mexico City, DF, 20 September 1985 – 11 October, 1985.
2. US Office of Foreign Disaster Assistance. "Situation Reports, Mexico City Earthquakes of 19–20 September, 1985". 20 September – 11 October, 1985.
3. Trained search and rescue teams from 8 nations – United States, France, Great Britain, West Germany, Switzerland, USSR, Italy, and Spain – participated in disaster operations in an humanitarian effort to assist the stricken nation.
4. US Office of Foreign Disaster Assistance. "Situation Reports, Mexico City Earthquakes of 19–20 September, 1985". 20 September – 11 October, 1985.
5. Professional observation, Mexico City, 25 September – 12 October, 1985.
6. George Natanson, CBS News, Interview, Mexico City, 29 September, 1985.
7. Professional observation and *Excelsior*, Mexico City, DF, 29 September, 1985.
8. *Excelsior*, Mexico City, DF, 28 September, 1985.
9. Professional observation, Community meeting, Colonia Morelos, Mexico City, 7 October, 1985.
10. Editorial, *Excelsior*, Mexico City, DF, 30 September, 1985. This observation was also confirmed by interviews with informed Mexican social scientists who were closely following the governmental policy process after the earthquake.
11. These findings are reported from a joint survey conducted by researchers from the University of Pittsburgh and the Instituto Tecnologico Autonomo de Mexico, Mexico City, DF, November, 1985. Survey supported by funds from research centers at the University of Pittsburgh and a Quick Response Grant from the National Science Foundation.
12. This section draws heavily upon a previously published article, L.K. Comfort, "International Disaster Assistance in the Mexico City Earthquake," *New World*, Vol. 1, No. 2, Fall 1986: 12–43.

13. Local magnitude, Ml, was reported as 7.2. These calculations were reported by the Seismological Department, University of Costa Rica. EQE International, Inc. 1991. *The April 22, 1991 Valled de la Estrella Costa Rica Earthquake: A Quick Look Report.* (May): p.3.
14. National Emergency Plan. National Emergency Committee. San Jose, Costa Rica, 1991:4.1. Portions of the plan were translated and made available by Dr. Teofilo Sarkis as part of his report to the United Nations Interregional Seminar, Jakarta, Indonesia, December 13–18, 1993. Dr. Sarkis played an active role in Red Cross medical response in Limon during the disaster operations, April 22–28, 1991.
15. This system, the Emergency Information System, was purchased with funds from the International Development Agency of Canada. Interview, Luis Diego Morales, Director of Planning, CNE, April 26, 1991.
16. *La Nacion*, April 25, 1991: pp. 4A, 6A, 8A, 11A.
17. Press conference conducted by Humberto Trejos, M.D., President, CNE, April 25, 1991, 6:00 p.m., San Jose, Costa Rica.
18. Reporter for *La Nacion*. Interview, Limon, Costa Rica, April 27, 1991.
19. Professional observation and interviews with operations staff, Limon Airport, April 27, 1991.
20. US Office of Foreign Disaster Assistance. "Situation Report, Costa Rica Earthquake of April 22, 1991," No. 4, Washington, DC, April 30, 1991: p.4.
21. I acknowledge, with thanks and appreciation, the work of Leslie Mohr and Joseph Narkevic, who assisted me with the content analysis of news reports from *La Nacion*.
22. We had first used two newspapers for the content analysis, *La Nacion* and *La Republica*, both published in San Jose. In calculating the number of mentions, we found that there was considerable duplication in the stories reported in the two newspapers. To avoid double counting an organization's participation, we dropped the stories from *La Republica* in our analysis. Consequently, the frequencies reported all derive from news stories reported in *La Nacion*.
23. There is a substantial literature on nonlinear, adaptive systems that presents cogently the primary characteristics of these systems. See, for example, S. A. Kauffman. 1993. *Origins of Order: Self-Organization and Selection in Evolution.* New York: Oxford University Press; L.K. Comfort. 1994. "Self-Organization in Complex Systems." *Journal of Public Administration Research and Theory*, Vol. 4, No. 3 (July):393–410.
24. The pioneering work of Bruno R. Lima, M.D., is important in framing this problem. See B.R. Lima. 1989. "Disaster Severity and Emotional Disturbance: Implications for Primary Mental Health Care in Developing Countries." *Acta Psychiatrica Scandinavica* 79, 74. See B.R. Lima, ed. *Psicosociales consecuencias de desastre: La esperiencia Latinoamericana.* Chicago. Hispanic American Family Center.
25. American Psychiatric Association. 1987. *Diagnostic and Statistical Manual of Mental Disorders*, Third Edition, Revised. (DSM-III-R) Washington, DC: 11.
26. American Psychiatric Association. 1987. DSM-III-R.:247–248.
27. This survey was carried out by an international, interdisciplinary research team from the universities of Ege and Ankara, Turkey and Pittsburgh, USA. The team conducted research in Erzincan, Erzurum, and Ankara, from June 27 – July 10, 1997. Please see Appendix C for characteristics of the sample. As a member of this team, I acknowledge with thanks and appreciation the work of my colleagues on the team, and support from the three universities and the Turkish Government for the conduct of this research. Other publications, from other disciplinary perspectives, have resulted from this study.
28. F. Cuharcoglu, Interim Governor, Erzincan Province. Interview, Erzincan, Turkey, July 7, 1992.
29. The content analysis was limited to articles from *Cumhuriyet*, a national newspaper published in Ankara. Regrettably, the coverage of events in Erzincan, a province in Eastern Turkey, was not extensive. Consequently, data from other official reports and documents were used to extend and confirm the identification of organizational actions.

CHAPTER SEVEN

OPERATIVE ADAPTIVE SYSTEMS: WHITTIER NARROWS, CALIFORNIA; LOMA PRIETA, CALIFORNIA; AND MARATHWADA, INDIA

Operative adaptive systems respond reasonably well to the demands of a severe seismic event, but they are still primarily reactive. That is, these systems emerge in response to particular events, and function effectively in specific communities. Yet, the organizational participants in these systems do not necessarily learn from the experience of disaster, or modify their behavior because of it, or use it to initiate policy change to reduce future seismic risk in their communities or the region. Three of the eleven response systems included in this study fall into this category, two following earthquakes in California: Whittier Narrows, October 1, 1987; and Loma Prieta, October 17, 1989. The third response system evolved following the Marathwada Earthquake in Maharashtra State, India, September 30, 1993.

Operative, adaptive systems are those in which the technical structure, organizational flexibility, and cultural openness to new methods of perceiving and responding to risk are neither high nor low, but approximately medium. The unusual case in this set is the response system that evolved following the Marathwada Earthquake in India, in which the technical structure of the building stock was very low, with houses made of mud and stone, but that of the communications system was very high, with the Indian Satellite System. In the classic, if somewhat misleading, application of averages, this wide variance in technical structure balances out to "medium". More relevant, however, is the powerful impact that the satellite communications system had upon the evolution of the response system in the Marathwada region characterized by extreme poverty. In these communities, the capacity for response to a sudden, urgent event existed prior to the earthquake, and existing administrative systems were able to use the technical systems effectively to reallocate resources and mobilize response to meet the immediate needs of the event. These systems may not be optimal in all aspects of disaster operations, but they facilitate response operations in a more timely, coherent manner, involving local organizations and groups.

Cases in this category display information processes that move to a more immediate level of analysis and interactive communication during disaster operations. These processes establish a critical link to evolving response operations not seen in the first two categories of nonadaptive and emergent adaptive systems. In operative adaptive systems, the information search process actively engages local participants in determining needs of the damaged area, and uses that information as the basis for timely decisions on the allocation of resources and deployment of resources and personnel. The result is a response system that is more closely linked to the actual communities and conditions in which disaster operations are being conducted, with greater involvement of local public organizations as well as local nonprofit and private organizations. These response systems shift closer to the local level in the conduct of disaster operations, reflecting a basic characteristic of complex, adaptive systems in their sensitivity to initial conditions (Prigogine and Stengers 1984; Kauffman 1993; Gell-Mann 1994).

Operative adaptive systems have clear technical requirements to support the rapidly evolving expansion of organizational actors and functions in disaster operations. A basic information infrastructure needs to be in place prior to the disaster event, with personnel who have the training and capacity to use it, as well as a shared goal to protect life and property for their community. This information infrastructure is both technical and organizational. That is, the technical means of communication need to be robust and functional under extreme conditions, whether it is telephone, radio, satellite, or electronic networks. Equally important, personnel involved in disaster operations need to have the organizational training and experience to use these means of communication effectively. If these "initial conditions" are in place, they create a basis for more comprehensive, accurate and timely information processes in disaster operations that engage the immediate communities more actively in assessing their own needs for assistance, and utilize available resources more appropriately to meet those needs, moving operations from response to recovery more quickly. Guided by the overall goal of protecting life and property for the community, these processes enable the system to become more coherent in its operations.

The basic set of information processes that characterize disaster operations emerge as stronger, more consistent and more interactive in terms of matching external resources appropriately to the needs of the damaged communities. These processes, again, are: 1) information search; 2) information exchange; 3) organizational learning; and 4) adaptive behavior and/or self organization. These processes, as observed in the three cases in this category of response systems, are not without error, controversy, or variation. They tend to function more effectively within subsets of the system than the system as a whole. Each of the three cases began to generate interorganizational collaboration under the urgent pressure of response to the disaster, but proved largely unable to sustain it past the immediate response period. The role of information in stimulating interorganizational response emerges in different ways, with slightly different results, in each of the three response systems which will be described briefly in turn.

Whittier Narrows, California, October 1, 1987 (M = 5.9)

On Thursday, October 1, 1987 at 7:42 a.m., a moderate earthquake struck the suburbs of the San Gabriel Valley east of Los Angeles. The epicenter was located in south Rosemead at the northwest end of the Whittier Fault, about 10 miles northeast of downtown Los Angeles. Eight persons were killed, and significant property damage prompted the affected cities, Los Angeles and Orange Counties, and the State of California to declare a state of emergency. Three days later, on Sunday, October 4, a severe aftershock measuring 5.5 Richter shook the area, damaging further buildings that had been weakened by the main shock. The increased damage and added danger resulted in a Presidential declaration of a state of emergency on October 7, 1987, releasing federal resources to the damaged cities and counties. Total property losses for the area were estimated at $358 million.[1]

Although the Whittier Narrows Earthquake was a moderate event, it is useful to our study of response systems following earthquakes for three reasons. First, it is typical of the type of moderate earthquakes that occur with relative frequency (every 10 to 20 years) in the densely populated area of Southern California, with its underlying network of earthquake faults. Second, this earthquake and its severe aftershock were located on blind, low-angle thrust faults in areas previously unmapped by the U.S. Geological Survey, seven kilometers beyond the known Whittier Fault.[2] Thrust faults are more difficult to detect and represent hidden risk in residential areas. Third, the damage extended over a surprisingly wide area of Los Angeles and Orange Counties, estimated at 190 square miles,[3] illustrating the importance of interorganizational coordination in earthquake response. These initial conditions meant that many jurisdictions were affected by the same event and were experiencing similar kinds of problems in disaster operations. These conditions placed heavy demands upon the intra- and interorganizational characteristics of the response system, affecting especially the basic processes of information search, information exchange, organizational learning and adaptive behavior/self organization.

Information Search

The earthquake and its aftershocks damaged buildings and infrastructure in at least fifteen cities in Los Angeles County and several in northern Orange County[4], but there was no single coordinated response for the entire area. Approximately 10,500 business and residential structures were damaged, most within 10 miles of the two epicenters.[5] Rather, each local jurisdiction responded to damage within its area of responsibility. Individual cities made efforts to coordinate their actions with one another, and jurisdictions supported one another through "mutual aid" agreements in which cities extend assistance and resources to one another in times of need. Yet, primary responsibility for disaster response lay with each local jurisdiction, resulting in significant variations in response operations for the separate cities.

The initial information search process varied from city to city, depending upon the level of damage, awareness, training, and access to available leadership. Findings are presented from a study of interorganizational coordination in disaster management that included seven cities that suffered the heaviest losses from the earthquake – Alhambra, Los Angeles, Monterey Park, Pasadena, Rosemead, Santa Fe Springs and Whittier.[6] The six smaller cities were all located in a continuous stretch of the San Gabriel Valley, and were of roughly comparable size and exposure to the earthquake. Response operations undertaken by each city differed in perceived effectiveness by their respective citizens, based in large part upon the timeliness and accuracy of the information provided to them by the public agencies.

Los Angeles, ten miles west of the epicenter, immediately activated its response plan, as is legally required for any earthquake over $M = 5.0$ Richter. This plan requires the City Emergency Operations Bureau to send its fleet of helicopters to gather information and survey damage to the city from the air. The Engineering Department dispatched its Earthquake Damage Evaluation Teams into the field by 8:30 a.m. to check a pre-identified list of critical facilities – freeways, bridges, dams, water reservoirs, communications facilities, the Port, the airport – for damage.[7] The reconnaissance teams inspected these facilities, and reported back to the Emergency Operations Bureau within an hour. By 10:00 a.m., the City declared a state of emergency. The City reported moderate damage to buildings and infrastructure. Traffic lights were out and electrical lines were down; approximately 87 fires were triggered by the earthquake in various forms. Public buildings were cracked and damaged, residents were frightened and requested help, but the basic operations of the city continued. City functions were largely back to normal by noon, 12:00 m. on October 1, 1987, when the demands for emergency response on an ordinary day in Los Angeles displaced the earthquake in priority.[8]

Other cities collected information within their own framework of emergency action. The City of Alhambra had a young management staff that had recently attended a training session for disaster preparedness at the California State Training Institute in San Luis Obispo. With training recommendations fresh in their minds, they carried out an almost textbook response to the disaster. They mobilized a Policy Council, consisting of the fire, police, and public works chiefs, city manager, and major agency heads, used the regular police and fire services to assess damage in the city, set up a public relations office to prepare and transmit press reports to citizens, opened a hotline for citizen requests for assistance, and created, within their own jurisdiction, an interorganizational response. Information search relied primarily on established means of damage assessment by responsible agencies, but the integration of information and its timely distribution to all relevant agency heads improved the communication and coordination among the agencies.[9]

Other jurisdictions followed different patterns. Pasadena, to the north and at the outer edge of the damaged area, did not declare a local emergency, but rather allowed each city organization to respond to specific requests within its particular field of responsibility. Monterey Park, to the south, relied heavily on its Police Department to direct the emergency response, and followed traditional patterns of information search within its boundaries. Rosemead, a small city that contracted both police and fire services to Los Angeles County, relied on the information search capacity of the County police and fire stations in its area. Whittier, an older city and closest to the epicenter, had a Central Business District with many unreinforced masonry buildings. Approximately 30 buildings in the uptown business district collapsed during the earthquake, resulting in substantial damage to its commercial sector. Whittier contracted fire services to the County of Los Angeles, but had its own Police Department, and mobilized its emergency response plan within the city. The City of Whittier activated its emergency plan and opened its Emergency Operations Center at City Hall at 8:15 a.m., following its pre-defined procedures for information search and damage assessment.[10]

Whittier had virtually no communications with external agencies in the immediate hours after the earthquake. All city agencies were fully engaged in meeting their responsibilities for emergency response and damage assessment. They reported back to the City Manager and Emergency Coordinator, who integrated information from the reports submitted by separate city agencies into a profile for the city. Within the city, telephone communications were out and radio frequencies were jammed, cutting off even emergency services from outside agencies. The Los Angeles County Fire Captain who commanded the Whittier Station, unable to communicate with his County Operations Center, learned that the epicenter of the earthquake was near Whittier only after fire engines from Orange County arrived to offer mutual aid.[11]

Santa Fe Springs, south of Whittier, is a relatively new city with a high commercial investment but low residential population. The City activated its emergency plan, which had recently been developed with the cooperation and support of the City's Chamber of Commerce. Santa Fe Springs had its own Fire Department, but contracted its police services to the Los Angeles County Sheriff's Office. The information search for this city focused largely on damage to commercial properties, and the substantial risk of hazardous materials releases from storage tanks and underground pipelines, a secondary hazard likely to be triggered by earthquakes.

Emergency response operations following this earthquake were complicated. Three of the seven cities included in this study — Rosemead, Whittier, and Santa Fe Springs — contracted some or all of their emergency services to Los Angeles County. While the County fire and police stations were staffed by personnel who were fully competent in their professional responsibilities, they lacked the detailed knowledge of people and places in their areas of responsibility that is accumulated by full-time residents. This fact made them even more dependent upon adequate information technology to communicate with relevant information sources. Without adequate means of communication, the communities were isolated from important sources of knowledge and support. Given these complex arrangements, information search processes proved relatively effective within each city, but search processes were difficult and slow among the cities subjected to the same event.

Information Exchange

The modes of information search shaped the processes of information exchange, which occurred primarily within cities, and with difficulty, between cities. Each of the participating cities largely followed its own emergency plan, and relied primarily on its own resources in meeting the demands of the event. Los Angeles County, which was directly involved in providing police and fire services to its contract cities, encountered the difficulty of integrating information to provide a comprehensive profile of the event for the affected area and recognized the importance of information exchange. This difficulty was equally obvious to administrators for the Operational Areas of the California Office of Emergency Services.

Individual cities developed means of information exchange within their own communities, some more effective than others. Alhambra used its Public Relations Officer to prepare and disseminate timely status reports of earthquake damage and available resources to its citizens, easing citizens' concerns and enabling them to take prompt action to meet household or business needs. Whittier established a telephone hotline to accept damage reports and calls for assistance from citizens, and enlisted the cooperation of the City's Chamber of Commerce to assist with the severe damage to the City's Central Business District. Santa Fe Springs' city officials met with business leaders to assess the damage to their city and to plan response and recovery strategies. Rosemead, dependent upon County emergency services, relied upon County reports to inform its response strategies. The City of Los Angeles followed well-established procedures for emergency response through its Emergency Operations Bureau, but had relatively little interaction with other jurisdictions. Monterey Park relied primarily on its Police Department for information functions, led by a chief who had earned the confidence of other city department administrators and the community. Pasadena followed its routine procedures of information exchange with separate departments reporting to the Office of the City Manager. Although the California Office of Emergency Services formally had administrative arrangements for Operational Areas for emergency services at the local level, these arrangements existed largely on paper and were used primarily to file the required state reports. Given the moderate impact of the earthquake, the individual cities acted relatively independently in an area-wide response to the disaster.

Organizational Learning

Given the limited means of information exchange among the primary cities involved in disaster response, I searched for evidence of a coherent organizational system that evolved in response to this earthquake. Using the N-K methodology and content analyses of news reports from the *Los Angeles Times*, we identified a system of 189 organizations that were reported to be directly involved in response operations.[12] The large majority, 123, or 65%, were public organizations. Private organizations represented the next largest category, with 38, or 20.1%, and nonprofit organizations made up the remaining set, with 28, or 14.8%. This response system attracted a sizeable involvement from private organizations, reflecting the substantial damage to the commercial sector in Whittier and other cities.

In marked contrast to the nonadaptive and emergent types of organizations, the largest proportion of the 123 public organizations was municipal, with 81, or 65.9% of the total. Next largest was state, with 17, or 13.8%; county organizations were a close third, with 16 or 13%, and national organizations, with 9, or 7.3%, made up the smallest proportion of the total number of public organizations. The distribution for the response system is summarized in Table 42.

County and municipal organizations together constituted over half, 51.2%, of the total response system, with very small national participation and no international organizations involved. Disaster operations were clearly focused on the local level, with local emergency response agencies meeting the needs of their communities.

Table 43 presents the matrix of 562 reported transactions performed by the set of organizations involved in disaster response. The largest proportion of transactions, 15.3%, involved disaster relief – providing food, water, shelter, clothing to displaced persons. The second largest number of reported transactions involved the inspection of buildings to assess damage and ensure safety. Emergency response was third, with 8.9% of the transactions, followed by federal/state aid with 7.7%. Communications and coordination, combined, constituted 7.1%, with communications representing the largest portion, 5.3%. These figures document a relatively small, but distinctive role for communications and coordination activities in response operations, as reported in news accounts.

Table 42. Frequency Distribution: Disaster Response System by Funding Source and Jurisdiction, 1987 Whittier Narrows Earthquake

Public										Nonprofit		Private		Total	
Int'l		Nat'l		State		County		Munic'l							
N	%	N	%	N	%	N	%	N	%	N	%	N	%	N	%
0	0.0	9	4.7	17	9.0	16	8.5	81	42.9	28	14.8	38	20.1	189	100.0

Part II: Shared Risk in Practice

Table 43. Frequency Distribution: Types of Transactions in Disaster Response by Funding Source and Jurisdiction, Whittier Narrows, CA Earthquake, October 1-21, 1987

Type of Transaction	Public Organizations									Nonprofit Organizations		Private Organizations		TOTALS	
	National		State		County		City								
	T	N	T	N	T	N	T	N	T	N	T	N	T	N	%
Emergency Response	0	0	1	1	9	3	24	11	10	6	6	5	50	26	8.9
Communication	3	2	4	4	3	3	7	6	2	2	11	6	30	23	5.3
Coordination of Response	1	1	2	1	1	1	5	5	1	1	0	0	10	9	1.8
Emerg. Response Center Activities	0	0	1	1	6	5	19	10	0	0	1	1	27	17	4.8
Medical Care/Health	0	0	0	0	2	1	4	4	0	0	0	0	6	5	1.1
Damage/Needs Assessment	0	0	6	1	5	2	19	12	1	1	0	0	31	16	5.5
Certification of Deaths	0	0	0	0	2	2	4	3	0	0	0	0	6	5	1.1
State of Emergency Declarations	3	1	6	3	1	1	5	5	0	0	0	0	15	10	2.7
Evacuation	0	0	0	0	3	3	11	8	0	0	3	3	17	14	3.0
Earthquake Assessment/Research	6	2	13	3	0	0	0	0	0	0	0	0	19	5	3.4
Housing Issues	0	0	0	0	0	0	20	12	10	6	0	0	30	18	5.3
Disaster Relief (food, shelter, etc.)	0	0	0	0	6	2	37	18	43	6	0	0	86	26	15.3
Building Inspection	1	1	3	2	11	7	36	16	1	1	5	4	57	31	10.1
Repair/Restore Utilities	0	0	0	0	0	0	3	2	0	0	19	6	22	8	3.9
Repair/Reconstruction/Recovery	0	0	2	2	1	1	18	9	3	1	10	9	34	22	6.0
Transportation/Traffic Issues	1	1	15	2	6	2	1	1	0	0	2	2	25	8	4.4
Visit by Officials	3	1	6	2	0	0	3	3	0	0	0	0	12	6	2.1
Federal/State Aid to Victims	12	5	15	7	1	1	4	3	10	8	1	1	43	25	7.7
Assistance (money, support)	0	0	2	1	2	2	8	7	10	7	1	1	23	18	4.1
Insurance Related Activities	0	0	0	0	0	0	0	0	0	0	7	6	7	6	1.2
Investigation/Recommendations for improved response	0	0	2	2	1	1	9	9	0	0	0	0	12	12	2.1
TOTAL	30	14	78	32	60	37	237	144	91	39	66	44	562	310	100.0
%	5.3		13.9		10.7		42.2		16.2		11.7		100.0		

T = Number of transactions; N = Number of Actors; % = Percent of Transactions
Sources: *The Los Angeles Times* and the State Seismic Safety Commission Report

Chapter 7. Operative Adaptive Systems

Table 44. Frequency Distribution: Types of Interactions in Disaster Response by Funding Source and Jurisdiction, Whittier Narrows, CA Earthquake, October 1–21, 1987

	Public												Nonprofit			Private			Total		
	National			State			County			City											
	K	N	%	K	N	%	K	N	%	K	N	%	K	N	%	K	N	%	K	N	%
Public																				
National	1	2	0.6	4	6	2.5	0	0	0.0	1	2	0.6	8	8	5.1	2	3	1.3	16	21	10.1
State				5	7	3.2	3	6	1.9	17	17	10.8	2	3	1.3	1	2	0.6	28	35	17.7
County							4	6	2.5	12	17	7.6	5	3	3.2	2	4	1.3	23	30	14.6
City										30	33	19.0	39	30	24.7	13	16	8.2	82	79	51.9
Nonprofit													8	7	5.1	1	2	0.6	9	9	5.7
Private																0	0	0.0	0	0	0.0
Total	1	2	0.6	9	13	5.7	7	12	4.4	60	69	38.0	62	51	39.2	19	27	12.0	158	174	100.0

K = Number of Interactions; N = Number of Actors; % = Percentage of total number of interactions
Sources: *The Los Angeles Times* and the Californian Seismic Safety Commission Report

Table 44 presents the matrix of interactions that provide a measure of the density of the system. Of the six measures of an adaptive system used in this analysis, K indicates on a gross scale the number of interactions between types of organizations engaged in emergency response. The table shows the relatively low level of total interactions, with 174 organizations engaging in a total of 158 interactions, or collaborative efforts in disaster response. The largest number of actors, 69, or 39.6%, were public organizations engaged in interactions at the municipal level. The next largest group, 51, or 29.3%, was nonprofit organizations, and private organizations constituted 27, or 15.5%. National public organizations had the smallest number of reported actors, 2, or 1.1%, with state and county organizations representing 7.5% and 6.9%, respectively. While these measures document a response system performing a recognized set of tasks in disaster operations, all measures show a substantial shift to the local level for performance.

Adaptive Behavior and/or Self-Organization

The nonlinear analysis of change in daily performance of disaster response functions documents the relative independence of the organizations engaged in disaster operations. Figure 28 shows the phase plane plot for the variables, disaster relief and communication/coordination. The change reported for these two variables showed several chaotic swings, but moved to a pattern of rough similarity in the later period. Figure 29 presents the marginal history for the relationship between the two variables, in which velocity represents a rough measure of the degree of coordination between them. Apart from a noticeable drop in coordination on Day 4, the day of the aftershock, activities reported for these two variables operated in a reasonably coherent way.

168   Part II: Shared Risk in Practice

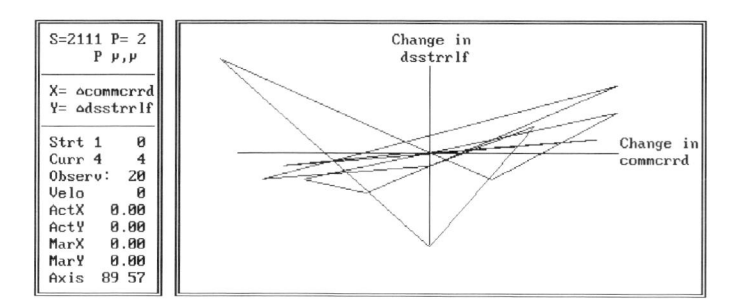

Figure 28. 1987 Whittier Narrows, CA Earthquake: Phase Plane Plot, Public Organizations, Emergency Response by Communication/Coordination

The logistic regression analysis, as expected, shows relatively weak performance across jurisdictions for public organizations. Regression equations were calculated on change data for the six variables identified in disaster response for all jurisdictional levels — national, state, county and city. For public organizations, three equations were found, and two produced statistically significant R^2 values at the 0.05 level. Figures 30 and 31 present the logistic equations for Damage Assessment ($R^2 = 0.46$; F = 16.35; k = 1.1) and Emergency Response ($R^2 = 0.30$; F = 8.11; k = 1.25). Change in actions by private organizations produced logistic equations for three variables in disaster response. These are: Communication/Coordination ($R^2 = 0.40$; F = 12.82; k = 1.25) shown in Figure 32; Recovery/Reconstruction ($R^2 = 0.25$; F = 6.26; k = 1.1); and Damage Assessment ($R^2 = 0.22$; F = 5.41; and k = 1.1). The R^2 values for all three equations are statistically significant at the 0.05 level. While the number of private actors is smaller, and the number of interactions in which these organizations were engaged is also smaller, they nonetheless adapted their performance on three of the main functions in disaster operations with which they were most directly involved and affected.

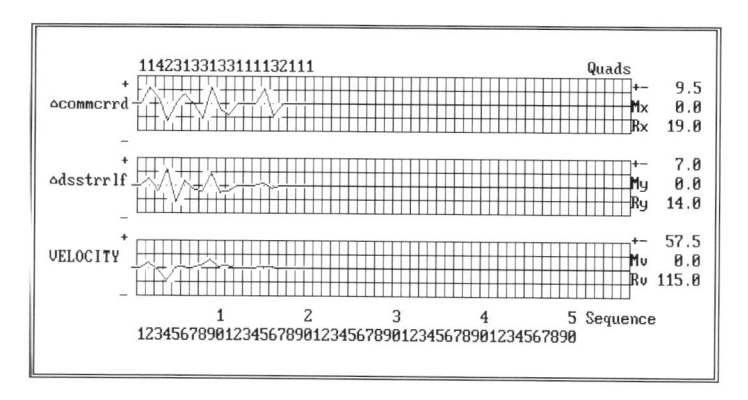

Figure 29. 1987 Whittier Narrows, CA Earthquake: Phase Plane Plot, Public Organizations, Disaster Relief by Communication/Coordination

Chapter 7. Operative Adaptive Systems   169

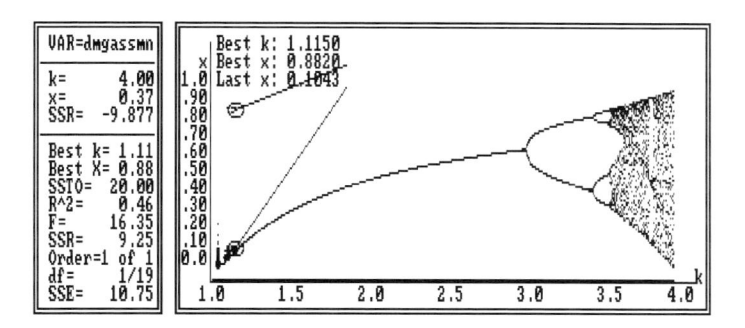

Figure 30. 1987 Whittier Narrows, CA Earthquake: Nonlinear Logistic Regression, Public Organizations, Damage Assessment

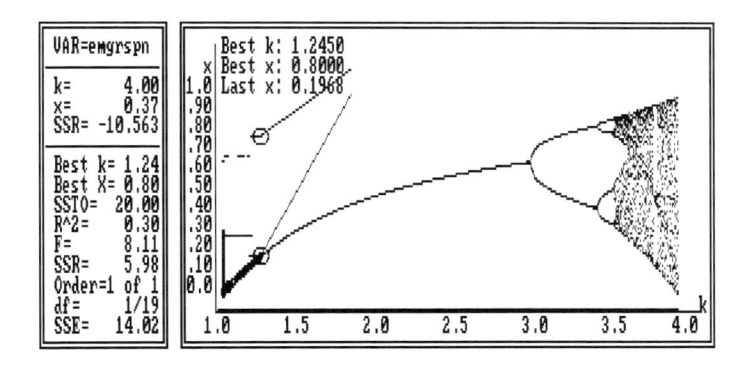

Figure 31. 1987 Whittier Narrows, CA Earthquake: Nonlinear Logistic Regression, Public Organizations, Emergency Response

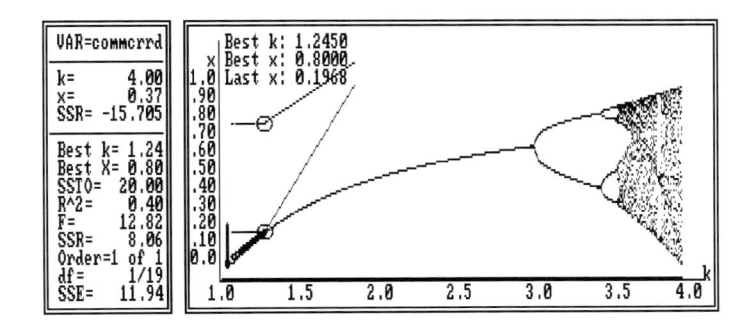

Figure 32. 1987 Whittier Narrows, CA Earthquake: Nonlinear Logistic Regression, Private Organizations, Communication/Coordination

The logistic equations, calculated with data including all public organizations — national, state, county and city — document the relatively low level of interaction across jurisdictional levels and identify the response functions that involved the most shared activity among public organizations: Emergency Response and Damage Assessment. Private organizations constitute a smaller, but more coherent group, and reflect the strong impact upon the private sector from damage incurred in the earthquake. Analysis of change data for the nonprofit sector produced no logistic equations, indicating the relatively low level of displacement of people and services for this moderate earthquake.

Loma Prieta, California, October 17, 1989 (M = 7.1)

At 5:04 p.m. on Monday, October 17, 1989, a powerful earthquake ruptured the southern segment of the San Andreas Fault in the San Francisco Bay Area in California. The earthquake's epicenter was located at Loma Prieta Mountain near Santa Cruz, with a reported magnitude of 7.1 on the Richter scale. Powerful ground shaking caused much damage to the historic business district of the city of Santa Cruz, where Victorian buildings of unreinforced masonry crumbled under the stress. In a pattern similar to that observed in the Mexico City Earthquake of 1985, the strong motion waves traveled considerable distances, becoming amplified as the force reached weak soil structures. This amplification of force caused considerable damage more than 60 miles away in San Francisco, with the liquefaction, collapse of houses, and ensuing fires in the Marina District, and more than 75 miles away in Oakland with the collapse of the I-880 freeway structure at Cypress Street and damage to buildings in the downtown area. The total cost of damages from this earthquake was estimated at $7.1 billion, much of it due to damaged infrastructure such as freeways, bridges, and electrical power stations and lines.

The damage pattern in the Loma Prieta Earthquake revealed the dangers and likely characteristics of a major earthquake in a metropolitan area, specifically one with widely varying soil structures as in most coastal cities. Although the primary effects of this powerful earthquake were in the relatively rural area of the Santa Cruz Mountains, the consequences stretched across the ten counties of the San Francisco Bay Area, snarling traffic with the collapse of the Bay Bridge, damaging other freeway structures, and interrupting business and other operations. Local emergency response agencies in each affected jurisdiction conducted their own operations in response to the event, but at least three primary centers of response operations emerged, reflecting the danger to life and property in their respective areas. These centers included: 1) the Pacific Garden Mall in downtown Santa Cruz, where 3 people were killed and several injured in collapsed buildings; 2) the Marina District in San Francisco, where 3 people were killed in a collapsed building and many others displaced from their homes by the ensuing fires; and 3) the Cypress Structure freeway collapse in Oakland, where 41 people were killed. Earthquake damage caused other deaths and displacement of people from their homes throughout the area, such as the spectacular collapse of the Bay Bridge in which one woman was killed, the crumbling of a brick wall in San Francisco onto a car in which 5 people were killed, and damage to residential and business buildings from Watsonville to Oakland.[13] Response operations focused on the areas which posed the greatest threat to life and affected the most people.

The disaster response system that emerged following this earthquake shared many of the characteristics of the Whitter Narrows response system, but on a much larger scale. California local emergency response agencies were mostly well-trained and equipped, and conducted disaster operations primarily within their own jurisdictions, with relatively little interaction or assistance from external agencies. Since the effects of the earthquake were scattered over a wide area, many emergency response organizations and jurisdictions — municipal, county, state, and federal — were involved. But this situation also created a complex set of interactions that resulted in different levels of performance in different areas, a condition that prompted considerable criticism from the affected groups. How and why these differences occurred in implementation of presumably the same federal and state policies for disaster management are issues of central importance to this study.

Information Search

Given the wide area of impact from this earthquake, information search was, for the most part, localized to the individual jurisdictions. This practice reflected the level of training, equipment and facilities available at each jurisdiction, and resulted in substantial unevenness in both the search for information and the reporting of events. In San Francisco, Oakland, Santa Cruz, and San Jose, the cities activated their Emergency Operating Centers and their first response agencies — police, fire, public works, emergency medical services — followed their emergency procedures and responded to calls. In virtually every jurisdiction, communications were overloaded or failed, resulting in delays, frustration, and confusion.

Innovative practices by the first response agencies enabled the local jurisdictions to carry out their response operations within their jurisdictions with substantial effectiveness, but not without problems. In San Francisco, the Fire and Police departments did not have helicopters available for aerial reconnaissance, but used standard police and fire practices of street reconnaissance and response to incoming calls. The fire dispatch failed, and dispatchers tracked ambulances by manual means. In Alameda County, the County's new computerized Emergency Communications System failed, and Fire and ambulance services responded to their own calls, with no central dispatching. Communications were out at Highland Hospital, the County's emergency medical receiving unit, and ambulances were not able to contact the hospital or provide information on incoming patients. Response agencies resorted to using runners to provide critical information or to establish critical contacts in organizing the response.

The City of Oakland had recently adopted an emergency plan, and had employed a three-person staff for its new Office of Emergency Services only months before the earthquake. This staff followed the new procedures to carry out a systematic search for damage in the city, relying on reports from Police and Fire personnel. In other cities and counties, information search procedures were carried out chiefly by using routine methods of fire and police patrols and response to 911 calls. As in the Whittier Narrows response, the greatest difficulty lay in obtaining a timely, coherent, accurate profile of the damage and response operations for the entire area.

Gaps in information at the agency and jurisdictional levels also posed difficulty in reporting the event accurately to the public, as the media focused on the most visible events in the most well-known areas, playing videotapes of the Bay Bridge collapse, the Marina District fires, and the Cypress Structure collapse over and over to national and international audiences. The equally damaged Santa Cruz business district and the serious loss of housing in the Santa Cruz and Watsonville areas were almost ignored in the national news during the first days of response.

Information Exchange[14]

According to the State's Emergency Plan, each city reports its status to the County Coordinator of Emergency Services, and the counties, in turn, report their status to the Regional[15] Coordinator. The Regional Operations Center for the California Office of Emergency Services has the responsibility for compiling a profile of damage and resource needs for the region and submitting it to the State Office in Sacramento. Following the October 17, 1987 earthquake, the Coordinator of Region II in Pleasant Hill waited for counties within the region to report their status, but the counties, in turn, were waiting for reports from their respective cities. The cities were so preoccupied with their immediate operations that they reported little information to any other level during the first hours and days of response. This difficulty was compounded by the failure of communications in many of the cities and counties, and the lack of any area-wide emergency communications capacity.

Creating and maintaining an area-wide profile of response operations was further hampered by the lack of staff and equipment to manage such an event at the Region II Office. The Regional Office did not have computers. The staff used the event, however, as evidence of the need for computers and requisitioned them from the State Office in Sacramento. The requisition was approved, but the delay involved in processing the requisition, obtaining the machines, programming them for operation, and entering data from the event meant they were unavailable for decision support in actual response operations for the October 17, 1989 earthquake. The machines would be in place for the recovery process and the next major disaster in the region.

Chapter 7. Operative Adaptive Systems 173

Limited by inadequate communications, information exchange occurred primarily within jurisdictions, rather than between jurisdictions. Interorganizational information exchange is most critical in a situation that requires multiorganizational and multijurisdictional response. Response to the collapse of the Interstate-880 freeway Cypress Street Viaduct in Oakland illustrated the urgency, complexity and interdependence of disaster operations that require timely information exchange. Such a situation leads to the emergence of a subsystem of organizations interacting to meet needs at a single site that could not be met by any single organization or jurisdiction. The evolving subsystem, in turn, becomes an important part of the larger response system for the whole disaster.

The collapse of the Cypress Street structure caused the greatest loss of life and placed the heaviest demands upon the interorganizational disaster response system. Each of the six disaster response functions was involved. Emergency response was immediate, as local personnel heard and saw the collapse. The Oakland Fire Department dispatched Engine Number 3 and Ladder Truck 3 to the scene, arriving within 2 minutes of the collapse. Allied Ambulance had an emergency vehicle in the area that immediately drove to the scene. As emergency personnel arrived and saw the destruction, they radioed to their respective dispatch centers, assessing the size and scope of the collapse and requesting reinforcements. The Oakland Police Department and the California Highway Patrol responded to the dispatch notification within minutes. Residents of the area, local business people, and workers at nearby industrial sites immediately began to assist victims to escape from the freeway, using fork lifts and heavy equipment volunteered by local industrial companies. Within ten minutes, approximately 50% of the emergency personnel who responded to the collapse were on scene and actively engaged in rescue operations.

Five primary organizations were involved in emergency search and rescue operations at the Cypress Structure: City of Oakland Fire Department, Police Department, and Allied Ambulance, a private ambulance company which provided contract emergency medical services to the City of Oakland; California Department of Transportation, responsible for the collapsed structure, and the California Highway Patrol, responsible for safety on the state's freeways. Each of the five organizations responded to the event through their respective notification systems. But communications failures hindered coordination among the agencies. The local telephone system was disrupted by the shock. Local radio frequencies were jammed. Organizational personnel had to rely on vehicle radios and improvisation during the first hours of the emergency. Pacific Telephone responded to emergency requests for assistance and by 6:30 p.m on October 17 had some emergency lines in operation. The company also provided cellular phones to emergency personnel to ease the communications burden. The problem of communication, and with it, information exchange, escalated as disaster operations progressed, when there was no simple accessible means of communicating information between agencies.

Information exchange was conducted primarily through the established mechanisms of disaster operations. At the Cypress Structure Command Post, the command structure included representatives from each of the five organizations involved in operations: City of Oakland Fire, Police, Allied Ambulance, California Department of Transportation and California Highway Patrol. Through interactive communication on site, these organizations worked together to develop the most effective strategies for response. Communications from local to state level have been described above, through the activities of the Region II Office of the California Office of Emergency Services. Information exchange between the state and federal agencies occurred at the Joint Federal/State Coordinating Office established in Mountain View, California. At this level, the federal agencies were primarily involved in providing disaster relief and financial assistance to the state agencies. The state agencies, in turn, were providing disaster relief and financial assistance to the counties, who were supporting the cities that had suffered damage and losses.

Although the structure for interorganizational information exchange and coordination of activities was in place, it operated primarily on demand and did not anticipate requests or actively facilitate the emergence of the response system.

Organizational Learning

The organizations participating in response operations, through their interactions with one another, formed a distinct response system, unified by the goal of reducing loss of life and threat of injury and protection of property. The content analysis of Bay Area newspapers, using the methodology of the N-K system, identified a system of 623 organizations that were actively engaged in response actions. Since the newspapers focused most heavily on the emergency response operations in their respective communities, four newspapers were used in this content analysis to cover the three primary sites. They are: *San Francisco Chronicle, San Francisco Examiner, Oakland Tribune,* and *San Jose Mercury.* Several characteristics made this a distinctive response system. First, it was large, by far the largest of the response systems identified in the entire set of 11 response systems. Second, the private sector played a significant role in disaster response, largely through donations to disaster relief. Third, the system included a sizeable number of international organizations, primarily private organizations that contributed to relief funds, but also public organizations that sent contributions and research teams. Finally, as in the Whittier Narrows earthquake, the largest proportion of public organizations was from the local level.

Of the total number of 623 organizations identified in the response system, 261, or 41.9%, were public. Private organizations had a slightly larger total, 264, or 42.4%. Nonprofit organizations made up the remaining portion, with 98, or 15.7%. Of the total number of public organizations, 94, or 36%, were city organizations; 51, or 19.6% were county organizations, 15, or 5.7% were special districts, e.g. water districts, transportation districts, port authorities. Together, these

Chapter 7. Operative Adaptive Systems

local organizations comprised 61.3% of the subset of 261 public organizations. State organizations were second, with 39, or 14.9%; national organizations were third, with 36, or 13.8%, and international organizations, with 26, or 10.0%, made a substantive contribution to the response system. A sizeable number of private international organizations, 62, or 23.5% of the subset of private organizations, made contributions to disaster relief. The significant proportion of international organizations participating in response reflects the high international profile of the San Francisco Bay Area, with its multicultural population, strong international trade relationships, and substantial investment from international firms. These figures are summarized in Table 45.

The unusual character of this response system is further shown by the analysis of transactions performed by organizations participating in response actions. Table 46 presents the data for a total of 1454 transactions involving 811 organizations, reported in the four Bay Area newspapers. Donations of money and goods claimed the largest number of reported transactions, 310, or 21.3%. Of that number, 121, or 39%, involved international actors, both public and private. The next largest category of transactions was disaster relief, with 140, or 9.6%; and third largest was transportation issues, with 119, or 8.2%. Emergency response was fourth, with 117 reported transactions, or 8.0%. Communication and coordination together constituted only 2.4% of the reported transactions, whereas governmental assistance ranked fifth with 7.2%.

The pattern of interactions, presented in Table 47, reveals that this large system was not tightly interconnected. That is, organizations participating in response were acting more independently than interdependently. Only 255 interactions were reported in total, and one-fourth of those involved private domestic organizations interacting primarily with nonprofit organizations.

Public-national and public-state organizations were second and third, both with 23.3% of the interactions, and public-city organizations were fourth, with 16.7% of the transactions. These findings confirm that disaster operations were carried out largely by local organizations at the local level, but with sizeable contributions from state, national, private and nonprofit organizations.

Table 45. Frequency Distribution: Disaster Response System by Funding Source and Jurisdiction, 1989 Loma Prieta Earthquake

Public							Nonprofit	Private	Total
Int'l	Nat'l	State	County	Sp. Dist.	City	Total			
N %	N %	N %	N %	N %	N %	N %	N %	N %	N %
26 4.2	36 5.8	39 6.3	51 8.2	15 2.4	94 15.0	261 41.9	98 15.7	264 42.4	623 100.0

Table 46. Frequency Distribution: Types of Transactions in Disaster Response by Funding Source and Jur-

Type of Transaction	Public Organizations														
	International			National			State			County			City		
	T	N	%	T	N	%	T	N	%	T	N	%	T	N	%
Emergency Response	0	0	0.0	10	5	0.7	15	5	1.0	23	11	1.6	46	25	3.2
Communication	0	0	0.0	0	0	0.0	1	1	0.1	8	5	0.6	0	0	0.0
Coordination of Response	0	0	0.0	3	2	0.2	0	0	0.0	2	2	0.1	3	3	0.2
Medical Care/Health	0	0	0.0	3	2	0.2	3	3	0.2	25	13	1.7	7	6	0.5
Damage/Needs Assessment	0	0	0.0	4	2	0.3	12	7	0.8	9	5	0.6	15	13	1.0
Certification of Deaths	0	0	0.0	0	0	0.0	6	3	0.4	17	4	1.2	2	1	0.1
State of Emergency Declaration	0	0	0.0	5	1	0.3	0	0	0.0	1	1	0.1	2	2	0.1
Evacuation	0	0	0.0	0	0	0.0	2	2	0.1	2	2	0.1	10	10	0.7
Earthquake Assessment/Research	0	0	0.0	19	2	1.3	11	6	0.8	0	0	0.0	0	0	0.0
Security Prevention of Looting	0	0	0.0	2	2	0.1	5	4	0.3	12	4	0.8	9	7	0.6
Housing Issues	0	0	0.0	2	1	0.1	0	0	0.0	7	6	0.5	0	0	0.0
Disaster Relief (food, shelter, etc.)	0	0	0.0	2	2	0.1	7	6	0.5	9	6	0.6	32	21	2.2
Donations (money, goods, etc.)	34	26	2.3	2	2	0.1	1	1	0.1	0	0	0.0	8	8	0.6
Building Inspection	0	0	0.0	4	3	0.3	4	3	0.3	26	11	1.8	27	15	1.9
Investigation into Road Failures	0	0	0.0	1	1	0.1	28	8	1.9	0	0	0.0	3	2	0.2
Repair of Roads/Bridges	0	0	0.0	4	3	0.3	52	6	3.6	0	0	0.0	6	4	0.4
Repair/Restore Utilities	0	0	0.0	2	1	0.1	0	0	0.0	3	2	0.2	1	1	0.1
Repair/Reconstruction/Recovery[a]	0	0	0.0	2	2	0.1	3	2	0.2	6	4	0.4	11	8	0.8
Transportation/Traffic Issues	1	1	0.1	3	2	0.2	31	7	2.1	4	3	0.3	3	3	0.2
Legal Issues (law suits)	0	0	0.0	0	0	0.0	7	3	0.5	0	0	0.0	0	0	0.0
Economic/Business Issues	0	0	0.0	3	3	0.2	1	1	0.1	0	0	0.0	5	4	0.3
Visit by Officials	0	0	0.0	28	7	1.9	4	1	0.3	1	1	0.1	3	1	0.2
Government Assistance	0	0	0.0	63	14	4.3	32	6	2.2	7	7	0.5	2	2	0.1
Insurance Related Issues	0	0	0.0	0	0	0.0	2	2	0.1	0	0	0.0	0	0	0.0
Economic Assistance to Indiv.	0	0	0.0	2	2	0.1	1	1	0.1	0	0	0.0	0	0	0.0
TOTAL	35	27	2.4	164	59	11.3	228	78	15.7	162	87	11.1	195	136	13.4

T = Number of Transactions; N = Number of Actors; % = Percent of Total Transactions
[a]Not including freeways, bridges, roads or utilities

Table 47. Frequency Distribution: Types of Interactions in Disaster Response by Funding Source and Juris-

	Public											
	International			National			State			Local		
	K	N	%	K	N	%	K	N	%	K	N	%
Public: International	0	0	0.0	1	2	0.5	0	0	0.0	0	0	0.0
Public: National				15	14	7.1	11	11	5.2	17	17	8.1
Public: State							20	14	9.5	20	28	9.5
Public: Local										17	27	8.1
Nonprofit: International												
Nonprofit: National												
Private: International												
Private: National												
Total	0	0	0.0	16	16	7.6	31	25	14.7	54	72	25.6

K = Interactions; N = Number of Cases; % = Percent of Total Interactions
Sources (Tables 46 & 47): *San Francisco Chronicle, Oakland Tribune, San Francisco Examiner, San Jose Mercury*

Chapter 7. Operative Adaptive Systems

isdiction, Loma Prieta, CA Earthquake, October 18 – November 7, 1989

| Special District ||| Nonprofit Organizations ||| Private Organizations |||||| TOTALS |||
| | | | | | | International ||| National ||| | | |
T	N	%	T	N	%	T	N	%	T	N	%	T	N	%
0	0	0.0	8	5	0.6	0	0	0.0	15	12	1.0	117	63	8.0
0	0	0.0	6	5	0.4	0	0	0.0	11	11	0.8	26	22	1.8
0	0	0.0	0	0	0.0	0	0	0.0	1	1	0.1	9	8	0.6
4	1	0.3	51	26	3.5	0	0	0.0	7	6	0.5	100	57	6.9
1	1	0.1	4	4	0.3	0	0	0.0	7	7	0.5	52	39	3.6
0	0	0.0	1	1	0.1	0	0	0.0	1	1	0.1	27	10	1.9
0	0	0.0	0	0	0.0	0	0	0.0	0	0	0.0	8	4	0.6
0	0	0.0	2	1	0.1	0	0	0.0	0	0	0.0	16	15	1.1
0	0	0.0	2	2	0.1	0	0	0.0	0	0	0.0	32	10	2.2
0	0	0.0	0	0	0.0	0	0	0.0	1	1	0.1	29	18	2.0
0	0	0.0	1	1	0.1	0	0	0.0	1	1	0.1	11	9	0.8
1	1	0.1	63	22	4.3	0	0	0.0	26	18	1.8	140	76	9.6
1	1	0.1	33	23	2.3	87	60	6.0	144	120	9.9	310	241	21.3
0	0	0.0	7	6	0.5	0	0	0.0	10	9	0.7	78	47	5.4
0	0	0.0	3	2	0.2	0	0	0.0	0	0	0.0	35	13	2.4
0	0	0.0	6	5	0.4	0	0	0.0	4	3	0.3	72	21	5.0
4	4	0.3	0	0	0.0	0	0	0.0	30	11	2.1	40	19	2.8
0	0	0.0	8	6	0.6	0	0	0.0	6	6	0.4	36	28	2.5
46	7	3.2	1	1	0.1	0	0	0.0	30	8	2.1	119	32	8.2
0	0	0.0	0	0	0.0	0	0	0.0	0	0	0.0	7	3	0.5
0	0	0.0	2	2	0.1	0	0	0.0	2	2	0.1	13	12	0.9
0	0	0.0	0	0	0.0	0	0	0.0	0	0	0.0	36	10	2.5
0	0	0.0	1	1	0.1	0	0	0.0	0	0	0.0	105	30	7.2
0	0	0.0	2	2	0.1	0	0	0.0	13	7	0.9	17	11	1.2
0	0	0.0	1	1	0.1	0	0	0.0	15	9	1.0	19	13	1.3
57	15	3.9	202	116	13.9	87	60	6.0	324	233	22.3	1454	811	100.0

diction, Loma Prieta, CA Earthquake, October 18 – November 7, 1989

| Nonprofit |||||| Private |||||| Total |||
| International ||| National ||| International ||| National ||| | | |
K	N	%	K	N	%	K	N	%	K	N	%	K	N	%
0	0	0.0	0	0	0.0	0	0	0.0	0	0	0.0	1	2	0.5
0	0	0.0	4	10	1.9	0	0	0.0	2	2	0.9	49	54	23.2
0	0	0.0	5	8	2.4	0	0	0.0	5	7	2.4	50	57	9.5
0	0	0.0	29	32	13.7	0	0	0.0	8	17	3.8	54	76	25.6
0	0	0.0	0	0	0.0	0	0	0.0	0	0	0.0	0	0	0.0
			3	6	1.4	1	2	0.5	42	44	19.9	46	52	21.8
						0	0	0.0	1	2	0.5	1	2	0.5
									10	14	4.7	10	14	4.7
0	0	0.0	41	56	19.4	1	2	0.5	68	86	32.2	211	257	100.0

178   Part II: Shared Risk in Practice

Adaptive Behavior and/or Self-Organization

The nonlinear analysis of change in the performance of the six primary disaster functions over a twenty-day period following the earthquake reveals a basic response system performing strongly on certain functions, more weakly on others, with private and nonprofit organizations supplementing the performance of public organizations. Again, all public organizations were grouped together across jurisdictional boundaries to identify the "public" sector of the system that had legal responsibility for the protection of life and property. Private organizations included both domestic and international organizations, and nonprofit included all charitable organizations.

Figure 33 shows the phase plane plot for change in reported emergency response activities by change in communication/coordination for public organizations. Except for Day Two in which communication/coordination was increasing and emergency response was decreasing, the two variables were strongly coordinated, falling quickly into a regular pattern. Figure 34 presents the marginal history which shows the daily change. In the last week of the response period, actions reported for emergency response decrease considerably, as the search and rescue activities were largely over.

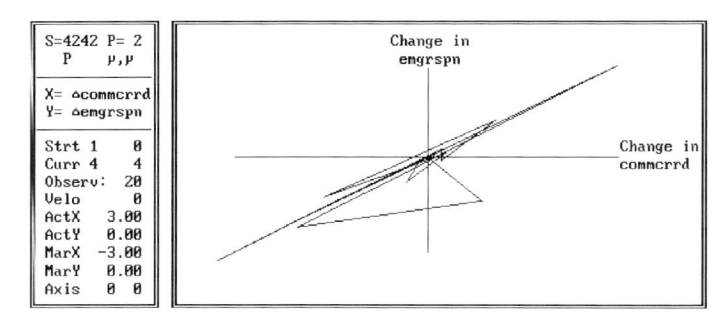

Figure 33. 1989 Loma Prieta, CA Earthquake: Phase Plane Plot, Public Organizations, Emergency Response by Communication/Coordination

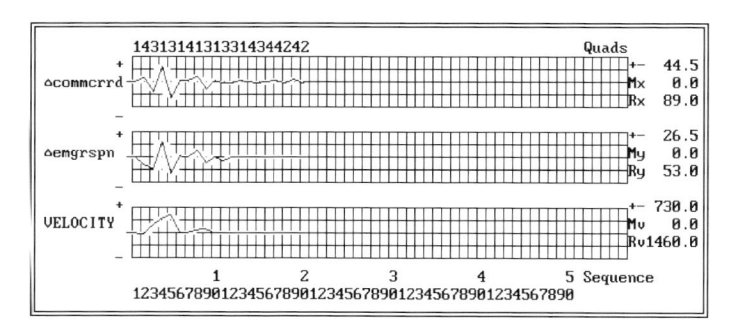

Figure 34. 1989 Loma Prieta, CA Earthquake: Marginal History, Public Organizations, Disaster Relief by Communication/Coordination

Chapter 7. Operative Adaptive Systems

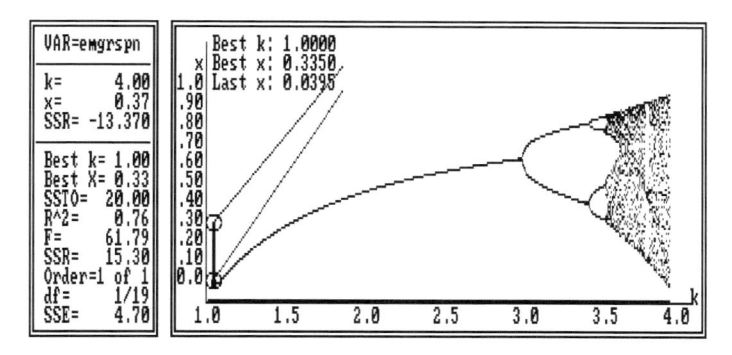

Figure 35. 1989 Loma Prieta, CA Earthquake: Nonlinear Logistic Regression, Public Organizations, Emergency Response

Figure 35 shows the logistic regression equation calculated for emergency response by public organizations, with an R^2 value of 0.76, F = 61.79, and k = 1.0. This finding indicates a strong emergency response that operated consistently with little variation in the system.

Figure 36 shows the logistic regression equation for Damage Assessment for public organizations, with an R^2 value of 0.60, F = 28.72, and k = 1.1. Figure 37 shows the significant, but lower value R^2 value of 0.42 for Communication/Coordination, with F = 13.57, and k = 1.1. A logistic regression equation was also found for Recovery/Reconstruction, but it was not statistically significant. These findings confirm the earlier profile presented by the distribution of transactions, in that the subsystem of public organizations performed well on their primary responsibilities of emergency response and damage assessment, less well on communication/coordination, and weakly on Recovery/Reconstruction.

The nonlinear analysis of change in performance for private organizations shows a relatively strong finding for Communication/Coordination, with an R^2

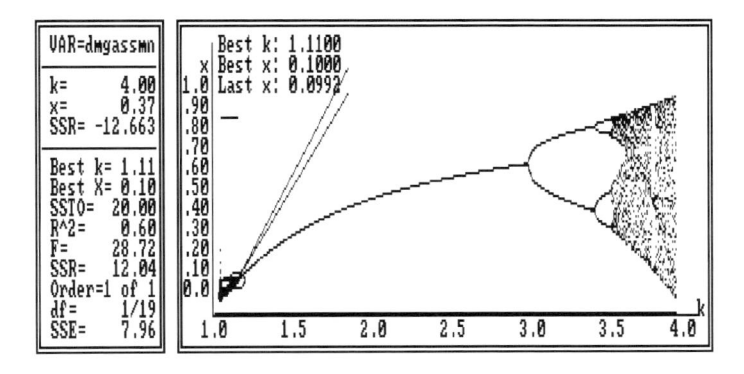

Figure 36. 1989 Loma Prieta, CA Earthquake: Nonlinear Logistic Regression, Public Organizations, Damage Assessment

180   *Part II: Shared Risk in Practice*

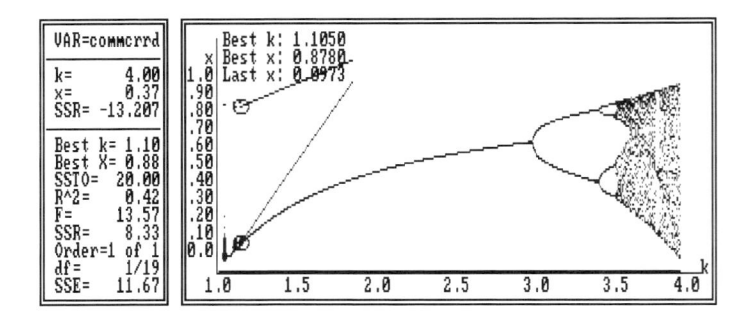

Figure 37. 1989 Loma Prieta, CA Earthquake: Nonlinear Logistic Regression, Public Organizations, Communication/Coordination

value of 0.58, F = 26.61, and k = 1.0, shown in Figure 38. Recovery/Reconstruction also shows a statistically significant value of R^2 = 0.45, F = 15.74, and k = 1.165. A logistic regression equation is found for Financial Assistance, but it is not significant. Nonprofit organizations show logistic regression equations for emergency response, R^2 = 0.40, F = 12.41, and k = 1.198, and Damage Assessment, R^2 = 0.21, F = 4.91, and k = 1.1. An equation is found for Disaster Relief, but the value of R^2 is not statistically significant.

These findings show that the large response system that evolved following the Loma Prieta Earthquake was weakly interconnected and would likely be vulnerable under greater stress.

Marathwada, India, September 30, 1993, M = 6.4[16]

On Thursday, September 30, 1993, a magnitude 6.4[17] earthquake struck the Marathwada region of Maharashtra State in Central India at 3:56 a.m. The epicenter of the earthquake was near the village of Killari in Latur District, with

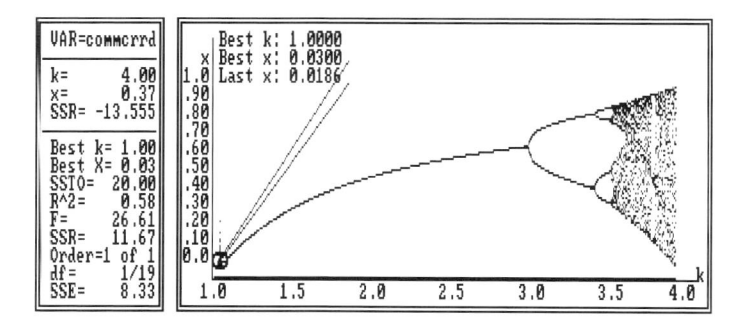

Figure 38. 1989 Loma Prieta, CA Earthquake: Nonlinear Logistic Regression, Private Organizations, Communication/Coordination

extensive damage reported throughout the District as well as in the adjoining district, Osmanabad. The initial social, economic, and technical conditions in the Latur and Osmanabad Districts prior to the earthquake shaped the dynamics of the evolving disaster response system. The two districts are located in an agricultural area that is moving gradually toward more productive, marketable crops and a higher standard of living for its inhabitants. Yet, most of the population lives in conditions of extreme poverty. Approximately 80% of the people in the area earn their living through agriculture, with more than 50% of the population earning less than $250 per year. The literacy rate is low, approximately 55% for men; 35% for women; 10% unreported (Census of India, 1991). The population is primarily Hindu, with a small proportion of Muslims in Osmanabad. Commerce is beginning to develop in the largest city, Latur, and signs of increasing literacy and economic development are also evident, but the economic and social needs of the area under normal times are great.[18]

Relative unawareness of seismic risk in this region of Central India heightened the social impact of this earthquake. Although India has known areas of severe seismic risk in its northern states near the Himalayas, this region in Central India had been classified as low risk by the Indian Meteorological Department (Chengappa with Menon 1993:54–55; Sharan, Gupta, and Sethi 1993:12; Gupta et al. 1993, IV:1–7). Consequently, residents of the area had not incorporated antiseismic criteria into their building codes nor developed local programs for seismic preparedness, despite tremors in the area recorded as recently as the preceding year (1992). Data now show that the Deccan plateau, which includes the Marathwada region of the Latur and Osmanabad districts, is seismically active and should be reclassified as vulnerable to major seismic risk. Outdated equipment, inadequate seismic monitoring, and underdeveloped knowledge of seismic risk for this region of India (Chengappa with Menon 1993, 56; Gupta et al. 1993, 7) contributed to the discrepancy between previous scientific reports and the actual geologic condition.

The combined lack of awareness of seismic risk and extreme poverty created conditions of serious vulnerability to damage from earthquakes in the region. Housing consisted largely of nonengineered structures, primarily built of stone, held together with mud. Wealthier homes had wooden beams that created a stronger structure for connecting ceilings to walls, but they also suffered heavy damage. Roads are rudimentary, with some of the villages connected only by dirt roads that turn to impassable mud during the rainy season. Further, India has no formal disaster management agency or plan that engages communities in risk assessment, preventive planning, and preparedness for disaster.

Three initial conditions – technical, organizational, and social – significantly affected the rapid evolution of a disaster response system following the Marathwada Earthquake. First, the Government of India had invested in a national satellite communications system in 1988, and had located downlinks in the offices of the State Ministers and District Collectors. Latur District, for example, maintains a branch of the National Informatics Centre (NIC) which provides basic information to support the district's administration. Two full-time Indian Administrative Service (IAS) officers and one trained person from each district department operate its NIC communications and information services. The satellite communications system allows multiway communications between the State of Maharashtra offices in Mumbai (formerly Bombay) and other district and subdistrict offices within the state, as well as among the 25 states in the nation and between state and national offices in New Delhi.[19] The satellite system serves as the base communications network, and computer links operate between the cities of the region: Solapur, Omerga, Latur, and Osmanabad. Within the cities, microwave links permit two-way communication among city offices. This technical capacity provides the mechanism for rapid, multiway communications among all agents responsible for conducting operations that require multi-agency coordination.

Second, the IAS has established a professional corps of educated public administrators who share a common background of professional training, accept a common set of responsibilities towards developing the capacities of the citizenry in their jurisdictions, and represent a strong presence of national government in state and local jurisdictions. Most officers have also had some experience with disaster response as part of their IAS training.[20] The IAS provides a national pool of trained professional administrators from which emergency assistance may be drawn during disaster operations. This national system provides an organizational structure which allows the rapid expansion of administrative capacity to meet urgent needs by integrating trained professional resources easily into an evolving response system, and reassigning them again after the needs have been met.

Finally, Indian society has a strong tradition of voluntary associations — religious, humanitarian, community, political, and professional — that are oriented to community activities and supported by deeply-held humanitarian values. The long-standing Hindu tradition provides a core set of widely shared beliefs that reinforces actions taken to help others. This societal norm of assisting those in need is powerfully activated by a sudden, destructive event such as an earthquake, and is reinforced by the working institutions of the nation.

This set of initial conditions — a strong technical capacity for communications, a strong organizational capacity for public organizations that allows flexible use of available resources, and a strong tradition of voluntary, humanitarian action — created an sociotechnical infrastructure that enabled a rapid transition from routine daily operations to dynamic disaster response.

Given these initial conditions, the Marathwada Earthquake caused extensive damage and loss of life. Measuring 6.4 on the Richter scale of magnitude, the earthquake was classified as moderate by geologists in other parts of the world. Its impact in rural Marathwada, however, was sobering. In Killari, for example, 1,220 persons out of a total population of 12,264 were killed; 1,282 were injured; and all 2,847 homes were destroyed.[21] Out of 936 villages in Latur District, 817 were damaged, as were 374 villages in the adjoining district, Osmanabad. Official reports listed a total of 7,582 dead, 21,849 injured, and 30,000 families or 175,000 people rendered homeless by the earthquake.[22] Total losses for the region were estimated at $36.6 billion.

Information Search

In this context of rural poverty and underdevelopment, a remarkably effective disaster response system evolved to meet the needs of the affected population following the earthquake. The system evolved in direct response to the search and exchange of information facilitated by the Indian Satellite System, accessible to public officials in this rural area. While the operations logs were not available from local disaster managers, this process of information search and exchange between administrative levels of government can be documented by an approximate record of response reconstructed from interviews with the District Collectors, members of the local Gram Panchayats or village councils, and other participants in the system who assumed emergency responsibilities.[23] The timing of response operations during the first critical hours and days following the disaster is reconstructed from data reported by public and medical administrators in charge of disaster operations. The times are necessarily approximate, as they were recalled from memory. Yet, the basic sequence of events was confirmed by multiple participants engaged in the response system,[24] as shown in Table 48.

Table 48. Log: Disaster Operations, Latur and Osmanabad Districts, 1993 Marathwada, India Earthquake

Village Level Operations, Killari, September 30, 1993

03:56 Earthquake occurred
04:00 Police patil transmitted news of event to District Police Control Room, Latur via Police wireless (radio)
04:01 Deputy Sarpanch (Deputy Mayor) starts to organize search and rescue operations by going house to house, asking assistance from those who are not injured
04:02 Police report that Killari hospitals are damaged; medical personnel are injured
04:45 First Police forces and medical teams from Latur arrive in Killari; District police teams support local search and rescue operations already underway
09:00 Chief Minister, Maharashtra State Government, Mumbai and Cabinet officers arrive in Killari to assess needs for disaster operations
10:00 Chief Minister, Maharashtra State, conducts urgent meeting of senior IAS officers and department heads of Latur, Osmanabad, and Solapur Districts; he designates Solapur, the largest city nearest the disaster area, as the headquarters for Marathwada Regional Disaster Operations and requests the District Collector of Solapur to serve as Regional Coordinating Officer for disaster operations; he grants the Regional Coordinating Officer an unlimited budget to support disaster operations
10:15 Voluntary organizations begin arrive in Killari and other villages to assist with search and rescue operations
10:30 Search and rescue operations continue through the day in Killari and other damaged villages

11:00 Voluntary organizations prepare and organize arrangements for cremation of the dead, distribution of food and water to survivors in area, assistance with medical care to the injured, working throughout the day

District Level Operations, Latur, September 30, 1993

04:00 District Collector's Office, Latur receives news of earthquake transmitted to Police Control Room via Police wireless (radio) from Killari
04:01 District Collector's Office reports event to Chief Minister's Office, State of Maharashtra, Mumbai via satellite link
04:02 District Control Room reports event to Government agencies in the district, and dispatches police force and medical team with supplies and ambulance to Killari
04:03 District Collector requests that hospitals in Latur and Solapur prepare to receive patients from the earthquake-affected area
04:04 District Collector requests Divisional Controller, State Transport, to send buses to transport injured persons
04:05 District Collector, Latur initiates inquiry regarding status of other damaged villages in district
05:00 District senior administrative and operations officers depart for Killari
08:00 District Collector requests assistance of voluntary organizations to supply food, water, clothing, and first aid to victims of earthquake

District Level Operations, Osmanabad, September 30, 1993

05:15 News of earthquake reached Collector's Office, Osmanabad District, approximately 102 km. from Omerga; damage reported from villages of Dalimb and Jewali
05:30 District Collector and senior operations officer started toward Dalimb, but received news of more extensive damage in Sastur, and changed direction to Sastur
06:30 District Collector immediately initiates search and rescue operations in damaged villages; deploys District resources for effort
09:00 District Minister, Osmanabad leaves for Killari to meet with Chief Minister, Maharashtra State regarding disaster operations

Marathwada Regional Level Operations, Solapur, September 30, 1993

04:04 District Collector, Solapur receives news of earthquake
04:05 District Collector, Solapur requests that hospitals in Solapur prepare to receive patients from the earthquake-affected area
04:08 Editor, Tarun Bharat, called reports of Earthquake to CNN, BBC
05:00 Twelve medical teams and 6 ambulances depart from Solapur for the disaster area
05:15 District Collector mobilized resources for disaster response: additional medical personnel, blood banks, doctors, worked on problem until 11:00 a.m.; sent medical students to EQ area in state transport buses
06:00 Solapur dispatched 2 fire engines and manpower to Killari
07:00 Chief Minister, Maharashtra State, and Cabinet Ministers arrive in Solapur and depart immediately for Killari, accompanied by District Collector, Solapur
08:30 Reporters from CNN, BBC arrive in Solapur; contact district collector, transmit news of EQ to world via international satellite
10:15 District Collector, now serving as the Marathwada Disaster Coordinating Officer, requests Army detachments from Pune, Hyderabad: 10,000 to 14,000 men to serve in disaster area
11:00 First team of volunteers started from Solapur to EQ area
13:00 Ambulances arrive at Solapur General Hospital with first patients injured in the earthquake

18:00   State transport buses arrive at N.M. Wadia Hospital with patients injured in earthquake;
22:00   More patients arrive at N.M. Wadia hospital from earthquake-damaged areas

State Level Operations, Mumbai, September 30, 1993

04:01   Chief Minister's Office, State of Maharashtra, Mumbai receives news of earthquake transmitted via satellite from District Collector's Office, Latur
04:04   Chief Minister's Office reports event to District Collectors' Offices in neighboring districts of Maharashtra State, to the Chief Minister's Offices in neighboring states, and to the Prime Minister's Office in New Delhi via satellite
04:05   Chief Secretary to Minister makes arrangements for Chief Minister, the Minister of Irrigation and Energy, and the Minister of Public Works to fly to Solapur
05:30   Chief Minister and Cabinet officers depart for Solapur
10:00   Chief Secretary coordinates disaster operations in Mumbai in accordance with direction from Chief Minister and staff, based in Solapur

This record of actions documents both the search for information and the process of communicating that information to public officials and administrators with the responsibility and authority to take action in response to this urgent event. The rapid mobilization of organizational response depended upon the technical facility of communicating information across substantial geographic and administrative distances.

Information Exchange

The availability of timely, accurate information enabled organizations and jurisdictions to exchange information and mobilize action in response to this event. Disaster operations continued at an intense rate with a rapidly expanding number of participants joining the interjurisdictional response system. For example, Medecins sans Frontiers, Netherlands, arrived the second morning, bringing a mobile dispensary for medicines and medical supplies. Engineers from the State Ministry of Public Works restored water the second day in some villages and by the fourth day in most others. Arrangements were made to truck in water via tankers for the remaining villages that suffered disruption in water distribution systems. Electricity was restored by the second day in most villages. Army detachments arrived on the morning of October 2, 1993, the third day following the disaster, to assume direction of the search and rescue operations and to assist in the cremation of the dead. The Army detachments succeeded in rescuing over 6,000 injured in Latur and over 3,000 injured in Osmanabad. In addition, over 6,000 dead were removed and cremated. Search and rescue operations were substantially completed by October 5, 1993, an extraordinary feat given the number of villages (67 villages severely affected; 1191 villages damaged) and breadth of the Marathwada region. Voluntary organizations contributed food, clothing, and assistance to search and rescue operations.[25] A distinct system of public, nonprofit, and private organizations engaged in response operations for the

Marathwada disaster evolved at a remarkably rapid rate. Nearly 12,000 nonmilitary personnel were engaged in rescue and relief operations. The system crossed jurisdictional boundaries from village to district to region to state to national to international levels.

Although the region experienced communications problems, e.g. microwave radio towers were dislocated and telephone lines were down, the Regional Disaster Coordinator attributed the rapid evolution of the disaster response system to their use of an extensive network of communications to coordinate disaster operations.[26] Disaster managers used computer links via the national satellite network to communicate among the regional cities. Within the cities, offices involved in disaster response communicated with one another via microwave radio. Villages, in turn, linked to city offices via wireless stations manned by volunteers. This technical capacity for multiway communications supported the organizational capacity to mobilize resources and personnel quickly to respond to urgent needs.

Organizational Learning

Using the methodology of the N-K system, we sought to define the major characteristics of this dynamic system. First, we identified 371 organizations that participated in disaster response organizations through a review of articles reported in English, Hindi, and Marathi newspapers.[27] While this list may not be comprehensive, it represents the major organizations, and types of organizations, that participated in disaster operations identified from public sources.

This response system revealed a high proportion of nonprofit organizations working with public organizations in carrying out disaster operations. Of the 371 organizations identified in the response system, 150, or 40.4%, were public organizations. Nonprofit organizations were nearly equal in number, 147, or 39.6%, of the total. Private organizations played a smaller, but significant role, with 74, or 19.9%, actively engaged in disaster operations. Of the 150 public organizations, the largest proportion, 55, or 36.6% were national. State organizations comprised the second largest proportion, with 44, or 29.3%. International organizations were a close third, with 34, or 22.7%, of the total, and local organizations represented the smallest proportion, with 17, or 11.3%. These findings are summarized in Table 49.

The large number of nonprofit organizations included an eclectic mix of charitable, religious, professional, business, trade union, and political party organizations. This group also included 30 international organizations, making up 20.4% of the total nonprofit organizations. Combined with the public subset of 34 international organizations and the private subset of 16, international organizations constituted 80, or 21.5% of the total response system. This relatively small proportion of international organizations provided vital access to external resources to aid in disaster relief and recovery, but it also indicates that the dominant response system for this earthquake was domestic, with Indian public, non-

Chapter 7. Operative Adaptive Systems

Table 49. Frequency Distribution: Disaster Response System by Funding Source and Jurisdiction, 1993 Marathwada, India Earthquake

Public					Total Public	Nonprofit				Private				Total	
Int'l		Nat'l		State		Mun'l			Int'l		Nat'l		Int'l		Nat'l		
N	%	N	%	N	%	N	%	N	%	N	%	N	%	N	%	N	%
34	9.2	55	14.8	44	11.8	17	4.6	150	40.4	30	8.1	117	31.5	16	4.3	58	15.6	371	100.0

profit and private organizations collaborating to meet the needs of the affected population.

Table 50 presents the distribution of 429 transactions performed in disaster response, as identified through newspaper reports. The largest proportion, 22.4%, involved donations of money and goods for disaster relief. The second largest proportion, 13.5%, involved actions taken to provide food, shelter and clothing to the people who had lost their homes and belongings.

Third, but representing a significant proportion, was the combined set of transactions involving communication and coordination, which made up 12.2% of the total transactions, and documented the sizeable effort placed by all levels of government in mobilizing their operations, respectively, in collaboration with other levels. Reports of fundraising and setting up special accounts for disaster relief represented 7.9% of the total transactions, and emergency response accounted for 6.3%.

Table 51 shows the distribution of interactions among the types of organizations engaged in disaster operations. The national government registered the highest proportion of interactions with all other types of organizations, at 29.6%. International organizations showed the next highest proportion of interactions, at 20.4%, and Maharashtra State Government was third, with 14.8%. These findings confirm the dominant role of the national government in mobilizing disaster response operations, but also the sizeable role of the state government in managing this disaster. International organizations played an important role, but less significant than for the response systems in the Nonadaptive and Emergent Adaptive categories. Nonprofit organizations played a significant role in this response system, in which national and international organizations combined accounted for more than one-fifth, or 22.3%, of the total number of interactions reported.

Adaptive Behavior and/or Self-Organization

The nonlinear analysis of change in basic disaster functions reveals patterns of dramatic shifts in performance over the three-week response period. Figure 39 shows the phase plane plot of change in emergency response by change in communication/coordination, as reported in newspaper accounts. After initial chaot-

188   Part II: Shared Risk in Practice

Table 50. Frequency Distribution of Types of Transactions Performed in Disaster Response Reported by

Type of Transaction	Public Organizations												
	International			National			Regional/State			Municipal		
	T	N	%	T	N	%	T	N	%	T	N	%
Emergency Response	2	2	0.5	11	12	2.6	5	6	1.2	1	1	0.2
Communication	1	1	0.2	9	10	2.1	8	9	1.9	3	3	0.7
Coordination of Response	2	2	0.5	9	12	2.1	9	12	2.1	1	1	0.2
Medical Care/Health	0	0	0.0	1	1	0.2	2	2	0.5	5	6	1.2
Damage/Needs Assessment	4	4	0.9	4	4	0.9	6	8	1.4	1	1	0.2
Certification of deaths	0	0	0.0	0	0	0.0	5	3	1.2	0	0	0.0
Earthquake Assessment/Research	4	4	0.9	15	16	3.5	1	1	0.2	0	0	0.0
Security/Prevention of Looting	0	0	0.0	0	0	0.0	0	0	0.0	0	0	0.0
Housing Issues	1	1	0.2	2	2	0.5	5	7	1.2	1	1	0.2
Disaster Relief (food, shelter, etc.)	7	9	1.6	9	9	2.1	4	4	0.9	2	2	0.5
Donations (money, goods, etc.)	15	24	3.5	4	6	0.9	6	6	1.4	1	1	0.2
Building Inspection	0	0	0.0	1	1	0.2	0	0	0.0	0	0	0.0
Building Code Issues	1	1	0.2	2	2	0.5	1	1	0.2	1	1	0.2
Repair of Freeways, Bridges, Roads	0	0	0.0	0	0	0.0	0	0	0.0	0	0	0.0
Repair/Restore Utilities	0	0	0.0	3	6	0.7	3	5	0.7	0	0	0.0
Repair/Reconstruction/Recovery[a]	0	0	0.0	4	5	0.9	7	10	1.6	1	2	0.2
Transportation/Traffic Issues	0	0	0.0	0	0	0.0	0	0	0.0	0	0	0.0
Hazardous Materials Releases	0	0	0.0	0	0	0.0	0	0	0.0	0	0	0.0
Legal/Enforcement/Fraud	0	0	0.0	0	0	0.0	1	1	0.2	0	0	0.0
Political Dialogue/Legislation	0	0	0.0	1	2	0.2	0	0	0.0	0	0	0.0
Business Recovery	0	0	0.0	0	0	0.0	1	1	0.2	0	0	0.0
Economic/Business Issues	0	0	0.0	1	1	0.2	0	0	0.0	0	0	0.0
Visits by Officials	0	0	0.0	2	2	0.5	2	2	0.5	0	0	0.0
Education Issues	0	0	0.0	0	0	0.0	1	1	0.2	1	1	0.2
Government Assistance	0	0	0.0	4	4	0.9	7	16	1.6	3	4	0.7
Insurance Related Issues	0	0	0.0	0	0	0.0	0	0	0.0	0	0	0.0
Loans (Private and International)	1	1	0.2	0	0	0.0	2	4	0.5	0	0	0.0
Psychological/Counseling Services	0	0	0.0	0	0	0.0	0	0	0.0	0	0	0.0
Fundraising/Account Setup	1	1	0.2	1	1	0.2	3	6	0.7	0	0	0.0
Volunteers	1	1	0.2	1	1	0.2	0	0	0.0	0	0	0.0
TOTAL	40	51	9.3	84	97	19.6	79	105	18.4	21	24	4.9

T = Number of Transactions; N = Number of Actors; % = Percent of Total Transactions
[a]Not including freeways, bridges, roads or utilities
Sources: *Times of India*, Bombay, October 1-21, 1993; *Times of India*, New Delhi; *The Hindustan Times*, New

ic swings in reported performance, accounts of emergency response operations taper off by Day 10, the task largely completed.

Figure 40 presents the marginal history of the interaction between the two variables, showing initial reports of communication/coordination to lead a sharp increase in activity in emergency response, with both dropping by Day 3, changing to a steady increase in reports on communication/coordination through Day 10, with roughly parallel performance by emergency response, tapering off as the activity was completed.

Primary Funding Source Maharashtra Earthquake, October 1–20, 1993

\multicolumn{6}{c}{Nonprofit Organizations}	\multicolumn{6}{c}{Private Organizations}	\multicolumn{3}{c}{TOTALS}

\multicolumn{3}{c}{National}	\multicolumn{3}{c}{International}	\multicolumn{3}{c}{National}	\multicolumn{3}{c}{International}			
T	N	%	T	N	%	T	N	%	T	N	%	T	N	%
4	5	0.9	4	4	0.9	0	0	0.0	0	0	0.0	27	30	6.3
4	4	0.9	1	1	0.2	0	0	0.0	0	0	0.0	26	28	6.1
3	3	0.7	1	1	0.2	1	1	0.2	0	0	0.0	26	32	6.1
5	6	1.2	1	2	0.2	3	4	0.7	0	0	0.0	17	21	4.0
1	1	0.2	0	0	0.0	0	0	0.0	0	0	0.0	16	18	3.7
0	0	0.0	0	0	0.0	0	0	0.0	0	0	0.0	5	3	1.2
1	1	0.2	0	0	0.0	0	0	0.0	0	0	0.0	21	22	4.9
1	1	0.2	0	0	0.0	0	0	0.0	0	0	0.0	1	1	0.2
4	19	0.9	1	1	0.2	2	2	0.5	0	0	0.0	16	33	3.7
25	26	5.8	4	5	0.9	6	8	1.4	1	1	0.2	58	64	13.5
37	61	8.6	7	7	1.6	19	30	4.4	7	8	1.6	96	143	22.4
2	2	0.5	0	0	0.0	0	0	0.0	0	0	0.0	3	3	0.7
3	3	0.7	0	0	0.0	0	0	0.0	0	0	0.0	8	8	1.9
0	0	0.0	0	0	0.0	0	0	0.0	0	0	0.0	0	0	0.0
0	0	0.0	0	0	0.0	2	3	0.5	1	3	0.2	9	17	2.1
1	1	0.2	0	0	0.0	4	4	0.9	0	0	0.0	17	22	4.0
0	0	0.0	0	0	0.0	0	0	0.0	0	0	0.0	0	0	0.0
0	0	0.0	0	0	0.0	0	0	0.0	0	0	0.0	0	0	0.0
1	1	0.2	0	0	0.0	0	0	0.0	0	0	0.0	2	2	0.5
8	9	1.9	0	0	0.0	0	0	0.0	0	0	0.0	9	11	2.1
1	2	0.2	0	0	0.0	0	0	0.0	0	0	0.0	2	3	0.5
0	0	0.0	0	0	0.0	0	0	0.0	0	0	0.0	1	1	0.2
0	0	0.0	0	0	0.0	0	0	0.0	0	0	0.0	4	4	0.9
1	1	0.2	0	0	0.0	0	0	0.0	0	0	0.0	3	3	0.7
0	0	0.0	0	0	0.0	0	0	0.0	0	0	0.0	14	24	3.3
0	0	0.0	0	0	0.0	0	0	0.0	0	0	0.0	0	0	0.0
0	0	0.0	0	0	0.0	0	0	0.0	0	0	0.0	3	5	0.7
0	0	0.0	0	0	0.0	1	1	0.2	0	0	0.0	1	1	0.2
17	18	4.0	8	12	1.9	4	4	0.9	0	0	0.0	34	42	7.9
7	8	1.6	1	1	0.2	0	0	0.0	0	0	0.0	10	11	2.3
126	172	29.4	28	34	6.5	42	57	9.8	9	12	2.1	429	552	100.0

Delhi, October 1, 2, 1993; *The Statesman*, Calcutta, October 1-3, 5-7, 10, 1993

Logistic regression analysis of performance by public organizations reported equations for four of the six disaster response functions that were statistically significant at the 0.05 level. Figures 41–44 presents the findings for Emergency Response, $R^2 = 0.69$, $F = 42.52$, and $k = 1.25$; Damage Assessment, $R^2 = 0.85$, $F = 105.98$, $k = 1.8$; Communication/Coordination, $R^2 = 0.64$, $F = 34.06$, $k = 1.2$; and Disaster Relief, $R^2 = 0.62$, $F = 31.5$, and $k = 1.2$. No equations were found for Recovery/ Reconstruction or Financial Assistance, using news reports.

190   *Part II: Shared Risk in Practice*

Table 51. Frequency Distribution: Types of Interactions in Disaster Response by Funding Source

	Public											
	International			National			State			Local[a]		
	K	N	%	K	N	%	K	N	%	K	N	%
Public: International	1	2	1.9	1	3	1.9	5	4	9.3	1	2	1.9
Public: National				3	6	5.6	9	17	16.7	2	6	3.7
Public: State							5	7	9.3	0	0	0.0
Public: Local[a]										3	6	5.6
Nonprofit: International												
Nonprofit: National												
Private: International												
Private: National												
Total	1	2	1.9	4	9	7.4	19	28	35.2	6	14	11.1

[a]Public-local includes sub-state jurisdictions: regional, district and municipal
K = Number of Interactions; N = Number of Cases; % = Row Total
Sources: *Times of India*, Bombay, October 1–21, 1993; *Times of India*, New Deli; *The Hindustan*

These findings document a rapidly evolving subsystem of public organizations adapting quickly to events in the disaster environment.

Within the sizeable subsystem of nonprofit organizations, logistic regression equations were found for Emergency Response, $R^2 = 0.21$; $F = 5.19$, and $k = 1.1$, and Disaster Relief, $R^2 = 0.44$, $F = 14.93$, and $k = 1.1$. These two functions represented the primary areas of activity for nonprofit organizations. No equations were found for Damage Assessment, Recovery/Reconstruction, or Financial Relief. Interestingly, an equation was found for Communication/Coordination that was not statistically significant, but slipped wholly into the realm of

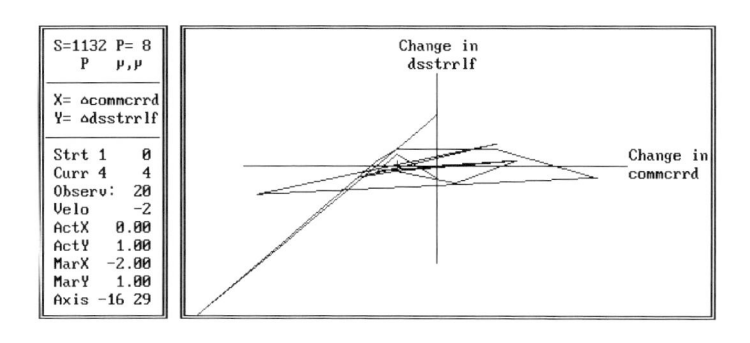

Figure 39. 1993 Marathwada, India Earthquake: Phase Plane Plot, Public Organizations, Disaster Relief by Communication/Coordination

and Jurisdiction, Marathwada, India Earthquake, October 1–21, 1993

Nonprofit						Private						Total		
International			National			International			National						
K	N	%	K	N	%	K	N	%	K	N	%	K	N	%
1	2	1.9	1	5	1.9	0	0	0.0	1	8	1.9	11	26	20.4
0	0	0.0	0	0	0.0	1	4	1.9	1	2	1.9	16	35	29.6
1	2	1.9	1	2	1.9	0	0	0.0	1	3	1.9	8	14	14.8
1	2	1.9	0	0	0.0	0	0	0.0	0	0	0.0	4	8	7.4
0	0	0.0	3	4	5.6	1	2	1.9	1	4	1.9	5	10	9.3
			7	14	13.0	0	0	0.0	0	0	0.0	7	14	13.0
						0	0	0.0	0	0	0.0	0	0	0.0
									3	8	5.6	3	8	5.6
3	6	5.6	12	25	22.2	2	6	3.7	7	25	13.0	54	115	100.0

Times, New Deli, October 1,2, 1993; and *The Statesman*, Calcutta, October 1–3, 5–7, 10, 1993

chaos with a k value of 4.0. This finding confirms reports, documented through on-site interviews with managers of nonprofit organizations, of lack of coordination among the many nonprofit organizations who contributed time, money and material goods to aid families who suffered losses from the earthquake.[28] Private organizations, which represented a smaller subsystem of the total response system, reported one active function, Disaster Relief ($R^2 = 0.12$; F = 2.55; k = 3.870). The value of R^2 is not statistically significant, but the interesting observation is that this subsystem slips wholly into chaotic behavior with a k value of 4.0.

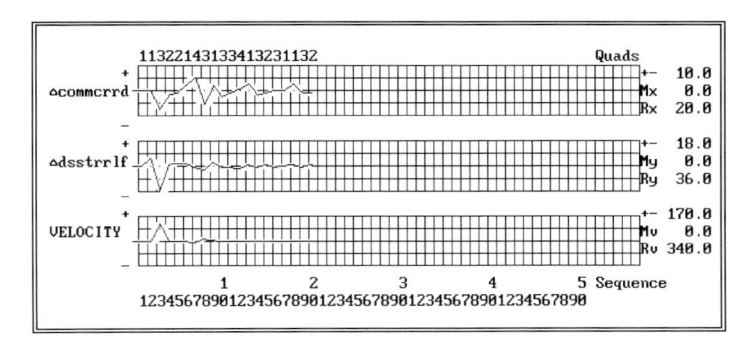

Figure 40. 1993 Marathwada, India Earthquake: Marginal History, Public Organizations, Disaster Relief by Communication/Coordination

192   Part II: Shared Risk in Practice

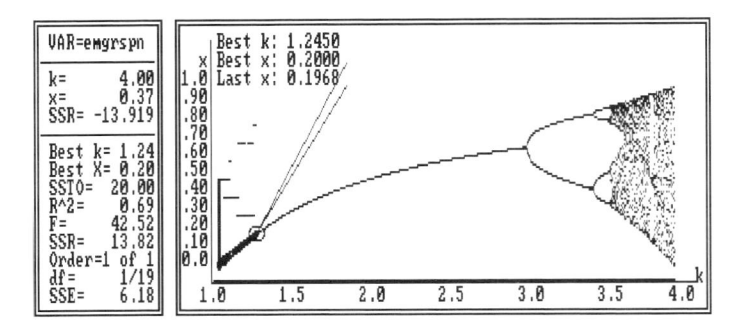

Figure 41. 1993 Marathwada, India Earthquake: Nonlinear Logistic Regression, Public Organizations, Emergency Response

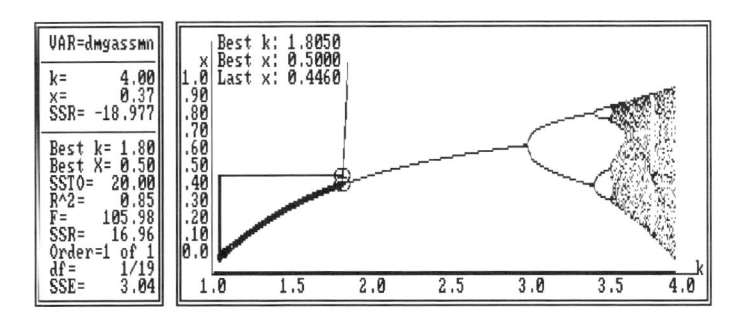

Figure 42. 1993 Marathwada, India Earthquake: Nonlinear Logistic Regression, Public Organizations, Damage Assessment

The data reported for the Marathwada Earthquake document the rapid evolution of a system of organizations engaged in interdependent response operations directed toward the goal of protection of life and property in the damaged communities. The system crossed jurisdictional lines as participants searched for the most appropriate and efficient means to meet urgent community needs. Most

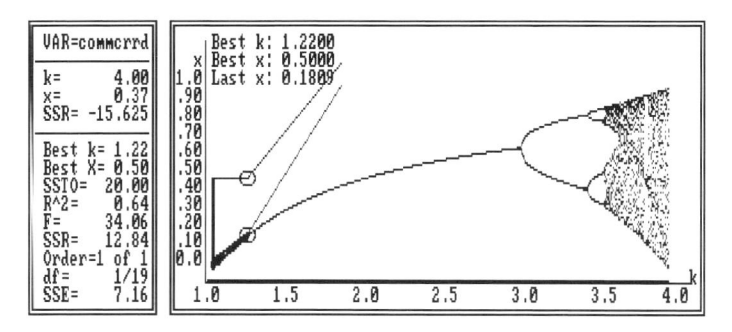

Figure 43. 1993 Marathwada, India Earthquake: Nonlinear Logistic Regression, Public Organizations, Communication/Coordination

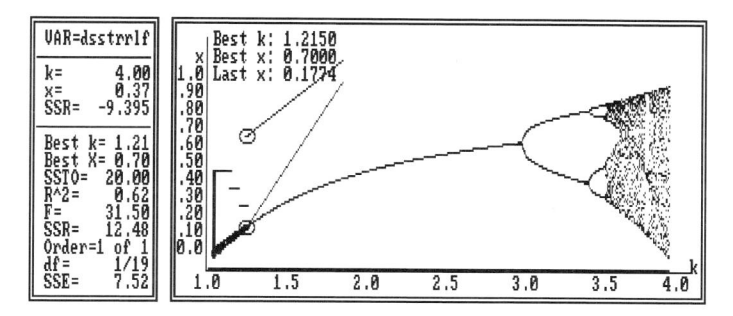

Figure 44. 1993 Marathwada, India Earthquake: Nonlinear Logistic Regression, Public Organizations, Disaster Relief

interesting, the disaster response system integrated public, private, and nonprofit organizations in the common effort of response to the critical needs of the disaster-affected population. The high degree of involvement by nonprofit organizations greatly supported the humanitarian needs of the victims, which could not be met by public organizations alone. This shared responsibility resulted in a high degree of cooperation among public and nonprofit organizations. Private organizations were also involved in disaster response, but to a lesser extent. Private companies frequently supported the voluntary contributions of time and wages by employees to the disaster response effort.

The rapid evolution of the disaster response system had marked consequences for the recovery of the damaged communities. First, since basic services were restored quickly and local administrators and village councils were directly involved in response operations, the affected communities moved relatively easily to re-engagement in reconstruction efforts. This fairly rapid transition from response to reconstruction resulted in a noticeably lower level of post traumatic stress among the earthquake-affected population than reported in previous disasters.[29] Many villages used the disaster as an opportunity to improve infrastructure and public facilities; their respective Village Councils pursued those goals vigorously in the design and reconstruction phases.

Second, other groups saw the reconstruction of housing and lifeline services as opportunities for employment training for young people, and allied these tasks with local technical colleges to provide supervised apprenticeships to local personnel. Third, the spontaneous response of voluntary and nongovernmental organizations to the humanitarian needs of the disaster-affected villages created a vital bridge of support and hope to earthquake victims that enabled them to find new sources of encouragement and strength from the wider national community. Given the shattering impact of the earthquake, the villagers moved relatively quickly to a new stage of collaborative interaction with other organizations in the interjurisdictional response system in the effort to rebuild their lives.

Finally, participants in the response process had a largely favorable perception of the interjurisdictional response effort, reflecting not only their participation

Table 52. Summary, Frequency Distributions, Operative Adaptive Systems by Funding Source and Jurisdiction

	Public								Total Public		Nonprofit		Private		Total	
	Int'		Nat'l		State		Local									
	N	%	N	%	N	%	N	%	N	%	N	%	N	%	N	%
Whittier Narrows	0	0.0	9	4.7	17	9.0	97	51.3	118	62.4	28	14.8	38	20.1	189	100.0
Loma Prieta	26	4.2	36	5.8	39	6.3	160	25.6	261	41.9	98	15.7	264	42.4	623	100.0
Marathwada, IN	34	22.7	55	46.0	44	23.3	17	8.0	150	40.3	147	39.5	75	20.2	371	100.0

in disaster operations but their candid assessment of the contributions of other groups. Of the 47 respondents included in the set of interviews for the field study, virtually all perceived the response positively. In the view of one retired lawyer who had volunteered his services, the government had "acted with rare promptness" to generate a framework for action that others could follow.

Reviewing the organizational distributions among the three response systems in this category of Operative Adaptive Systems, one can see characteristics that distinguish this category from the previous two, Nonadaptive and Emergent Adaptive Systems. Table 52 summarizes the distributions for all three systems. For purposes of comparison, the local government organizations are grouped together for each system.

In this subset of response systems, three striking characteristics emerge. First, the distribution of participating organizations shifts markedly away from the national and international level and much more toward state and local organizations. In the California response systems, the shift to the local level was evident among the public organizations. Second, private organizations played a significant role in these response systems, working in collaboration with public and nonprofit organizations. In the Loma Prieta response system, the number of private organizations contributing to disaster operations exceeded the number of public organizations involved by a slight margin. Third, nonprofit organizations participated actively in the response activities. In Marathwada, nonprofit organizations played a role nearly equivalent to that of the public organizations. Many of the nonprofit organizations were community-based and actively engaged community residents in activities related to disaster relief. In each of these response systems, active engagement from the community level characterized their operations.

NOTES

1. Tierney, K. 1988. "The Whittier Narrows, California Earthquake of 1987 — Social Aspects". *Earthquake Spectra*, Vol. 4, No. 1:14.
2. EQE Engineering. 1987. *Summary of the October 1, 1987 Whittier, California Earthquake*. San Francisco: EQE Engineering.

Chapter 7. Operative Adaptive Systems

3. French, S. and G. Rudholm. 1990. "Damage to Public Property in the Whittier Narrows Earthquake: Implications for Earthquake Insurance". *Earthquake Spectra*. Vol. 6, No. 1:105.
4. Leyendecker, E.V., L.M. Highland, M. Hopper, E.P. Arnold, P. Thenhaus, and P. Powers. 1988. "Early Results of Isoseismal Studies and Damage Surveys". *Earthquake Spectra*, Vol. 4. No. 1. *The Whittier Narrows, California Earthquake of October 1, 1987*: p. 1. See also City of Los Angeles. 1987. "Whittier Narrows Earthquakes, October 1 and 4, 1987: After Action Report". Los Angeles, CA: Emergency Operations Organization.
5. Web, F.H. 1987. "Whittier Narrows Earthquake: Los Angeles County". *California Geology*, Vol. 40, No. 12 (December):275–281.
6. Data collection for this section was supported by National Science Foundation Grant #CES 88-0425. "Interorganizational Coordination in Disaster Management: a Model for an Interactive Information System". Findings from this study are reported in the 1992 final report by the same title, submitted to the National Science Foundation.
7. Taghavi, M. 1987. "Summary of October, 1987 Earthquake Damage Reports". Structural Engineering Division, City of Los Angeles, October 19, 1987.
8. Lawrence Burk, Deputy City Engineer, Bureau of Engineering; Sam Matsumuru, Principal Assistant Division Engineer, Bureau of Engineering, Department of Public Works, City of Los Angeles. Interviews, Los Angeles, CA, May 21–23, 1988.
9. Alhambra city officials. Interviews, Alhambra, CA, May 18–25, 1988.
10. Thomas Mauk, City Manager, and other Whittier city officials. Interviews, Whittier, CA, June 2–10, 1988.
11. Jim Rounds, Los Angeles County Fire Captain. Interview, Whittier Station, Whittier, CA, June 17, 1988.
12. I am grateful to Bernadette Palumbo, Claremont Graduate School, Patricia Campbell, California State University, Los Angeles, and Keun Namkoong, Graduate School of Public and International Affairs, University of Pittsburgh, for their assistance with the data collection and administration of this study. Leslie Mohr and Susan Wade, Graduate School of Public and International Affairs, University of Pittsburgh, assisted with the content analysis of the data. Giovanni Reyes, University of Pittsburgh, assisted with the chaos analysis of the data.
13. The number of dead are taken from news reports in Bay Area newspapers, October 18–21, 1989, but this information is summarized in a table presented in Chapter 11: "Socioeconomic Impacts and Emergency Response", *Loma Prieta Earthquake Reconnaissance Report*, L. Benuska, Technical Editor, Earthquake Spectra, Supplement to Vol. 6, May 1990:396.
14. This section draws heavily upon the report that I wrote as a member of the Reconnaissance Team for the National Research Council following the Loma Prieta Earthquake. My assignment for the team was to investigate "Interagency and Intergovernmental Coordination" in emergency response. The report was substantially incorporated into Chapter 11: "Socioeconomic Impacts and Emergency Response", *Loma Prieta Earthquake Reconnaissance Report*, L. Benuska, Technical Editor, Earthquake Spectra, Supplement to Vol. 6, May 1990:407–412.
15. Coordinator, Region II Office, California Office of Emergency Services. Interview, Lafayette, CA. October 22, 1989.
16. This section draws heavily upon sections of a paper, L.K. Comfort, "Self-Organization in Disaster Response: Global Strategies to Support Local Action", presented at a Conference on the United Nations and International Crisis Management, sponsored by the Center for International Studies at the University of California, Berkeley, November 7–8, 1995.
17. The magnitude of the earthquake was estimated at $M_b = 6.3$ and $M_s = 6.4$ Richter scale by the US Geological Survey. The earthquake was reported as M = 6.5 in the press. *India Today*, October 11, 1993:54.
18. Tables 2 and 3 (see Appendix D) present data on primary administrative, social, and occupational characteristics of the Latur and Osmanabad Districts (Census of India, 1991).
19. Praveen Pardeshi, District Collector, Latur. Interview, December 22, 1993.
20. Dineshkumar Jain, District Collector, Solapur. Interview, December 22, 1993.

21. Tata Institute of Social Sciences. 1994. "Survey of People Affected by the Earthquake in the Latur and Osmanabad Districts (1993)". Bombay, India: Joint Action Group of Institutions for Social Work Education. Final Report, February:142.
22. A Preliminary Report by the Government of Maharashtra. Bombay, 1993. Table 4 (see Appendix D) summarizes the damage caused by the earthquake to lives and private property in the districts of Latur and Osmanabad.
23. I acknowledge, with thanks and appreciation, the assistance of Dr. Sharayu Anantaram, SNDT University, Bombay, India in arranging and conducting these interviews in the Solapur, Latur, Osmanabad Districts in India, December 19–24, 1993.
24. The schedule of activities for the first hours of the disaster is drawn from reports of disaster operations from the Maharashtra State Government and the District Collectors' Offices of Latur and Osmanabad, as well as interviews with the following managers:
    1. Volunteer, Jan Kalyan Samiti and RSS
    2. District Collector, Solapur, who served at the Governor's request as Coordinator of Disaster Operations for the Marathwada Region
    3. Chief Surgeon, General Hospital, Solapur
    4. Cardiologist, N.M. Wadia Charitable Hospital, Solapur
    5. Director, Red Cross Branch, Solapur
    6. Subeditor, *Tarun Bharat* [*Young India*]
    7. District Collector, Latur
    8. District Collector, Osmanabad
    9. Deputy Sarpanch, Killari
    10. Member, Gram Panchayat, Killari
25. Regional Coordinating Officer, Marathwada Disaster Operations. Interview, Solapur, India, December 22, 1993.
26. Regional Disaster Coordinator and District Collector, Indian Administrative Service. Interview, Solapur, India, December 22, 1993.
27. These newspapers included the following English language publications: *The Times of India, Bombay*; *The Times of India, Calcutta*, The Hindustan Times, New Delhi, *The Statesman*, Calcutta; *The Business Standard*, Calcutta; *Indian Express*, New Delhi; *Economic Times*, Bombay; *The Observer of Business and Politics*; New Delhi. In addition, the following Marathi newspapers were reviewed to identify local organizations: *Lokasatta*, [People's Power], Bombay; *Maharashtra Times*, Bombay; *Tarun Bharat*, Solapur; *Marathwada*, Aurangabad; *Dainik Lokamat*, [Public Opinion Daily], Latur; *Dainik Ekamat*, [Unanimity Daily], Latur; *Bhukamp Varta*, [Earthquake News] fortnightly paper, edited by Shaila Lohia; *Lokprabha*, Bombay, weekly paper.
28. Nalin Sheth, Coordinator of Nonprofit Organizations, Latur District. Interview, Latur, India, December 23, 1993. Mr. Sheth was asked by the Latur District Collector, Praveensingh Pardeshi, to organize the operations of the nonprofit organizations which, while well intentioned, were nonetheless creating difficulties for the smooth operations of disaster relief.
29. See, for example, the findings on post traumatic stress reported for a sample of the population from Erzincan, Turkey, following the earthquake of March 13, 1992 in L. Comfort, A. Tekin, E. Pretto, B. Kirimli, D. Angus and Other Members of the International, Interdisciplinary Disaster Research Group. 1998. "Time, Knowledge, and Action: The Effect of Trauma upon Community Capacity for Action". *International Journal of Mass Emergencies and Disasters*, Vol. 16, No. 1: 73–91.

CHAPTER EIGHT

AUTO ADAPTIVE SYSTEMS: SELF-ORGANIZATION OR DYSFUNCTION IN NORTHRIDGE, CALIFORNIA AND HANSHIN, JAPAN

The 'Edge of Chaos'

At the 'edge of chaos', complex, organizational systems have the potential for creative adaptation or stunning failure in response to sudden, major changes in their operating environments. The difference lies in the information available to the participants in the system, and the extent to which they are able to absorb the information and act on it, in timely manner. Creative response, as outlined briefly in Chapter 4, requires a system that is high on technical structure, high on organizational flexibility, and high on cultural openness to new information and new methods of action.

In a complex society with a large population and many organizations holding diverse responsibilities for the sustenance of the community, a 'sociotechnical' system is able to use the capacity of information technology to search for, analyze, and disseminate information to support interorganizational decision-making on public issues requiring collaborative action. Such a system also uses this technology in the rapid implementation of those decisions, evaluation of actions taken, and timely feedback to the many participants on the changing state of the system. The timely flow of information through the system enables participants with different responsibilities operating in different locations to adjust more quickly to the changing environment and adapt their performance in reciprocal actions to achieve their shared goal.

In his discussion of the 'edge of chaos', Kauffman (1993) notes that a critical characteristic of this state for any system is the degree of uncertainty under which action is taken and the strategies that actors use to reduce that uncertainty. In "inquiring systems" (Churchman 1971, Comfort 1993, 1997), uncertainty is the condition that prompts information search as a means to achieve the basic goal shared by members of the system. Information search becomes both a strategy for reducing uncertainty and the first step in the process of adapting to change in the environment. In this inherently volatile state, the amount, quality, and timeliness of information available to decision-makers operating at different positions in the system affects the capacity of the system to perform effectively in an environment undergoing change.

Timely, accurate information available to key members simultaneously is likely to increase the system's efficiency in operation. Inadequate or delayed information at multiple points within the system, conversely, decreases its efficiency. In a complex, interdependent system, successive gaps or delays in information compound into almost certain failure. Operating in a dynamic environment, the system has the potential either for creative innovation in adapting to change, or chaotic failure resulting from its inability to adapt effectively or in time. The difference between creativity and chaos lies in the quality and timeliness of information available to the principal actors in the system and their ability to absorb and act on it.

Sociotechnical systems are necessarily interdisciplinary, and consequently are difficult to design, build and maintain. The technical components require advanced knowledge and skills in engineering, computer science, and, for seismic policy, seismology, geophysics and geology. The social components require an understanding of organizational design, public policy, sociology and communications. Sociotechnical systems require a "team approach" for effective operation, since no one person can master all of the knowledge and skills required to manage its complex tasks. Rather, a set of experienced and able managers, each with in-depth knowledge and skills in a particular field, but with sufficient understanding of the complementary fields, is best able to guide and maintain a sociotechnical system. Since these systems are interdependent and function on the basis of mutual understanding, effective communications skills are requisite for each member participating in management decisions.

Two response systems in the set of eleven included in this study showed initial potential for operating at the 'edge of chaos', but revealed very different results in practice. The systems that evolved following the Northridge, California Earthquake in 1994 and the Hanshin, Japan Earthquake in 1995 reveal the volatility and capacity for both creativity and failure in the midst of sudden, destructive change in the built and social environments. The two response systems are particularly interesting because the earthquakes occurred in metropolitan areas of the advanced technological societies of the U.S. and Japan. Both metropolitan regions had developed complex organizational systems and advanced technical systems to support the daily delivery of services to populations of several millions. How effectively could these respective communities adapt their performance and reallocate resources and energy to meet the needs of sudden, urgent, change following a severe earthquake? Response systems evolved in each community, but at different rates and with different consequences for their populations. Each case will be outlined briefly below.

Northridge, California, January 17, 1994 (M = 6.7)

At 4:31 a.m. on January 17, 1994, an earthquake measuring 6.7 on the Richter scale struck the communities of Northridge, Reseda, and Granada Hills in the San Fernando Valley, a section of the City of Los Angeles. The earthquake was the largest to occur in a modern, heavily-populated urban area in California, affecting directly or indirectly approximately three million people in parts of Los Angeles and adjacent cities. The timing of the event, early in the morning on a holiday weekend, contributed to a low death toll and minimized the damage that would have been likely in this area under normal daytime activities. Fifty-nine people died in earthquake-related circumstances, which included 19 deaths from heart attacks. Approximately 33 deaths were the direct result of collapsed buildings. Thousands of persons reported injuries, ranging from cuts and bruises to serious trauma requiring hospitalization. Area hospitals reported treating over 2,800 injured persons within 72 hours following the earthquake, admitting 530 patients for hospital treatment.[1] Less traumatic, but equally urgent were the shelter and welfare needs of nearly 33,000 people who suffered damage to their homes. The large scale of this disaster was mitigated only by the knowledge that it could have been much worse, except for the fortuitous timing of the event.

Response operations were activated immediately by the earthquake,[2] and carried out largely by experienced, well-trained, local emergency service organizations. State and federal organizations responded promptly to requests for assistance, and mobilized back-up resources to support the local efforts. The first response, including urban search and rescue teams engaged in life-saving activities and emergency response teams engaged in identifying and stabilizing life-threatening conditions, was completed within 36 hours. From that point, the needs of the community turned to restoring basic public services and meeting the human needs generated by the significant loss of housing, property, jobs, transportation, and access to other services such as medical care and nutrition.[3] The costs of this disaster, in terms of lost public infrastructure, damage to housing, businesses, schools, hospitals, and the costs of services provided to those rendered homeless and jobless were estimated at $25.7 billion, close to the losses suffered in the massively destructive Hurricane Andrew in South Florida and Louisiana in August, 1992. With economic losses of this magnitude, the Northridge Earthquake was clearly a national disaster, as reserves and resources from the entire nation were directed toward re-establishing the economic, social, and infrastructure systems of the Los Angeles Basin.

Information Search

Years of training, practice and experience with actual disasters governed the information search processes used by public agencies and jurisdictions following the Northridge Earthquake. These organizations largely followed their pre-existing emergency plans, which called for immediate activation of their Emergency

Operations Centers following an earthquake of M = 5.0 or above. Public agencies and jurisdictions had reconnaissance strategies written into their emergency plans, which specified an immediate strategy of visual damage assessment. This strategy included systematic observation of critical facilities and neighborhoods by local fire engine companies, police patrols, and engineering damage evaluation teams, who reported directly to their respective stations. These reports were then summarized by each organization and forwarded to the Emergency Operations Center to create a comprehensive assessment for the jurisdiction.

The damage assessments were used as the basis for allocating resources and personnel and determining how much and what kind of additional resources would be needed for response operations in each jurisdiction. For the most part, each public agency and jurisdiction followed its pre-planned strategy in gathering information as the basis for operations. In some instances, where freeway structures were down or transportation constraints made it impossible for emergency personnel to report to their assigned stations, they reported to the nearest station in their vicinity, contributing their skills and observations to the operations initiated at that location.

Information search strategies outlined in the respective emergency plans allowed nearly simultaneous assessments of damage by different organizations and jurisdictions. For example, as a procedure specified in its emergency plan, the City of Los Angeles activated its helicopter team at daybreak for overflights to assess damage within the City's boundaries. Simultaneously, NASA activated its planes at first light for aerial reconnaissance of the entire earthquake-affected region, as part of the State emergency plan that specified collaboration between state and federal agencies on disaster response. Each strategy contributed to the information search at the respective jurisdictional levels, which then were corroborated by information from other levels of government to provide a comprehensive assessment of the damage in the region. The state also used a computerized loss estimation model developed by EQE, Inc., an Irvine-based engineering firm, which estimated the losses to property in the Los Angeles region, based on the magnitude, location of the epicenter, and intensity of the ground shaking triggered by the earthquake. The combined set of information search processes produced credible damage estimates very quickly, and led to a remarkably rapid series of disaster declarations: City of Los Angeles at 5:45 a.m.; County of Los Angeles at 6:00 a.m.; State of California at 9:05 a.m.; Presidential declaration at 2:08 p.m. on January 17, 1994.

Information Exchange

Since the earthquake occurred in the Metropolitan Los Angeles Region, a significant technical information infrastructure was already in place through the local public jurisdictions, private telecommunications companies, and state investment to support information exchange among organizations participating in disaster operations. In addition, the Federal Emergency Management Agency

(FEMA) made a substantial investment in information technology to support dissemination of public information and interorganizational decision processes, much larger than for previous disasters. For example, in a "normal" disaster, FEMA spends approximately $100,000 for communications equipment. For Hurricane Iniki in 1992, FEMA spent $240,000. Three weeks after the Northridge Earthquake, FEMA had already spent $1.5 million on telecommunications, communications and related information technology, with continuing costs estimated at $3—3.5 million.[4] In terms of personnel, the Joint Public Information Office had a staff of 140 working at the Disaster Field Office, 30 assigned by the State of California, 110 employed by FEMA.

Given the critical need for timely, accurate information in the mobilization of response operations for disaster, the technical capacity to support information exchange governs the extent to which it can actually be achieved. Seven different mechanisms of communication were used in the Northridge disaster response: 1) satellite communications provided by the five Mobile Emergency Radio System (MERS) units brought to Northridge by FEMA from other states; 2) Operational Area Satellite Information System (OASIS), California's state-sponsored satellite system, partly operational at that time; 3) Caltech-USGS Broadcast of Earthquakes/Rapid Earthquake Data Integration (CUBE/REDI) system; 4) Emergency Digital Information System (EDIS); 5) Emergency Broadcast System (EBS); 6) Geographic Information System (GIS), quickly organized by California's Office of Emergency Services (OES); and 7) Recovery Channel, a special television channel initiated by FEMA to provide information about disaster response and recovery operations to residents affected by the earthquake.

In addition, routine technologies were used in new ways to support the intense demand for communication in disaster operations. Participating organizations made a major shift to cellular telephones for intra-agency communication. California OES, for example, issued 3,800 cellular phones to employees and volunteers working in disaster response and recovery. The Los Angeles County Fire Department had recently opened a computerized communications center that received approximately 200 calls per hour, or 3.3 calls per minute[5] in the first hours after the earthquake.

The varied set of communications mechanisms used in response operations indicate a major investment in information technology by FEMA, supplemented by substantial investments by the State of California, Los Angeles County and local organizations. This substantial technical information infrastructure facilitated greatly the exchange of information among the public organizations involved in disaster operations located at various sites, and the jurisdictional levels participating in disaster response. Again, specific communications processes were not without delay or failure, but multiple mechanisms created many avenues of transmission of timely information and opportunities for access to many groups in the disaster-affected region. This strong technical capacity made possible frequent and substantive exchanges of information among the organizational participants in the disaster response process, creating a 'sociotechnical system'. Gaps still existed, but the rate, frequency, and content of information exchange reported by managers of public organizations played a major role in the mobilization of the disaster response system. An abbreviated log of disaster operations across jurisdictional levels documents the practiced response at each level which drove the disaster declaration process. Table 53 presents, briefly, the timing of response actions.

Table 53. Log: Multijurisdictional Operations, Northridge, CA Earthquake

City of Los Angeles, January 17, 1994

04:31 Earthquake occurs, 6.7 Richter scale
04:32 Emergency procedures activated for all City departments; damage survey initiated
04:38 Emergency Operations Center activated
05:30 Mayor of Los Angeles arrives at EOC
05:44 Departmental field reports of damage submitted to Mayor
05:45 Mayor declares State of Emergency for City of Los Angeles, activating emergency powers
06:30 Directors of departments arrive at EOC to coordinate disaster operations
08:00 Planning Department ordered chemical toilets for shelters
09:00 Red Cross opened shelters at pre-designated locations

Los Angeles County, January 17, 1994

04:31 Earthquake occurs
04:32 Emergency procedures activated
04:45 Emergency Operations Center opened for 24 hour duty
05:00 Fire trucks dispatched to check critical areas
05:01 Fire Department personnel began reporting to their respective stations, reporting observations of damage as they arrived
05:31 Departmental damage reports are collected and summarized
05:46 Damage report for the County is submitted to the County Administrator and Board of Commissioners
06:00 Chairman, Board of Commissioners declares a State of Emergency for Los Angeles County, and requests assistance from the Governor's Office of Emergency Services
07:30 Full complement of personnel on duty
08:00 Fire Department organized Urban Search and Rescue operation; activated specially trained USAR team

California Governor's Office of Emergency Services

04:31 Earthquake occurs
04:32 Director of OES receives electronic notification of Magnitude 6.7 earthquake in Northridge Area
04:33 Office of Emergency Services initiates emergency procedures
05:50 Director of OES calls Director of FEMA, requests USAR teams on stand-by notice
06:01 OES receives report of disaster declaration by Los Angeles County
06:30 California Office of Emergency Services activates its Emergency Operations Center
06:31 Staff from Caltrans, CHP, CDF, National Guard and Health report to the EOC to determine level of support needed by local jurisdictions

09:05 Governor Wilson declares a State of Emergency for Los Angeles, Orange, and Ventura Counties and requests federal assistance

Federal Emergency Management Agency, January 17, 1994, Region IX, San Francisco, CA

05:45 Received call to activate Regional Operations Center (ROC)
05:46 NASA activates overflight of earthquake-damaged region in accordance with State plan
09:00 Meeting of Emergency Support Function managers

FEMA Headquarters, January 17, 1994, Washington, DC, PST

05:50 James Lee Witt, Director of FEMA, receives call from Director of California OES for USAR teams
09:05 Governor Wilson of California requests federal assistance in response to earthquake damage in Los Angeles Metro Region
11:00 J.L. Witt prepares to leave for Los Angeles
14:08 President Clinton declares a State of Emergency for Los Angeles, Ventura, and Orange Counties

---

Due to near-simultaneous actions according to pre-planned emergency procedures, the local, state and national jurisdictions mobilized a full-scale multijurisdictional response to the earthquake in a little over eight hours. Prompt action at each level initiated complementary actions by nonprofit and private organizations to provide assistance to dislocated residents of the damaged region.

Equally dramatic was the increase in federal personnel involved in response operations at the joint federal/state Disaster Field Office in Pasadena, California. On January 17, 1997, 125 employees were assigned to the Disaster Field Office to conduct follow-up work from the October, 1993 Malibu Fires.[6] Two weeks later, the number of federal employees and volunteers working in disaster operations had increased to 9,200 (Situation Report, Federal Emergency Management Agency, 31 January, 1994). This figure represented staff from 16 federal agencies, and did not include state, county, or municipal employees, or volunteers from organizations other than the Red Cross. Each of these employees or volunteers needed to be integrated into the response system; that is, briefed on the situation, assigned tasks, equipment and reporting instructions, and debriefed when tasks were completed. Eventually, staff from eleven additional federal agencies joined the response, bringing the number of federal agencies involved in disaster operations to a total of twenty-seven.

In addition, the response system included personnel from 55 of the 88 cities in Los Angeles County, three counties in the Los Angeles Region, five adminstrative departments of the State of California, and 304 business, religious, educational, and voluntary organizations from the Southern California region.[7] These figures, taken together, document the emergence of a multiorganizational, multijurisdictional response system that evolved within hours of the earthquake and continued in operation at high levels of effort for approximately three weeks. Some organizations continued their efforts for months following the earthquake, while others returned to normal operations in early February.

Organizational Learning

A strong sociotechnical infrastructure supports organizational learning both within and among the organizations and jurisdictions participating in disaster response. Disaster operations involve sets of interdependent problems. As the situation changes at one location, it affects needs or resources at other sites. Monitoring the effects of actions taken by one organization in relation to actions taken by other agencies or groups creates the feedback needed for organizational learning. Active management involves anticipating problems and acting to reduce them before they become serious issues for the welfare of the community. This process can be carried out most effectively through multiway communication among members of the group.

The sociotechnical infrastructure for the Northridge Earthquake produced a large disaster response system. Using the N-K methodology and content analyses of news reports, we identified a system of 378 organizations that engaged in disaster operations, as reported in the *Los Angeles Times*. This set of organizations is summarized in Table 54, by funding type and jurisdictional level.

The distribution reveals a plurality of organizations from the public sector, at 38.4%, but substantial participation from both the nonprofit sector, at 27.8%, and a full third of the organizations from the private sector, at 33.9%. Of the 145 public organizations, more than half, or 53.8%, were from the local level: city, county, and regional jurisdictions. State organizations were the next largest group, representing 24.8% of the public organizations. National organizations were third, at 20.73%, and one international organization, the City of Berlin, made a contribution to disaster relief.

Table 55 presents the matrix of 859 transactions performed in disaster operations following the Northridge Earthquake, identified through the content analysis of news reports to the *Los Angeles Times*. The largest proportion, 11.8%, involved communication among participating organizations in the response system, and between the response system and the community. The second largest proportion, 10.8%, involved actions taken to provide disaster assistance to those

Table 54. Frequency Distribution: Disaster Response System by Funding Source and Jurisdiction, 1994 Northridge Earthquake

Public									Nonprofit		Private		Total Public, Private, NPO	
Int'l		Nat'l		State		Regional/ County		City		Total					
N	%	N	%	N	%	N	%	N	%	N	%	N	%	N	%	N	%
1	0.3	30	7.9	36	9.5	22	5.8	56	14.8	145	38.3	105	27.8	128	33.9	378	100.0

who experienced loss. Coordination followed as a close third, with 9.5% of the transactions. Combining communication and coordination, the two functions represented 21.3% of the transactions, and served as the driving force for the evolving response system. Other types of transactions revealed important characteristics of the response system. Donations, largely from the private sector, constituted 8.6% of the transactions. Combined with disaster relief, the two types accounted for 19.4% of the total transactions.

Damage/needs assessment, at 7.3%, represented a sizeable proportion of the response effort, while transportation/traffic issues and repair of the freeways, bridges and roads together made up 8.4% of the transactions, documenting transportation as a serious issue in recovery. Transactions involving education issues at 5.0%, business recovery and economic issues at 4.4%, and housing at 2.3% constituted important subcomponents of the response system. The strong findings on communication and coordination document these functions as the driving force which integrates the actions of the member organizations into a coherent response system.

In validation of the premise that local organizations govern disaster response, the highest proportion of transactions, 182 or 21.2%, involved public city organizations. The next highest proportion, 175, or 20.4%, were performed by non-profit organizations, and private organizations performed 166, or 19.3%, of the transactions. A large proportion of the private transactions were single, one-time donations to disaster relief, rather than the sustained involvement over time by some public organizations.

Table 56 presents the matrix of interactions among the types of agencies by funding source and jurisdiction. While federal agencies were involved in the largest number of reported interactions, the next largest number involved municipal organizations, and the combined set of local organizations represented 31.9%. Interactions reported for local and state organizations together, at 47.4%, exceeded the federal share. The findings in this matrix document the density of interactions among the organizations participating in the disaster response system.

Adaptive Behavior and/or Self-Organization

The nonlinear analysis of change in disaster response functions over the three-week period immediately following the earthquake reveals a strong immediate response by local organizations, and rapid evolution of the disaster response system.

Complex development in organizations depends upon the capacity of many individuals to communicate and process large amounts of information quickly. During the Northridge Earthquake disaster operations, these processes were enhanced by advanced information technology. For example, in order to facilitate the registration and information processes to obtain disaster assistance, FEMA introduced two mechanisms that employed information technology. First,

Table 55. Frequency Distribution: Types of Transactions in Disaster Response by Funding Source and

| Type of Transaction | Public Organizations ||||||||||||
| | Federal ||| State ||| Regional ||| County |||
	T	N	%	T	N	%	T	N	%	T	N	%
Emergency Response	7	5	0.8	4	2	0.5	0	0	0.0	5	3	0.6
Communication	20	8	2.3	24	11	2.8	0	0	0.0	4	3	0.5
Coordination of Response	31	12	3.6	6	4	0.7	0	0	0.0	1	1	0.1
Medical Care/Health	2	2	0.2	4	4	0.5	0	0	0.0	2	2	0.2
Damage/Needs Assessment	6	5	0.7	12	10	1.4	1	1	0.1	4	4	0.5
Certification of Deaths	0	0	0.0	1	1	0.1	0	0	0.0	2	2	0.2
Earthquake Assessment/Research	14	3	1.6	4	4	0.5	0	0	0.0	0	0	0.0
Security Prevention of Looting	1	1	0.1	5	1	0.6	0	0	0.0	4	1	0.5
Housing Issues	10	2	1.2	0	0	0.0	0	0	0.0	0	0	0.0
Disaster Relief (food, shelter, etc.)	7	3	0.8	1	1	0.1	0	0	0.0	6	4	0.7
Donations (money, goods, etc.)	0	0	0.0	0	0	0.0	0	0	0.0	1	1	0.1
Building Inspection	1	1	0.1	2	2	0.2	0	0	0.0	1	1	0.1
Building Codes Issues	0	0	0.0	3	3	0.3	0	0	0.0	0	0	0.0
Repair of Freeways, Bridges, Roads	3	2	0.3	12	3	1.4	0	0	0.0	0	0	0.0
Repair/Restore Utilities	1	1	0.1	0	0	0.0	0	0	0.0	3	3	0.3
Repair/Reconstruction/Recovery[a]	4	4	0.5	4	4	0.5	1	1	0.1	1	1	0.1
Transportation/Traffic Issues	2	2	0.2	13	3	1.5	9	1	1.0	13	4	1.5
Hazardous Materials Releases	1	1	0.1	3	2	0.3	0	0	0.0	0	0	0.0
Legal/Enforcement/Fraud	4	8	0.5	3	3	0.3	0	0	0.0	3	3	0.3
Legislation/Legislative Processes	4	1	0.5	2	2	0.2	0	0	0.0	0	0	0.0
Business Recovery	1	1	0.1	0	0	0.0	1	1	0.1	3	3	0.3
Economic/Business Issues	0	5	0.0	6	3	0.7	0	0	0.0	3	1	0.3
Visits by Officials	5	1	0.6	1	1	0.1	0	0	0.0	0	0	0.0
Education Issues	3	8	0.3	1	1	0.1	0	0	0.0	3	1	0.3
Government Assistance	22	0	2.6	0	0	0.0	0	0	0.0	0	0	0.0
Insurance Related Issues	0	0	0.0	2	1	0.2	0	0	0.0	0	0	0.0
Loans by Private Banks	0	0	0.0	0	0	0.0	0	0	0.0	0	0	0.0
Psychological/Counseling Services	0	0	0.0	2	2	0.2	0	0	0.0	1	1	0.1
TOTAL	149	76	17.3	115	68	13.4	12	4	1.4	60	39	7.0

T = Number of Transactions; N = Number of Actors; % = Percent of Total Transactions
[a]Not including freeways, bridges, roads or utilities
Note: One international transaction (donation), by one actor not included in table
Sources: *The Los Angeles Times* and *The Daily News*

FEMA established a "hotline", an information center with a toll-free telephone line which callers could use to obtain current, accurate information about the disaster assistance process. The number of calls to the "hotline" reveals the interaction among different components in the response system. As calls to the hotline increased, documenting the number of individuals requesting help, the number of persons living in shelters declined. People affected by the earthquake sought information regarding assistance, and acted on that information, when they received it.

Jurisdiction, Northridge Earthquake, January 18 – February 6, 1994

			Nonprofit Organizations			Private Organizations			TOTALS		
City											
T	N	%	T	N	%	T	N	%	T	N	%
14	4	1.6	0	0	0.0	1	1	0.1	31	15	3.6
23	9	2.7	15	15	1.7	15	16	1.7	101	62	11.8
22	14	2.6	11	14	1.3	11	10	1.3	82	55	9.5
0	0	0.0	20	19	2.3	0	0	0.0	28	27	3.3
26	19	3.0	6	6	0.7	8	8	0.9	63	53	7.3
0	0	0.0	0	0	0.0	0	0	0.0	3	3	0.3
0	0	0.0	4	1	0.5	16	2	1.9	38	10	4.4
8	1	0.9	0	0	0.0	1	2	0.1	19	6	2.2
3	1	0.3	7	7	0.8	0	0	0.0	20	10	2.3
8	7	0.9	70	27	8.1	1	1	0.1	93	43	10.8
1	1	0.1	14	13	1.6	58	47	6.8	74	62	8.6
9	3	1.0	3	3	0.3	0	0	0.0	16	10	1.9
6	3	0.7	2	4	0.2	0	0	0.0	11	10	1.3
0	0	0.0	0	0	0.0	4	4	0.5	19	9	2.2
5	2	0.6	0	0	0.0	7	4	0.8	16	10	1.9
9	9	1.0	2	2	0.2	4	4	0.5	25	25	2.9
10	7	1.2	0	0	0.0	6	5	0.7	53	22	6.2
2	2	0.2	0	0	0.0	1	1	0.1	7	6	0.8
3	2	0.3	1	1	0.1	1	1	0.1	15	18	1.7
1	1	0.1	0	0	0.0	0	0	0.0	7	4	0.8
0	0	0.0	1	1	0.1	1	1	0.1	7	7	0.8
2	1	0.2	1	1	0.1	17	20	2.0	29	31	3.4
2	3	0.2	1	1	0.1	0	0	0.0	9	6	1.0
24	12	2.8	9	9	1.0	3	2	0.3	43	33	5.0
0	0	0.0	0	0	0.0.	0	0	0.0	22	0	2.6
0	0	0.0	0	0	0.0	5	5	0.6	7	6	0.8
0	0	0.0	0	0	0.0	6	5	0.7	6	5	0.7
4	4	0.5	8	8	0.9	0	0	0.0 1	15	15	1.7
182	105	21.2	175	132	20.4	166	139	9.3	859	563	100.0

Second, FEMA augmented the registration process, traditionally done in person at Disaster Assistance Centers (DACs) established for this purpose, by establishing a National Teleregistration Center (NTC) with toll-free lines, where residents seeking assistance could call 24 hours a day. The number of registrations for disaster assistance, both at the NTC and at the DACs set record highs for prompt service to disaster victims. By January 31, 1994, two weeks after the event, fixed DACs reported 58,314 registrations for individual assistance, mobile DACs reported 4,366, and the NTC reported 153,218, for a total of 215,898

Table 56. Frequency Distribution: Types of Interactions in Disaster Response by Funding Source and Jurisdiction, Northridge Earthquake, January 18 – February 6, 1994

| Type of Interaction | Public Organizations ||||||||||| Nonprofit || Private || TOTAL ||
| | Federal || State || County || City || School District || | | | | | |
	K	N	%	K	N	%	K	N	%	K	N	%	K	N	%	K	N	%	K	N	%	K	N	%
Public: Federal	29	15	11.8	20	18	8.2	9	14	3.7	20	17	8.2	8	5	3.3	11	13	4.5	5	3	2.0	102	85	41.6
Public: State				12	10	4.9	5	9	2.0	14	13	5.7	1	2	0.4	4	8	1.6	2	4	0.8	38	46	15.5
Public: County							2	4	0.8	15	14	6.1	1	2	0.4	2	4	0.8	1	2	0.4	21	26	8.6
Public: City										16	14	6.5	9	6	3.7	16	19	6.5	8	13	3.3	49	52	20.0
Public: School District													0	0	0.0	5	7	2.0	3	4	1.2	8	11	3.3
Nonprofit																18	25	7.3	7	8	2.9	25	33	10.2
Private																			2	4	0.8	2	4	0.8
TOTAL	29	15	11.8	32	28	13.1	16	27	6.5	65	58	26.5	19	15	7.8	56	76	22.9	28	38	11.4	245	257	100.0

K = Number of Interactions; N = Number of Actors; % = Percent of Interactions Row Total
Sources: *Los Angeles Times* and the *Daily News*

Chapter 8: Auto-Adaptive Systems 209

registrations. Each mechanism enhanced use of the other, and contributed to a record number of registrations, 310,000 by February 10, 1994.

Other measures show the rapid evolution of the Northridge response system. First, a total of 33,000 people initially sought shelter because their homes were destroyed or damaged, requiring 42 Red Cross shelters at peak demand. The number of persons in shelters dropped to less than 3,000 by February 6, 1994, as "reassurance teams" of clergy, social workers, and public agency representatives offered assistance to frightened residents. The situation also showed collaborative work in which public agencies from multiple jurisdictions, private housing agents and nonprofit organizations formed a "Housing Network" that matched families seeking housing with available units.

With the escalation of response operations, new sites of operation and requirements for coordination among them emerged as critical needs for maintaining a coherent response process. This multiplicative aspect is seen most vividly in the rapid expansion of disaster assistance centers in the different neighborhoods of the city, with a consequent increase in communication between each center and headquarters, as well as among the different centers. This increase in communication was essential to coordinate services among the set of DACs in order to ensure quality service delivery at each one. Disaster Field Office (DFO) staff opened 11 Disaster Assistance Centers on January 20, 1994, three days after the earthquake. This number increased almost daily for the next ten days when it peaked at 20 fixed site DACs and 14 mobile DACs on January 30, 1994. That number remained until February 8, when one more DAC was added for a total of 21 fixed DACs. Then Centers were combined and converted to Community Recovery Centers, as needs shifted from immediate individual assistance to recovery and reconstruction from losses.[8] The Recovery Channel, supported by the MERS units, and thousands of cellular phones provided critical communication linkages between the DACs and the DFO, federal, state and local agencies, and among the 35 operating DACs.

The nonlinear analysis of frequency data derived from the content analysis of news reports documents the change in response actions over the first three weeks following the earthquakes. It also reveals a surprising vulnerability in the integration of the response system among the types of organizations participating in response operations. Figure 45 presents a phase plane plot of the daily change in the relationship between the number of reported actions in emergency response and the number of reported actions in communication/coordination for public organizations. Figure 46 shows the marginal history of the reported daily change between the two variables for the first twenty-one days following the earthquake for public organizations. The marginal history confirms the pattern portrayed graphically in the phase plane plot. Change in reported emergency response activities for public organizations appears to move relatively independently of change in reported communication/coordination activities. This pattern reflects the intensive professional training for public emergency response organizations in California that produces a rapid and reliable pattern of emer-

210   Part II: Shared Risk in Practice

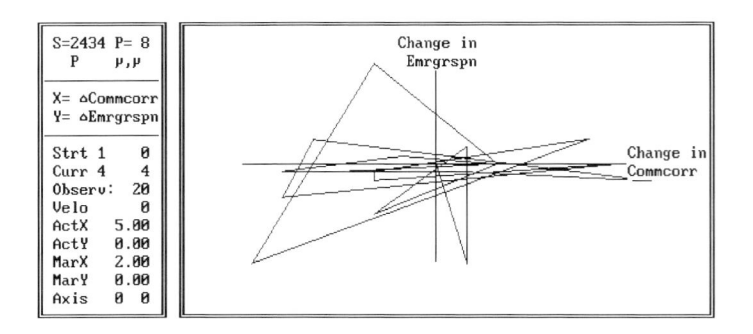

Figure 45. 1994 Northridge, CA Earthquake: Phase Plane Plot, Public Organizations, Emergency Response by Communication/Coordination

gency response, but one that operates largely independently from other response functions or community organizations. This pattern, which may function well in moderate disasters, could prove vulnerable to heavier demands and a more urgent time line in a catastrophic event.

The nonlinear logistic regression analysis reports equations for all six response functions for public organizations. The equations reveal variations in both strength and stability among the six functions, which indicate potential vulnerability in the community response system. The functions of emergency response and damage assessment show high values for R^2 and low values for k, indicating stability in performance. Figure 47 reports the nonlinear logistic equation for emergency response, showing a value of $R^2 = 0.65$, $F = 34.6$, $p < 0.001$, and $k = 1.00$. Damage assessment reports an R^2 value of 0.50, $F = 18.9$; $p < 0.001$, $k = 1.00$. Communication/coordination shows an R^2 value of 0.33, $p < 0.01$, and $k = 1.0$.

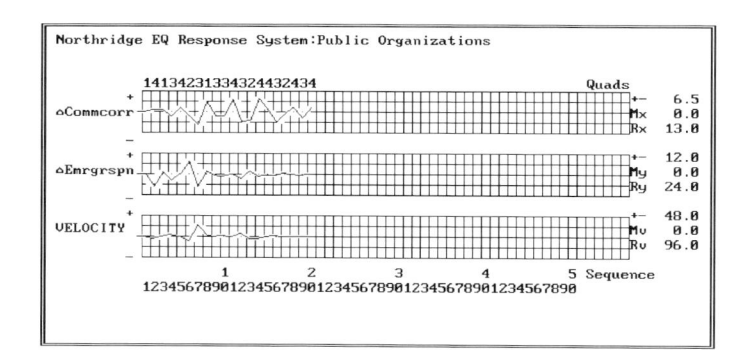

Figure 46. 1994 Northridge, CA Earthquake: Marginal History, Public Organizations, Emergency Response by Communication/Coordination

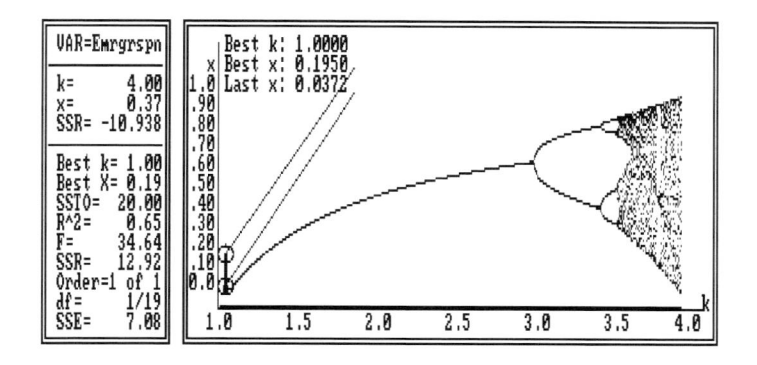

Figure 47. 1994 Northridge, CA Earthquake: Logistic Regression Analysis, Public Organizations, Emergency Response

But the logistic regression equations calculated for the remaining three response functions performed by public organizations reveal surprising instability and low values for explained variation in the rate of change in performance. The functions of disaster relief and financial assistance, both vital to resilience of the community in disaster, report R^2 values below 0.20 which is not statistically significant. The function of recovery/reconstruction reported $R^2 = 0.20$, which is statistically significant, but a k value of 3.95 which indicates that change in performance had slipped into the chaotic range, above 3.6. Financial assistance also reported a k value in the chaotic range. These findings are interesting because they reflect the substantial interaction of public organizations with private organizations in recovery/reconstruction and financial assistance and with nonprofit organizations in disaster relief. Yet, these interactions appear to have been carried out without the same level of training, preparedness and investment in information infrastructure as was available to emergency response.

The high levels of performance in emergency response, damage assessment, and communication/coordination among public organizations combine with more irregular performance in the other three functions to produce an R^2 value that explains nearly two-thirds of the variance in performance of the subsystem for public organizations. However, grouping frequency data from public, private and nonprofit organizations to represent the entire Northridge disaster response system produces a logistic equation with an R^2 value of 0.38, $F = 11.6$, $p < 0.05$, and $k = 1.0$. This finding reveals greater variance in performance than reported for the subsystem of public organizations, which is not surprising, given findings of chaotic performance on response functions in the nonprofit and private sectors.

These findings indicate that the extensive investment in response training for public organizations in California has not been carried over into the private and nonprofit sectors, which also play important roles in community response and recovery. The response system that evolved following the Northridge Earthquake was stable, but more vulnerable than expected. Without equivalent preparation and full integration of the private and nonprofit sectors into a rapidly evolving response system, the Los Angeles community is likely to be vulnerable in an earthquake of larger magnitude that would create heavier demands upon the entire community than the Northridge event. Table 57 summarizes findings from the nonlinear regression analysis for public organizations, and shows the contrast with the total system of public, private and nonprofit organizations.

These findings demonstrate that organizations experienced in disaster response and damage assessment, and supported by strong performance in communication/coordination, are able to stabilize the response system, even with chaotic performance in the functions of recovery/reconstruction and financial assistance, and near chaotic performance in disaster relief.

The evidence supports the emergence of self organizing, creative activities within and between jurisdictions to form a coherent response to the disaster. Disaster operations were not perfect, but remarkably effective given the rapid increase in complexity and the intensity of the disaster environment. The second case in this set of response systems, characterized by a similar potential for reduction of seismic risk, demonstrated a different response to an actual earthquake.

Hanshin, Japan, January 17, 1995 (M = 7.2)

A severe earthquake struck the Hanshin region of Japan at 5:46 a.m. on January 17, 1995, registering 7.2 on the Richter scale of magnitude. The epicenter was located on northern Awaji Island, just off shore from Kobe, a city of 1.5 million population. The rupture registered strong ground motion directly through downtown Kobe and northward to the neighboring cities of Nishinomiya, Ashiya, Itami City, Amagasaki, Takarazuka, and other towns in Southern Hyogo Prefecture. A disaster response system evolved following this event, revealing significant aspects of the process of self organization in dynamic, uncertain environments.

Table 57. Summary, Logistic Regression Analysis of Disaster Response Functions, Northridge Earthquake, January 18 – February 6, 1994

Response Function	R^2	F	p	k
Emergency Response	0.65	34.6	0.001	1.00
Damage Assessment	0.50	18.9	0.001	1.46
Communication/Coordination	0.33	9.7	0.001	1.00
Disaster Relief	0.09	1.9	NS	3.31
Recovery/reconstruction	0.20	4.7	0.05	*3.95*
Financial assistance	0.12	2.6	NS	*3.92*
Subsystem, Public Organizations	0.65	35.1	0.001	1.00
System, Public, Private, Nonprofit Organizations	0.38	11.6	0.01	1.00

Figures in italics indicate that the value of k registers above 3.66, within the chaotic range

The initial conditions prevailing in the Southern Hyogo Prefecture of Japan in January, 1995 shaped in significant ways the response system that evolved following this disaster. The technical, organizational, and social conditions of this metropolitan region were those of an advanced industrial society. Kobe, the principal city in the Hanshin region, is located in the south central section of Honshu, the main island of Japan. Geographically, the city stretches 30 kilometers east to west along Osaka Bay, with the Rokko Mountains rising steeply to the north. Kobe is a modern city, with interdependent systems of transportation, industry, trade, banking, education, and medical care linking the city to others in the region. The transportation system, for example, is an advanced mix of highspeed rail transport, local railways, city bus lines, and expressways, connected to international transport via a major new regional airport and a busy international shipping port, the sixth largest in the world. Extensive networks of telecommunications, electrical, gas, and sewer lines provide efficient, modern service to this metropolitan region of over 10 million people. Building structures represent a mix of types, with sophisticated seismic engineering in high-rise buildings interspersed with old style wooden houses with heavy tile roofs. The technical profile of the region is generally strong and, prior to the earthquake, was a matter of pride for residents of the region.

Organizationally, the area was not well prepared for seismic risk. Although the islands of Japan are located at the juncture of three tectonic plates and seismic risk is well known in the nation, residents generally believed the Hanshin region, which had last experienced a moderate earthquake (M = 6.1 Richter scale) in 1916, was relatively stable in contrast to the Tokyo Region, which had suffered a major earthquake with heavy losses in 1923. Consequently, relatively little investment had been made in earthquake preparedness, either by public organizations or residents. While cities in the region had emergency plans, their preparation had been oriented toward small, local disasters of fires and floods.

Private utility companies, such as Kansai Electric Co. and Osaka Gas Co., demonstrated substantial investment in seismic mitigation efforts to protect their interests, but were not directly linked to the public agencies. Socially, there existed little tradition of voluntary organizations or community self help associations. Most people focused their lives on their work associations and their families. Although the initial technical systems were strong, there was little interorganizational capacity to reallocate resources and action in timely response when these interdependent systems failed under the severe shock of the unanticipated earthquake.

In the densely populated, complex urban environment of the Hanshin region, a Magnitude 7.2 earthquake set off a cascading effect in the area's network of interdependent systems at 5:46 a.m. on January 17, 1995. Failure in one system triggered failure in another which triggered further failure in a third, each failure compounding the damage and leading to full-scale disaster, affecting approximately 4 million people in the metropolitan region.

The damage was extensive. The death toll has climbed past 6,300 in recent reports (National Land Agency 1995)[9] and the number of injured totaled 41, 648 in the April 25, 1995 report. The total losses in housing were 101, 233 homes totally destroyed, 107, 269 homes half destroyed, and 182,190 homes partially destroyed, for a total of 390,692 damaged homes. A total of 3,669 public buildings were damaged or destroyed, as cited by the National Land Agency.

The dynamics of the destruction were sobering. The strong vertical ground motion ruptured underground gas and water mains, causing leaks and disrupting service throughout the region. An estimated 4,500 km. of gas lines were heavily damaged, and 1,200,000 houses were left without water. Electrical facilities were also damaged, cutting off sources of electrical power to 850,000 city departments, businesses, and households. The total cost of the disaster is estimated at over US $200 billion. As the gas mains ruptured, fires broke out. With no water available for fire suppression, the fires raged largely unchecked through seriously damaged sections of the city. In Kobe, 60 fires broke out before 6:00 a.m. on January 17, 1995, and burned simultaneously. Before 9:00 a.m., the number of fires burning simultaneously had increased to 85, with a total of 109 fires reported for the city of Kobe, and a total of 294 fires for the entire earthquake-affected area. The major cause of the fires was broken gas mains. Debris from collapsed buildings blocked the streets, preventing fire trucks from getting through. The National Land Agency reported over 9,403 blockages in roads for the area.

These conditions proved overwhelming for the Kobe Fire Department which had primary responsibility for emergency response, but a total of 11 fire stations in the city, 176 engines, and 305 personnel on duty when the earthquake occurred. Three of the 11 stations were damaged in the earthquake, and even with emergency call-out procedures, only 663 personnel were able to report for duty within the first two hours. The actual destruction was beyond any training scenario for municipal emergency response.

Information Search

Interdependent emergency response organizations were unable to make a rapid transition to an emergency response system vital to saving lives in the first hours following the earthquake. Under the urgent conditions of disaster, communications capability was critical. The Kobe Fire Department had just installed an advanced computerized dispatch system with video monitors in December, 1994, but it was not yet operational and was not used in disaster operations. Telephone lines were out of order during the first day in large sections of the region, while others were overloaded. The 119 dispatch logs for January 17, 1995 showed that over 1800 emergency calls had been made on 118 emergency circuits, at roughly 100 calls per hour or 1.7 calls per minute. Yet, these were only the calls that could get through. The number of calls attempted, but not completed, cannot be estimated. Fire departments had their own radio systems, but could not communicate with other departments. Communications capability proved very limited in the first critical hours following the earthquake. The basic information infrastructure needed to support the search for, and exchange of, information in the dynamic disaster environment was either not available or not functioning.

The business sector had invested in information technology that performed well within its limited range, but business organizations did not have clear, effective communication linkages with public sector agencies responsible for life and property. Public sector investments in information technology either were not fully operational, e.g. Kobe Fire Department's GIS and computerized dispatch system, or failed, e.g. Hyogo Prefecture's satellite communication system, to support decision making in disaster operations.

Information Exchange

The damaged communications infrastructure severely restricted information exchange in response operations during the first critical hours following the Hanshin Earthquake. During this time, the fires broke out of control and spread rapidly throughout the city. Valiant efforts were made to suppress the fires, but the combination of simultaneous ignition, lack of water, lack of electrical power for pumping water, the direction of the winds, and the number of wooden buildings fueled the fires and completely overwhelmed the local fire resources. Only hours after the initial outbreak of fires did prefectural and national response agencies learn of the severe conditions in Kobe, late, almost too late, to provide much needed support. The operations logs from municipal, prefectural and national fire agencies, presented in Table 58, reveal the near absence of information exchange among the jurisdictional levels during this period, and its consequent effects upon response operations.

Table 58. Log: Multijurisdictional Operations, Hanshin, Japan Earthquake

Kobe City Fire Department, January 17, 1995

05:46 Earthquake occurred; almost all of the 119 emergency lines were occupied; emergency summons issued to personnel
05:53 First fire report and three others followed; at least 60 fires were burning simultaneously
06:15 Chief, Fire Department arrived at Kita Suma branch office, called the control center, and received reports of the disaster situation and rescue operations
06:25 Chief, Fire Department left Kita Suma branch office for the Fire Department. On the way, he observed the disaster situation
06:40 Fire Chief ordered a pump truck team at Tarumizu Fire Station sent to the Nagata area
06:50 Center control room was established; Mayor arrived at control room
07:00 Kobe City Disaster Operations Center was established
07:10 Chief, Fire Department arrived at the Operations Center; vice head, Operations Center tried to call prefecture to report disaster, but could not get through
07:20 Chief ordered two pump truck teams at Kita Fire Station sent to Hyogo area
07:30 Chief, Operations Center reported disaster and prevention activities to mayor

08:00 Chief ordered a pump truck team at Tarumizu Fire Station sent to Nagata area
08:30 Chief ordered a pump truck team at Kita Fire Station sent to Nagata area
09:00 Vice Chief, Operations Center briefed prefectural government on the disaster
09:20 Operations Chief ordered a Fire Defense Mobile Unit helicopter to gather information on status of disaster in the entire city
09:30 Chiefs of Fire Departments of Kyoto City and Osaka City offered support. Asked the prefectural government for the possible mobilization of Self-Defense Force (Planning Adjustment Department)
09:40 Received a report from the Fire Defense Mobile Unit helicopter. At least 20 additional fires were reported, and building collapses were observed all over the city, especially in the eastern part
09:50 Chief, Fire Department advised the mayor to request a wide area fire fighting support and mobilization of Self-Defense Force; suggested that Fire Departments deal with fires and Police and Self-Defense Force carry out rescue operations. The mayor requested the governor of Hyogo Prefecture to send wide area fire fighting support
10:00 Mayor of Kobe requested the governor of Hyogo Prefecture to mobilize the Self-Defense Force. The Minister of Fire Defense Agency, the Ministry of Home Affairs accepted the request. The Governor of Hyogo Prefecture reported that relevant governors had received the order

Source: "Hanshin – Awaji Daishinsai (Kobe Shiiki) ni okeru Shobokatsudo no Kiroku", Kobe City Fire Department, Kobe, Japan, March, 1995

Hyogo Prefecture, January 17, 1995

09:20 Helicopters of Kobe Fire Department were activated, and officials gave a disaster report to the Operations Center by radio; operations were delayed due to liquefaction at heliport. In the afternoon, the Fire Defense heliport was moved to Hiyodori Dai
09:50 Governor of Hyogo Prefecture receives request from Mayor of Kobe for wide area fire fighting support
10:00 Governor of Hyogo Prefecture receives request from Mayor of Kobe for mobilization of Self-Defense Force from National Fire Defense Agency in Tokyo
10:01 Governor of Hyogo Prefecture requests wide area fire fighting support and mobilization of Self-Defense Force from National Fire Defense Agency in Tokyo
10:30 Disaster Prevention Center organized seven special teams to carry out mission, with 6 personnel to a team. The first medium team (three small teams, 18 personnel) was mobilized in Nagata area. It carried out fire fighting and rescue operations, securing water from fire fighting ships, etc.
11:10 Fire brigades from Mita City (north of Kobe) arrived at Nagata-ku
13:15 Self-Defense Force, the Third Division, Himeji Special Regiment arrived with 216 members
13:40 Ten fire fighting teams from Osaka City arrived. Thereafter, fire brigades arrived one after another. Tokyo Fire Defense Agency, Nagoya City Fire Department, and Hiroshima City Fire Department responded with support teams. Yokohama City Fire Department, Kawasaki City Fire Department, Kyoto City Fire Department sent helicopters
24:00 Reinforcements arrived: 182 pump truck teams with 860 personnel, 9 helicopters with 52 personnel, and 2,562 Self-Defense Force members to assist in fire fighting operations.

Source: "Hanshin – Awaji Daishinsai (Kobe Shiiki) ni okeru Shobokatsudo no Kiroku". Kobe City Fire Department, Hyogo Prefecture, Kobe, Japan, March, 1995

National Fire Defense Agency, Tokyo, January 17, 1995, Director's Report

06:30 Awakened at home at usual time; turned on television; learned of earthquake from news report. Did not receive any calls; planned to go to office at usual time, 9:30 a.m.

07:30 At breakfast, watched the news, saw the photos of the shinkansen collapse. Realized that the earthquake was serious; but did not know scale of damage
08:40 Arrived at office, earlier than usual. Telephone communications were out between Tokyo and Kobe. Tokyo Fire Department called to ask the status of Kobe. Without knowing the damage, they were preparing to send a support team and two helicopters to Kobe. In Fire Department, protocol is not to send assistance unless requested
09:00 Established communication with Kobe; established a support team
10:01 First report from Kobe — they requested support — request came from Kobe City Mayor through the governor of Hyogo Prefecture via telephone
10:02 Called Fire Defense Agencies that had helicopters, e.g. Hiroshima; there are 12 Fire Defense Agencies with helicopters; some helicopters couldn't fly, they under inspection. Mobilized response to Kobe

Source: Director, Ambulance and Rescue Service Division, Fire Defense Agency, Ministry of Home Affairs. Interview, Tokyo 100, Japan, Tuesday, May 16, 1995

Information exchange did occur among the participating organizations and jurisdictions, but frequently late or inadequate. The cumulative effect of inadequate information disrupted efforts to mobilize response and triggered further failure in other sectors of disaster reponse.

Organizational Learning

Operating under the urgent, stressful conditions of disaster, participating response organizations had little time for reflection and less opportunity for learning new methods of coping with their dynamic environment. Using the N-K methodology and content analysis of news reports published in the *Japan Times*, we identified a response system of 391organizations that participated in disaster operations. Of this total, public organizations made up the largest proportion, with 42.2%. Private organizations composed the next largest category, at 35.5%, and nonprofit organizations made up 22.3% of the total response system. Examining the breakdown for the 165 public organizations, the largest proportion was municipal, with 71, or 43% of the total number. Prefectural organizations constituted 15.8%; national organizations represented 23%, and international organizations made up 18.2% of the public organizations contributing to disaster response. Table 59 summarizes these findings.

The content analysis identified a large number of organizations contributing to disaster operations. However, Table 60 shows a relatively small number of reported transactions, 136 performed by 197 organizations.

Many organizations were listed as contributors to disaster relief, but were not actively engaged in disaster operations. The largest proportion, 35, or 25.7%, involved provision of disaster relief. Emergency response accounted for 20, or 14.7%, of the total transactions reported. Medical/health care to the injured was third, with 13, or 9.6%, and communication/coordination combined represented 11.7% of the total. These findings must be interpreted carefully, however, given the low number of actual transactions. The number of reported transactions is

Table 59. Frequency Distribution: Disaster Response System by Funding Source and Jurisdiction, 1995 Hanshin, Japan Earthquake

Public										Nonprofit		Private		Total Public, Private, NPR	
Int'l		Nat'l		Prefectural		Municipal		Total							
N	%	N	%	N	%	N	%	N	%	N	%	N	%	N	%
30	7.7	38	9.7	26	6.6	71	18.2	165	42.2	87	22.3	139	35.5	391	100.0

Table 60. Frequency Distribution: Types of Transactions in Disaster Response by Funding Source and

Type of Transaction	Public Organizations												
	International			National			Regional/State			Municipal			
	T	N	%	T	N	%	T	N	%	T	N	%
Emergency Response	1	2	0.7	11	11	8.1	4	4	2.9	4	8	2.9
Communication	1	1	0.7	1	1	0.7	0	0	0.0	0	0	0.0
Coordination of Response	1	3	0.7	4	6	2.9	3	3	2.2	2	2	1.5
Medical Care/Health	0	0	0.0	1	1	0.7	0	0	0.0	6	8	4.4
Damage/Needs Assessment	0	0	0.0	2	3	1.5	0	0	0.0	2	2	1.5
Certification of Deaths	0	0	0.0	0	0	0.0	0	0	0.0	0	0	0.0
Earthquake Assessment/Research	0	0	0.0	0	0	0.0	0	0	0.0	0	0	0.0
Security Prevention of Looting	0	0	0.0	0	0	0.0	0	0	0.0	0	0	0.0
Housing Issues	0	0	0.0	0	0	0.0	0	0	0.0	0	0	0.0
Disaster Relief (food, shelter, etc.)	2	2	1.5	11	15	8.1	6	7	4.4	8	16	5.9
Donations (money, goods, etc.)	0	0	0.0	0	0	0.0	0	0	0.0	0	0	0.0
Building Inspection	0	0	0.0	0	0	0.0	0	0	0.0	0	0	0.0
Building Code Issues	0	0	0.0	0	0	0.0	0	0	0.0	0	0	0.0
Repair of Freeways, Bridges, Roads	0	0	0.0	0	0	0.0	0	0	0.0	0	0	0.0
Repair/Restore Utilities	0	0	0.0	0	0	0.0	0	0	0.0	0	0	0.0
Repair/Reconstruction/Recovery[a]	0	0	0.0	2	4	1.5	1	1	0.7	3	3	2.2
Transportation/Traffic Issues	0	0	0.0	0	0	0.0	0	0	0.0	0	0	0.0
Hazardous Materials Releases	0	0	0.0	0	0	0.0	0	0	0.0	0	0	0.0
Legal/Enforcement/Fraud	0	0	0.0	0	0	0.0	0	0	0.0	0	0	0.0
Political Dialogue/Legislation	0	0	0.0	1	1	0.7	1	1	0.7	0	0	0.0
Business Recovery	0	0	0.0	2	2	1.5	0	0	0.0	1	1	0.7
Economic/Business Issues	0	0	0.0	1	2	0.7	0	0	0.0	0	0	0.0
Visits by Officials	0	0	0.0	0	0	0.0	0	0	0.0	0	0	0.0
Education Issues	0	0	0.0	0	0	0.0	0	0	0.0	0	0	0.0
Government Assistance	0	0	0.0	1	2	0.7	0	0	0.0	0	0	0.0
Insurance Related Issues	0	0	0.0	0	0	0.0	0	0	0.0	0	0	0.0
Loans (Private and International)	0	0	0.0	2	3	1.5	0	0	0.0	0	0	0.0
Psychological/Counseling Services	1	1	0.7	0	0	0.0	0	0	0.0	0	0	0.0
Fundraising/Account Setup	3	3	2.2	0	0	0.0	0	0	0.0	0	0	0.0
Volunteers	0	0	0.0	0	0	0.0	0	0	0.0	0	0	0.0
TOTAL	9	12	6.6	39	51	28.7	15	16	11.0	26	40	19.1

T = Number of Transactions; N = Number of Actors; % = Percent of Total Transactions
[a]Not including freeways, bridges, roads or utilities
Note: One international transaction (donation), by one actor not included in table
Source: *The Japan Times*

limited for the size of the identified response system, which is nearly triple the number of organizations actively participating in disaster operations.

Table 61 presents the matrix of 53 reported interactions performed by 120 organizations involved in disaster operations. This low number of interactions indicates a response system in which participating organizations were operating largely independently, with relatively little shared responsibility for interdependent tasks. Much of the content of these reported interactions involved contributions by municipalities outside of Kobe to assistance in the delivery of inter-

Jurisdiction, Hanshin, Japan Earthquake, January 18 – February 6, 1995

Nonprofit Organizations							Private Organizations							TOTALS		
International			National			International			National					
T	N	%	T	N	%	T	N	%	T	N	%	T	N	%
0	0	0.0	0	0	0.0	0	0	0.0	0	0	0.0	20	25	14.7
1	1	0.7	1	3	0.7	0	0	0.0	0	0	0.0	4	6	2.9
0	0	0.0	2	2	1.5	0	0	0.0	0	0	0.0	12	16	8.8
1	1	0.7	3	5	2.2	0	0	0.0	2	4	1.5	13	19	9.6
0	0	0.0	0	0	0.0	0	0	0.0	2	3	1.5	6	8	4.4
0	0	0.0	0	0	0.0	0	0	0.0	0	0	0.0	0	0	0.0
0	0	0.0	0	0	0.0	0	0	0.0	0	0	0.0	0	0	0.0
0	0	0.0	0	0	0.0	0	0	0.0	0	0	0.0	0	0	0.0
0	0	0.0	0	0	0.0	0	0	0.0	0	0	0.0	0	0	0.0
0	0	0.0	3	7	2.2	0	0	0.0	5	6	3.7	35	53	25.7
0	0	0.0	0	0	0.0	1	3	0.7	0	0	0.0	1	3	0.7
0	0	0.0	0	0	0.0	0	0	0.0	0	0	0.0	0	0	0.0
0	0	0.0	0	0	0.0	0	0	0.0	0	0	0.0	0	0	0.0
0	0	0.0	0	0	0.0	0	0	0.0	0	0	0.0	0	0	0.0
0	0	0.0	1	1	0.7	0	0	0.0	1	1	0.7	8	10	5.9
0	0	0.0	0	0	0.0	0	0	0.0	0	0	0.0	0	0	0.0
0	0	0.0	0	0	0.0	0	0	0.0	0	0	0.0	0	0	0.0
0	0	0.0	0	0	0.0	0	0	0.0	0	0	0.0	0	0	0.0
0	0	0.0	0	0	0.0	0	0	0.0	0	0	0.0	2	2	1.5
0	0	0.0	0	0	0.0	0	0	0.0	0	0	0.0	3	3	2.2
0	0	0.0	0	0	0.0	0	0	0.0	0	0	0.0	1	2	0.7
0	0	0.0	0	0	0.0	0	0	0.0	0	0	0.0	0	0	0.0
0	0	0.0	0	0	0.0	0	0	0.0	0	0	0.0	0	0	0.0
0	0	0.0	0	0	0.0	0	0	0.0	0	0	0.0	1	2	0.7
0	0	0.0	0	0	0.0	0	0	0.0	0	0	0.0	0	0	0.0
0	0	0.0	0	0	0.0	0	0	0.0	1	1	0.7	3	4	2.2
0	0	0.0	2	4	1.5	0	0	0.0	0	0	0.0	3	5	2.2
2	2	1.5	7	8	5.1	6	17	4.4	6	9	4.4	24	39	17.6
0	0	0.0	0	0	0.0	0	0	0.0	0	0	0.0	0	0	0.0
4	4	2.9	19	30	14	7	20	5.2	17	24	12.5	136	197	100.0

Table 61. Frequency Distribution: Types of Interactions in Disaster Response by Funding Source

Type of Interaction	Public Organizations											
	International			National			Prefectural			Municipal		
	K	N	%	K	N	%	K	N	%	K	N	%
Public: International	0	0	0.0	4	5	7.5	1	2	1.9	0	0	0.0
Public: National				4	6	7.5	5	5	9.4	11	29	20.8
Public: Prefectural							0	0	0.0	2	4	3.8
Public: Municipal										1	2	1.9
Nonprofit: International												
Nonprofit: National												
Private: International												
Private: National												
TOTAL	0	0	0.0	8	11	15.1	6	7	11.3	14	35	26.4

K = Number of Interactions; N = Number of Actors; % = Percent of Total Interactions
Source: *The Japan Times*

rupted services such as water distribution. Of the set of 53 interactions, the largest proportion were performed by national organizations, 24, or 45.3%. International and municipal organizations each were identified in 6 interactions, or 11.3%, respectively. National nonprofit organizations accounted for 8, or 15.1% of the reported interactions. Again, the low number of reported interactions may reflect lack of access or awareness by the journalists who are observing disaster operations, but it is a striking comparison to the number reported for other disaster response systems.

These findings indicate a large response system that is very loosely interconnected, with many organizations performing single functions rather than a set of organizations integrating their efforts to accomplish multiple functions directed toward the same goal.

Adaptive Behavior and/or Self-Organization

Self-organization did occur, but later and more sporadically in the response period. Instances of innovative behavior characterized the response, for example, in the efforts made by firemen as they sought to halt the destructive force of fire. Without electrical power in damaged wards of Kobe, there was no water pressure in the mains. Fire companies connected long hoses and ran them for several kilometers to pump water from Osaka Bay to suppress fires in the most severely affected wards. After this destructive event, the Kobe Fire Department, working in conjunction with a computer scientist at a local university, has modeled the spread of the fire to study the dynamic conditions of its rapid escalation in order to mitigate risk of fire in future earthquakes.[10] The challenge is to build upon

Chapter 8: Auto-Adaptive Systems

and Jurisdiction, Hanshin, Japan Earthquake, January 18 – February 6, 1995

Nonprofit Organizations						Private Organizations						TOTAL		
International			National			International			National					
K	N	%	K	N	%	K	N	%	K	N	%	K	N	%
0	0	0.0	1	2	1.9	0	0	0.0	0	0	0.0	6	9	11.3
0	0	0.0	0	0	0.0	0	0	0.0	4	12	7.5	24	52	45.3
0	0	0.0	0	0	0.0	0	0	0.0	2	4	3.8	4	8	7.5
0	0	0.0	3	4	5.7	0	0	0.0	2	6	3.8	6	12	11.3
0	0	0.0	2	4	3.8	1	3	1.9	2	9	3.8	5	16	9.5
			4	10	7.5	2	9	3.8	2	4	3.8	8	23	15.1
						0	0	0.0	0	0	0.0	0	0	0.0
									0	0	0.0	0	0	0.0
0	0	0.0	10	20	18.9	3	12	5.7	12	35	22.6	53	120	100.0

this spontaneous base of interest and experience to foster a continuing exchange of information, knowledge, and skills in the mitigation of risk in Japan and other nations.

The nonlinear analysis of change in performance of basic emergency functions documents the actual rates of change among the subsets of public, private and nonprofit organizations. Figure 48 presents the phase plane plot for change in emergency response by change in communication/ coordination. Figure 49 presents the marginal history for that change over the twenty-day period immediately following the earthquake. Reports of emergency response and communica-

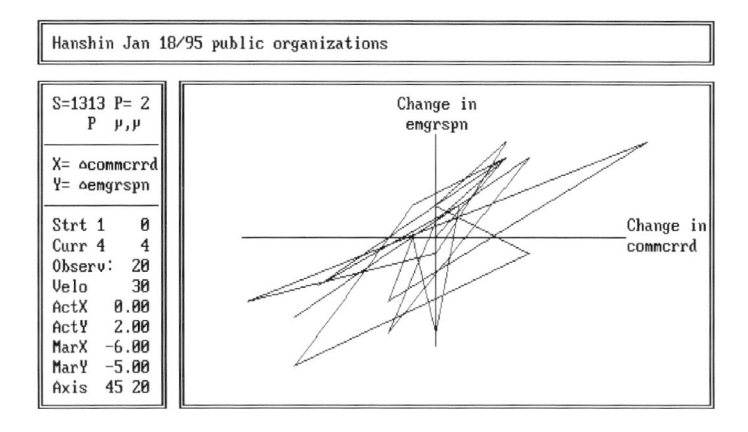

Figure 48. 1995 Hanshin, Japan Earthquake: Phase Plane Plot, Public Organizations, Emergency Response by Communication/Coordination

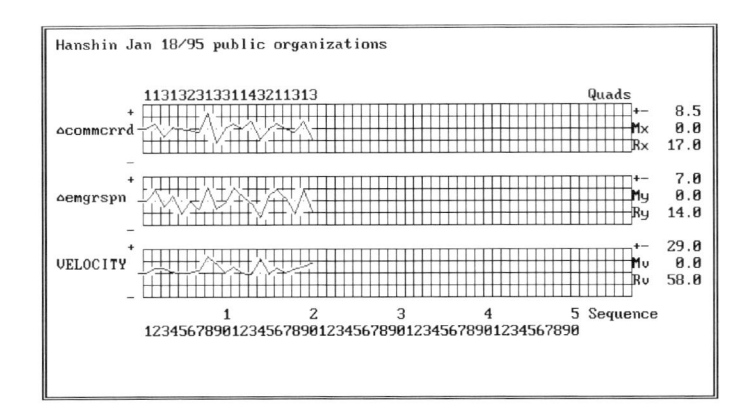

```
Hanshin Jan 18/95 public organizations
                                                            Quads
          11313231331143211313
                                                        +−    8.5
  ΔcommcrrdMx    0.0
                                                        Rx   17.0

                                                        +−    7.0
  ΔemgrspnMy    0.0
                                                        Ry   14.0

                                                        +−   29.0
  VELOCITYMv    0.0
                                                        Rv   58.0

            1         2         3         4        5 Sequence
          12345678901234567890123456789012345678901234567890
```

Figure 49. 1995 Hanshin, Japan Earthquake: Marginal History, Public Organizations, Emergency Response by Communication/Coordination

tion/coordination activities appear to increase and decrease with some regularity, with reports of emergency response activities showing a steep increase on Day 7, leading a sharp peak in reports of communication/coordination on Day 8, late in the response period. This decline in communication/coordination is followed by an increase in reported emergency response activities again on Day 11, as the two functions settle into a more regular pattern. Interestingly, reported activities in emergency response appear to lead reports on communication/coordination, rather than the reverse.

This may reflect the substantial protest expressed in news reports regarding the lateness of the actions taken by government agencies in emergency response and the lack of coordination among the levels of government in delivery of services.[11]

Table 62 summarizes the nonlinear logistic regression analysis for the Hanshin response system. The findings indicate that specific disaster response functions were operating, but that virtually the entire system was operating in high chaos, with k values above 3.66, as reported in the content analysis of news reports. Only two of the six disaster response functions were operating on a level of stability below chaos, with damage assessment showing a fair degree of disorder at k = 2.1 and financial assistance relatively stable at k = 1.4. Financial assistance was clearly the strongest, and most stable, of the six response functions. No equation was found for the response function of Recovery/reconstruction, which likely meant that these activities were not fully underway by the end of the response period, or the first 21 days following the earthquake. The subsystem of public organizations reported an R^2 of 0.33, which is statistically significant at the 0.01 level, but explains only one-third of the variance in the system's operations. A logistic regression equation was found for the whole system, including public, private, and nonprofit organizations, but the F value was not statistically significant. As shown, both the subsystem of public organizations and the total

Table 62. Summary, Logistic Regression Analysis of Disaster Response Functions, Hanshin Earthquake, January 18 – February 6, 1995

Response Function	R^2	F	p	k
Emergency Response	0.23	5.7	0.0	*3.95*
Damage Assessment	0.55	23.6	0.001	2.09
Communication/Coordination	0.56	24.0	0.001	*3.95*
Disaster Relief	0.01	0.2	NS	*3.95*
Recovery/reconstruction	—	—	—	—
Financial Assistance	0.53	21.2	0.001	1.42
Subsystem, Public Organizations	0.33	9.3	0.01	*3.95*
System: Public, Private, Nonprofit	0.11	2.3	NS	*3.95*

Italics indicate k values that fall in the chaotic range, above 3.66
— indicates no equation was found

response system of public, private and nonprofit organizations were registering high chaos, with k values = 3.95. These findings indicate a chaotic, unstable response system, but nonetheless, a response system in which four primary disaster response functions revealed statistically significant R^2 values.

Reviewing the outcomes of the two response systems, several striking differences emerge. Table 63 presents a comparison of the two systems by funding source and jurisdictional level. First, the two response systems involved approximately the same number of organizations, 378 for Northridge in comparison to 391 for Hanshin. Second, the distribution of organizations among types of organizations was fairly similar in terms of jurisdictional level and funding source. Notable exceptions were the larger number of nonprofit organizations participating in the Northridge response, and the larger number of private organizations participating in the Hanshin response. International organizations played a larger role in the Hanshin response system, 7.6% of the organizations participating, as opposed to 0.3% for Northridge.

Given the general similarity in size and distribution of the two response systems, the most striking finding is the difference in the number of transactions and interactions reported for the two response systems. While the content analy-

Table 63. Summary, Frequency Distributions, Auto-Adaptive Systems, by Funding Source and Jurisdiction

	Public								Total Public		Nonprofit		Private		Total	
	Int'l		Nat'l		State		Local									
	N	%	N	%	N	%	N	%	N	%	N	%	N	%	N	%
Northridge	1	0.3	30	7.9	36	9.5	76	20.1	145	38.4	105	27.8	128	33.9	378	100.0
Hanshin	30	7.6	38	9.7	26	6.6	71	18.2	165	42.2	87	22.3	139	35.5	391	100.0

sis for the Northridge system reported 971 transactions performed by organizations in disaster response, the content analysis for the Hanshin system reported 130 transactions. The difference in the number of interactions for the two systems is also large. The number of interactions reported for the Northridge system was 218, in contrast to 53 interactions for the Hanshin response system. The Northridge response system, more densely interconnected than the Hanshin system, engaged many organizations in timely, collaborative actions that moved the disaster operations more quickly into recovery, reducing both the time spent by disaster victims in traumatic conditions and the cost in lives and damages. These measures reveal differences in coordination and communication among the two response systems that resulted in strikingly different consequences for their respective communities.

Interpretation of Findings

The reported findings lead to the following interpretation:
Northridge:
- Jurisdictions acted simultaneously in mobilizing response to the event; local agencies responded immediately, but state and federal jurisdictions also sent in advance teams to provide support to local agencies and to assess the requirements for state and federal assistance
- Major investment in communications and information technology by the federal government created a disaster information infrastructure that enabled different agencies and jurisdictions to communicate more easily with one another, enabling the response system to function in a coherent, adaptive manner
- While this information infrastructure served important functions of interorganizational coordination, many of the principal managers in the response system saw the system as a hurried response to an urgent need, and believed they could have functioned at a much more professional level, had the technology been in place prior to the event. This was especially true of the GIS capacity which was hastily put in place, but nearly too late to provide substantive decision support to practicing managers.
- Information technology introduced into organizations that already had high levels of training and experience in disaster operations maximized the performance of these organizations. Introduced into organizations without familiarity with the technology, e.g. GIS, the effort was less effective. Therefore, the initial organizational and environmental conditions governed the outcome of the investment in information technology. Addressing those conditions prior to disaster increases the effectiveness of system performance during actual disaster operations.

Hanshin:
- Jurisdictions followed their existing emergency plans, which required lower jurisdictions to request assistance from higher jurisdictions before aid would

be sent. In the first urgent, critical hours of disaster, local jurisdictions were overwhelmed and could not communicate with the higher jurisdictions. Consequently, external assistance did not arrive to any substantial degree until eighteen hours after the event, and much needed aid in medical care, food, water and shelter began to arrive only four days after the event.
- Although the Hanshin area had a sound basic information infrastructure, its application in emergency management was not fully realized. Information technology for local governments was either not sufficiently developed or inoperable.
- Practicing disaster managers, nearly unanimously, indicated a lack of adequate communication not only within their own organizations, but especially between jurisdictions with emergency reponsibilities.
- Only the private organizations in the Hanshin region appeared to have adequate, functioning communications systems. The public organizations revealed a serious under-investment in information technology to support interorganizational decision making.
- The obvious lack of coordination in disaster response appeared largely due to lack of communication. When communication was re-established among jurisdictions and the full profile of the disaster consequences was communicated to responsible agencies, the nation responded to the needs of the affected area with discipline and compassion.

Noting the differences and similarities between the Northridge and Hanshin disaster response systems, this analysis confirms important characteristics of a nonlinear model of disaster response. First, disaster creates a "symmetry-shattering event" (Kiel 1994) that both disrupts established patterns of thought and action and creates the opportunity to redesign an emergency response system that 'fits' the environment more effectively. Second, the critical function of aggregating units from different levels of an intergovernmental disaster response system easily into a wider system of response underscores both the difficulty of this task under linear models of organization and the interdependence of these units in a massive, large-scale disaster. This function and its capacity to mobilize resources — personnel, equipment, supplies, skills, and knowledge — requires a mechanism of information exchange to achieve a shared system-wide goal: protection of life and property. This function appears to be performed more effectively in rapidly changing disaster environments by a nonlinear, dynamic system that is able to coordinate diverse resources, materials, and personnel across previously established organizational and jurisdictional boundaries. It is essentially an organizational system operating in parallel, supported by a strong, distributed information system. Third, the goal of the disaster response system serves as an 'internal model' for self organizing processes. This goal allows participants from diverse perspectives, experience, and resources to adjust their actions and contributions to that of other participants in the system. Finally, an 'epistemic community' (Haas 1990) of knowledgeable people from diverse backgrounds, experience, and organizations that focuses on the shared problem and articulates a common goal is vital to formulating strategies of risk reduction and collective response which can be communicated to a wider set of responsible actors.

Creating a community of knowledgeable people is essential to the development of 'resonance' or willingness to support shared action when necessary to sustain the goal of a responsible, civil society. Such a community was clearly evident in the Northridge response system, forged from the crucible of previous disasters. There are some indications that the formation of such a self-organizing community of knowledgeable people is already taking place in Japan, for example, through the Pan Pacific Forum in Kobe that is meeting regularly to form policy recommendations for the reconstruction process. The group includes about 40 responsible community leaders from education, business, medical care, publishing, and voluntary organizations.[12]

Conclusion

Several conclusions can be drawn from the analysis of the Northridge and Hanshin response systems. The two systems were essentially the same size, and evolved from metropolitan regions of similar economic and technical development and seismic risk. Yet, the two systems achieved very different results under the impact of severe earthquakes. The Northridge response system proved essentially stable, with some chaotic functions, while the Hanshin response system proved essentially chaotic, with some stable functions. The difference in results depends upon the available capacity for interorganizational communication and coordination, which in turn depends upon ready access to information technology and a current, viable community knowledge base regarding seismic risk.

NOTES

1. Earthquake Engineering Research Institute. 1994. *Northridge Earthquake, January 17, 1994, Preliminary Reconnaissance Report*, Chapter 9, Social Impacts and Emergency Response: 86–89.
2. Emergency plans for both the City and County of Los Angeles activate immediately in event of an earthquake of 5.0 Richter scale or more. All emergency service personnel report immediately to their stations, or to the nearest station if conditions do not permit travel to their assigned duty base. Los Angeles County Emergency Operations Plan. Los Angeles County Emergency Operations Center Briefing, Los Angeles, CA, February 4, 1994.
3. Detailed accounts of damage resulting from the earthquake and the numbers of individuals, households, and businesses affected are presented in a number of sources. These include the daily coverage of the event in the *Los Angeles Times*, the situation reports prepared by the Federal Emergency Management Agency and the California Office of Emergency Services, the transcript of the California Seismic Safety Commission's hearings on the response to the earthquake, and reports of professional organizations such as the Earthquake Engineering Research Institute, the Earthquake Engineering Research Center of the University of California, Berkeley, and EQE International, an engineering firm with offices in San Francisco and Irvine, California.
4. Schiffrin. L. Manager, Emergency Support Function #2: Communications. Interview, Disaster Field Office, Pasadena, CA, February 2, 1994.

5. Chief, Communications, Los Angeles County Fire Department. Interview, Los Angeles, February 2, 1994.
6. Chief, Finance, Northridge Earthquake Disaster Operations, FEMA. Interview, Federal/State Disaster Field Office, Pasadena, CA, April 9, 1994.
7. Federal Emergency Management Agency. "Situation Reports, Northridge Earthquake, January 17, 1994". Pasadena, CA: Federal/State Disaster Field Office, January 18 — February 6, 1994.
8. Federal Emergency Management Agency. "Situation Reports, Northridge Earthquake, January 17, 1994". Pasadena, CA: January 18 — February 6, 1994.
9. Summary of Reports from Ministries regarding Status of Hanshin-Awaji Disaster Operations, National Land Agency, Tokyo, Japan, April 23, 1995.
10. Captain, Kobe Fire Department. Interview and demonstration, May 22, 1996, Kobe, Japan.
11. See *Japan Times*, January 18 — February 6, 1995.
12. Dean, Graduate School of International Cooperation Studies, Kobe University. Interview, Kobe, Japan, May 14, 1995. Other fora also serve this capacity, such as the Pan Pacific Conference held by Canada as its contribution to the United Nations International Decade for Natural Hazard Reduction in Vancouver, Canada in July, 1996.

Part III
Future Strategies:
Managing Risk in Complex, Adaptive Systems

CHAPTER NINE

ADAPTATION TO DISASTER: EVOLVING PATTERNS OF RESPONSE

Response Systems in Practice: A Comparison of Types of Adaptation to Disaster

The analysis of field studies presented in the preceding chapters documents four types of adaptation to the sudden, shattering strike of an earthquake — nonadaptive, emergent adaptive, operative adaptive and auto adaptive systems. Each type of adaptation represents a response to some combination of failure in existing technical as well as organizational systems, which produced varying patterns of action, inaction, innovation, and determination to restore community functions. These types reflect different states of a community's capacity to adapt to sudden change, reallocate its resources and energies effectively in response to major threat, and manage different degrees of dependence upon external resources for recovery without losing its basic capacity for performance. In each state of adaptation, the community's functional operating system is reconfigured, but at widely varying costs in lives, property, time, and resources.

In Chapter Four, I outlined a preliminary model of inter-organizational transition in dynamic environments, informed by the literature on organizational change. In this chapter, I return to that model and compare it to findings from the four subsets of disaster response systems observed in practice. The question is whether actual practice, documented by the field studies, confirms or modifies the theoretical explanation of an evolving disaster response system.

The model identified five stages in the evolution of a response system following disaster: initial conditions, information search, information exchange, intra- and inter-organizational learning, and adaptive behavior. The first stage characterized the initial conditions existing in the community prior to the earthquake. These conditions included not only the physical characteristics of the community — its location, geographic, geologic, and engineered features — but also its social, economic and political conditions which contribute to its level of awareness of risk and organizational capacity for informed, collective action to reduce that risk. The earthquake clarifies the immediate goal for the community, and activates a set of information processes — information search and information exchange — that serve as a basis for collective action. These interactive processes lead to intra-organizational learning within separate agencies, which contributes to inter-organizational learning among the set of organizations engaged in response operations. Since all organizations in the community share the risk of disaster, all are responsible for taking appropriate action to reduce that risk.

The three dynamic phases — information search, information exchange, and intra/inter-organizational learning — culminate in adaptive behavior, which represents an accommodation to the disaster event and achieves some level of coordinated action that can be distinguished as a response system. The degree of coherence that is achieved by the response system reflects the extent to which the community has been able to shift its performance from routine tasks to meet urgent demands from the impact of the earthquake. This coherence is measured by the number and type of transactions performed by organizations engaged in response operations, the number and density of interactions among the participating organizations, and the amount of variation explained in the six basic functions of disaster response by the nonlinear logistic regression analysis for each system.

The eleven field studies present cumulative evidence of the evolution of disaster response systems in practice. Table 64 summarizes the distribution of organizations participating in the response systems included in this analysis. The systems range in size from 171 organizations identified in Erzincan, Turkey to 623 organizations identified in Loma Prieta, California, and vary in their distribution among public, private and nonprofit sources of funding.

The total number of organizations identified for the response systems roughly reflects the magnitude of the earthquakes or the extent of area affected. The two exceptions are San Salvador, with a relatively large response system of 375 organizations for an earthquake of M = 5.4, and Northern Armenia, with a small response system of 198 organizations for an earthquake of M = 6.9. In both cases, the size of the response system likely reflected the political situations at the time. San Salvador was the capital city of a small nation torn by civil strife, echoing the tensions of the Cold War. Armenia was a small republic within the larger Soviet Union, with all response functions legally performed by public organizations.

More striking is the distribution of organizations participating in response among public organizations. In six of the eleven response systems, international organizations played a substantive role, ranging from 16.2% of total public organizations participating in response operations in San Salvador, to nearly one-third, or 31%, in Erzincan, Turkey. International participation was important in two other systems, Marathwada, India, with 9.2%, and Hanshin, Japan, with 7.6%. The near absence of international participation in the three California earthquakes is striking. In the Loma Prieta response system, 26 international

Chapter 9: Adaptation to Disaster 233

Table 64. Summary, Frequency Distributions, Organizational Disaster Response Systems by Type, Funding Source and Jurisdiction

	Public Organizations							Total Public		Nonprofit Organizations						Private Organizations				Total	
	Internat'l		National		State		Municipal				Internat'l		National		Internat'l		National			
	N	%	N	%	N	%	N	%	N	%	N	%	N	%	N	%	N	%	N	%
Nonadaptive																				
San Salvador	61	16.3	96	25.7	0	0.0	48	12.8	206	54.9	23	6.1	80	21.4	12	3.2	54	14.4	374	100.0
Ecuador	54	22.4	79	32.8	6	2.5	15	6.2	154	63.9	12	5.0	59	24.5	6	2.5	10	4.1	241	100.0
Armenia	33	16.6	39	19.7	40	20.2	15	7.6	127	64.1	34	17.1	24	12.1	13	6.7	0	0.0	198	100.0
Emergent Adaptive																				
Mexico City	90	18.0	117	23.4	38	7.6	19	3.8	264	52.8	59	11.8	100	20.0	28	5.6	49	9.8	500	100.0
Costa Rica	52	21.2	74	30.2	6	2.4	10	4.1	142	57.9	15	6.1	35	14.2	16	6.5	37	15.1	245	100.0
Erzincan	53	31.0	40	23.4	13	7.6	28	16.4	134	78.4	25	14.6	4	2.3	1	0.5	7	4.0	171	100.0
Operative Adaptive																				
Whittier Narrows	0	0.0	9	4.7	17	9.0	98	51.9	118	62.4	0	0.0	27	14.8	0	0.0	38	20.1	189	100.0
Loma Prieta	26	4.2	36	5.8	39	6.3	160	25.6	261	41.9	0	0.0	98	15.7	0	0.0	264	42.4	623	100.0
Marathwada	34	9.2	55	14.8	44	11.9	17	4.6	150	40.4	30	8.1	117	31.5	16	4.3	58	15.6	371	100.0
Auto Adaptive																				
Northridge	1	0.3	30	7.9	36	9.5	78	20.6	145	38.4	0	0.0	105	27.8	0	0.0	128	33.9	378	100.0
Hanshin	31	7.9	38	9.7	27	6.9	71	18.2	165	42.2	16	4.1	71	18.1	26	6.6	87	22.3	391	100.0

organizations — largely international businesses with branch offices in San Francisco — contributed donations to earthquake recovery, out of a total of 623 organizations participating in response operations. One international participant — the City of Berlin — out of 378 organizations made a contribution to the Northridge response system, and no international participants were identified in the Whittier Narrows response system. The international community bears the highest burden in providing disaster assistance for those systems that show not only the greatest humanitarian need, but also the lowest capacity for adaptation.

The California cases also illustrate the importance of first response at the local level. As shown in Table 64, participation of public organizations from the municipal level is low in the Nonadaptive and Emergent Adaptive cases. National public organizations clearly are the most dominant in the respective systems. This condition shifts markedly in the Operative Adaptive cases, as public municipal organizations compose over half of the organizations participating in response operations for the Whittier Narrows system, and one-fourth of the operations in the Loma Prieta response system. In the Marathwada system, national organizations are important, but state organizations play a substantial role, with municipal organizations playing a lesser role in this developing country. Although there is participation from international public organizations, they play a significantly smaller role than in the Nonadaptive and Emergent Adaptive cases. These findings demonstrate a major tenet of complex, adaptive systems, that adaptive behavior is governed by the local level. There is also a noticeable increase in the participation of nonprofit and private organizations in both the Operative Adaptive and Auto Adaptive cases. The response systems in these cases are engaging the organizations from the wider society more fully in disaster operations.

Since the general context of each community prior to the earthquake was discussed in Chapter 4, I will focus in this section on the type, kind and quality of the information infrastructure that was available to disaster managers prior to the earthquake, and the ways in which this infrastructure affected the subsequent development of the response systems. The evolution of the response systems marks a transition from a state of routine, single-agency operations to a state of collective operations engaging many organizations from the community and wider society.

Initial Conditions

In the preliminary model, four conditions were identified as necessary, but not sufficient, to initiate inter-organizational transition. They are as follows:
1. Articulation of commonly understood meanings among participating organizations and key audiences in the wider environment
2. Sufficient trust among leaders, organizations, and citizens to enable them to accept a shared goal
3. Sufficient resonance between the organizations seeking change and their potential sources of support for action

4. Sufficient capacity and resources to sustain collective action among participating organizations in order to achieve a shared goal

In practice, the four conditions are interdependent. That is, the absence of, or a weakness in, any one condition inhibits interactions among the system's organizational components, thus weakening its ability to share resources collaboratively or to coordinate action on multiple levels of responsibility. The critical condition for adaptive performance is the fourth, which in practice relies upon a technical and organizational information structure. The evolving response system's effectiveness increases with its ability to hold and exchange information to support rapid mobilization of complex disaster operations. To the extent that this capacity is not present, not developed, or not utilized, the response system is limited in its ability to meet the needs of the community under duress.

Evidence from the response systems included in the nonadaptive and emergent systems subgroups supports this requirement of information to support adaptive behavior. In each of the cases of nonadaptive response — San Salvador, Ecuador and Armenia — the information infrastructure was not sufficiently developed to hold and exchange information among the levels of government and types of organizations participating in the system. Further, the political situations in each of these countries at the time of the respective earthquakes were characterized by wary relationships between government and people, verging on mistrust.

In the set of emergent adaptive systems — Mexico City, Costa Rica, and Erzincan — elements of a sociotechnical information infrastructure were present, but operated either inconsistently or without sufficient connection to all components to support the sustained operation of a response system. In Mexico City, news of the earthquake broadcast globally through satellite communications elicited an outpouring of disaster assistance, but the federal government had little capacity to transfer this assistance to the neighborhood level. In Costa Rica, the breakdown of communications between the central government's agencies in San Jose and the provincial city of Limon, coupled with the near absence of communications with the outlying villages damaged by the earthquake, seriously inhibited timely, effective response operations to the outlying areas. In Erzincan, Turkey, a sophisticated damage assessment model run by the national government in Ankara proved insufficient to guide disaster operations without the active participation of local responders. Without continuing support for inter-organizational decision making, bold initiatives taken in each nation eventually dissipated into routine performance.

In the subset of Operative Adaptive response systems — Whittier Narrows, California, Loma Prieta, California, and Maharashtra, India — information infrastructures did exist. However, they were not well linked to the organizational systems of disaster response, and therefore, supported some, but not all, functions of disaster operations. In the subset of potentially Auto Adaptive systems — Northridge, California and Hanshin, Japan — information systems were critical to both the successes and failures in each response system, with timing making the crucial difference between order and chaos in response operations.

Reviewing the set of field studies, I found conditions and characteristics that both facilitate and hinder the evolution of response systems in practice. While the systems evolved in very different ways, there were critical points in the development of each response system when choices made in practice moved the system either toward chaos or order. The choices reflected the quality of information available to responsible decision-makers at the time, but the ensuing consequences for the respective communities were strikingly different. The differences in outcomes among the set of communities reflect, in substantive ways, the accuracy, timeliness, and comprehensiveness of that information and the interactions generated in response to it under the stressful, urgent environment of disaster. They also reflect the capacity of the system to absorb new information and integrate it successfully to support many kinds of action at different levels of responsibility within the system. These differences are shown in a comparison of functions in the dynamic phases of the model, presented in turn.

Information Search

The initial stage of disaster response involves information search and the level within the evolving organizational system at which this function occurs. Information search includes those processes which an organization undertakes to determine the nature of the event, its impact upon the community and its infrastructure, the resources currently available for response, and the likely consequences for further damage or disruption. This information serves as the basis upon which the organization mobilizes its response to the event and bases its request to other organizations for assistance, if necessary. Information search is carried out through the function of communication, which makes possible the function of coordination. Table 65 shows the proportion of reported transactions in communication and coordination for the set of eleven response systems.

The data show a somewhat mixed pattern in the proportion of total transactions that involve communication and coordination among the different types of response systems, but a general increase corresponding to the level of adaptation which the respective systems achieve. The two California response systems — Whittier Narrows and Loma Prieta — present striking exceptions. Both events affected multiple cities, each of which had its own emergency response plan. In both cases, emergency response operations were performed largely by local emergency response organizations following their own emergency procedures. These procedures are considered standard, and were not likely to attract the attention of the media.

News reports, therefore, may not have identified fully what I observed in the field, which were intensive efforts to communicate within each city, but relatively little attention given to establishing communications with other cities or other levels of jurisdiction.

The importance of communication and coordination for effective disaster response is clearly shown in the cases of Auto Adaptive systems. Northridge,

Chapter 9: Adaptation to Disaster 237

Table 65. Summary, Communication and Coordination Functions by Type of Response System, Jurisdiction and Funding Source

Type of Response System	Public Organizations									Nonprofit Organizations						Private Organizations				Total Function		Total System	
	International		National		Provincial		Municipal		International		National		International		National					
	T	%	T	%	T	%	T	%	T	%	T	%	T	%	T	%	T	%	T	%
Nonadaptive																					
San Salvador																					
Communication	2	0.3	5	0.7	0	0.0	1	0.1	0	0.0	4	0.6	0	0.0	1	0.1	13	1.9	685	100.0
Coordination	2	0.3	12	1.8	0	0.0	1	0.1	1	0.1	5	0.7	0	0.0	0	0.0	21	3.1	685	100.0
Ecuador																					
Communication	0	0.0	6	0.9	0	0.0	0	0.0	0	0.0	0	0.0	0	0.0	4	0.6	10	1.6	633	100.0
Coordination	3	0.5	15	2.4	0	0.0	1	0.2	0	0.0	3	0.5	0	0.0	0	0.0	22	3.5	633	100.0
Armenia																					
Communication	5	1.9	17	6.5	5	1.9	0	0.0	0	0.0	1	0.4	0	0.0	0	0.0	28	10.7	262	100.0
Coordination	0	0.0	17	6.5	3	1.1	0	0.0	0	0.0	0	0.0	0	0.0	0	0.0	20	7.6	262	100.0
Emergent Adaptive																					
Mexico City																					
Communication	3	0.5	18	2.8	7	1.1	2	0.3	0	0.0	11	1.7	1	0.2	1	0.2	43	6.7	641	100.0
Coordination	6	0.9	27	4.2	1	0.2	2	0.3	1	0.2	12	1.9	0	0.0	0	0.0	49	7.6	641	100.0
Costa Rica																					
Communication	2	0.5	31	8.2	0	0.0	1	0.3	3	0.8	5	1.3	0	0.0	0	0.0	42	11.1	378	100.0
Coordination	7	1.9	23	6.1	0	0.0	0	0.0	5	1.3	5	1.3	0	0.0	0	0.0	40	10.6	378	100.0
Erzincan, Turkey																					
Communication	1	0.3	2	0.6	1	0.3	1	0.3	0	0.0	0	0.0	0	0.0	1	0.3	6	1.7	348	100.0
Coordination	8	2.3	10	2.9	3	0.9	2	0.6	0	0.0	0	0.0	0	0.0	0	0.0	23	6.6	348	100.0
Operative Adaptive																					
Whittier Narrows, CA																					
Communication	0	0.0	3	0.5	4	0.7	10	1.8	0	0.0	2	0.4	0	0.0	11	2.0	30	5.3	562	100.0
Coordination	0	0.0	1	0.2	2	0.4	6	1.1	0	0.0	1	0.2	0	0.0	0	0.0	10	1.8	562	100.0
Loma Prieta, CA																					
Communication	0	0.0	0	0.0	1	0.1	8	0.6	0	0.0	6	0.4	0	0.0	11	0.8	26	1.8	1454	100.0
Coordination	0	0.0	3	0.2	0	0.0	5	0.3	0	0.0	0	0.0	0	0.0	1	0.1	9	0.6	1454	100.0
Marathwada, India																					
Communication	1	0.2	9	2.1	8	1.9	3	0.7	1	0.2	4	0.9	0	0.0	0	0.0	26	6.1	429	100.0
Coordination	2	0.5	9	2.1	9	2.1	1	0.2	1	0.2	3	0.7	0	0.0	1	0.2	26	6.1	429	100.0
Auto Adaptive																					
Northridge, CA																					
Communication	0	0.0	20	2.3	24	2.8	27	3.1	0	0.0	15	1.7	0	0.0	15	1.7	101	11.8	859	100.0
Coordination	0	0.0	31	3.6	6	0.7	23	2.7	0	0.0	11	1.3	0	0.0	11	1.3	82	9.5	859	100.0
Hanshin, Japan																					
Communication	1	0.7	1	0.7	0	0.0	0	0.0	1	0.7	1	0.7	0	0.0	0	0.0	4	2.9	136	100.0
Coordination	1	0.7	4	2.9	3	2.2	2	1.5	0	0.0	2	1.5	0	0.0	0	0.0	12	8.8	136	100.0

which achieved a high level of adaptation in response to the earthquake, registered a combined subset of 183 transactions in communication and coordination, or 21.3% of the total number of transactions identified through the content analysis. Communication, with 11.8% of the total transactions, created a base of common knowledge that could support coordination, which represented 9.5% of the total transactions. Only Costa Rica, with a set of transactions half the size, matched this percentage with 82, or 21.7% of the transactions involved in communication and coordination. The news reports in Costa Rica were colored by the dispute between the President and the Comision Nacional de Emergencia, which had legal responsibility for disaster operations, over timely response. A significant amount of the news coverage on the disaster reported the lack of communication and coordination among the relevant parties in disaster response, and actions taken to correct this condition. The broad categories of the content analysis pick up the negative as well as positive reports on communication and coordination, and may have reflected the general concern for timely communication and coordination, as well as actual steps taken to produce it. Data for Hanshin, Japan, show a relatively small number of transactions, 136, for a severe earthquake, but 11.7% of the total were involved in communication and coordination.

The type of information infrastructure available in a given community and its accessibility to groups participating in disaster response operations determines the extent and quality of information search processes performed by participating organizations. Each of the three response systems that fell in the "nonemergent" subset were characterized by limited or disrupted or controlled processes of information search. The Government of El Salvador, operating in the midst of civil war, had stringent controls on information search. Civilian organizations, supportive of disaster response, were nonetheless constrained in their activities due to the risk of escalating the hostilities.

Ecuador, experiencing tensions between its elected president and the military chiefs, had to focus attention on maintaining the stability of its government as well as to respond to the needs of the disaster-affected communities. Since the consequences from the earthquake were spread over three geographic zones, the communications between the outlying areas and the central government in Quito were both limited and damaged by the event. Communications under normal times are difficult in this stunningly beautiful, but still developing nation. Given the destruction to roads, telephone lines, and the limited availability of radio communications in this Andean nation, information search was delayed and incomplete at best. In Soviet Armenia, emergency communications were controlled by the Red Army in December, 1988, with little to no information search processes carried out with participation and exchange.

The type of information infrastructure available to support information search processes in disaster operations and the degree of accessibility to this infrastructure by local managers generally improved for each subsequent set of cases, with Northridge, an Auto Adaptive system, observed to have the most extensive access to information. The extent and timeliness of information search processes carried out within each response system, in turn, created the basis for information exchange.

Information Exchange

Information exchange is essential to marshaling coordinated action that is crucial to effective disaster response. It is also the point at which gaps in performance most frequently occur under the urgent stress of disaster. Tables 66–69 present a comparative profile of the distribution of interactions among the four sets of response systems. These interactions represent actions taken by two or more organizations to carry out tasks vital to the protection of life and property of the stricken communities. The level at which these interactions occur within the response system indicate the level of engagement and adaptiveness of the wider community to the threat posed by the earthquake.

In Table 66, Nonadaptive Systems, the largest number of interactions occur at the national level, with a substantial number of interactions occurring at the international level. For San Salvador, the combined set of interactions for the international and national levels of jurisdiction are 77.5%, more than three-fourths of the total interactions in the response system. For Ecuador, the proportion of combined interactions at the international and national levels is even higher, at 86.7%. For Armenia, nearly two-thirds, or 65.8% of the total interactions occurred at these levels of jurisdiction. All three response systems revealed some evidence of interactions occurring with nonprofit organizations, both international and national, but the private sectors showed minimal participation. In Soviet Armenia, there was participation only by the international private sector, as no formal private sector existed.

The pattern of national jurisdictions leading the participation of other organizations in disaster operations, with strong support from the international organizations, continued in the second set, Emergent Adaptive Systems, Table 67. Data for Mexico City show that 78.5% of the reported interactions involved national and international organizations. Data for Costa Rica show that 89.2% of the reported interactions involved national and international organizations, with nearly one-third, 31.3%, representing international organizations. Erzincan, Turkey showed a similar pattern, with 87.8% of the reported interactions involving national and international organizations. In Erzincan, one-third of the relatively small number of reported interactions involved international agencies.

The pattern of interactions shows a near reversal in the next set of cases, Operative Adaptive Systems, Table 68. In the Whittier Narrows response system, a relatively small system for a moderate earthquake, two-thirds, or 66.5%, of the interactions occurred at the local/municipal level, with an additional 17.7% of the interactions occurring at the state level. The overwhelming proportion of interactions, 84.2%, in this response system involved sub-national public agencies. In the California cases, multiple jurisdictions at the local level – county,

Table 66. Summary, Frequency Distribution: Types of Interactions in Disaster Response by Funding

Type of Interaction	Public Organizations											
	International			National[a]			Provincial[b]			Municipal		
	K	N	%	K	N	%	K	N	%	K	N	%
San Salvador												
Public: International	9	18	5.8	12	21	7.7	0	0	0.0	0	0	0.0
Public: National				38	53	24.4	0	0	0.0	12	37	7.7
Public: Provincial							0	0	0.0	0	0	0.0
Public: Municipal										2	4	1.3
Nonprofit: International												
Nonprofit: National												
Private: International												
Private: National												
TOTAL	9	18	5.8	50	74	32.1	0	0	0.0	14	41	9.0
Napo Province, Ecuador												
Public: International	2	5	1.3	31	46	20.5	0	0	0.0	0	0	0.0
Public: National				58	58	38.4	5	11	3.3	5	10	3.3
Public: Provincial							0	0	0.0	1	4	0.7
Public: Municipal										0	0	0.0
Nonprofit: International												
Nonprofit: National												
Private: International												
Private: National												
TOTAL	2	5	1.3	89	104	58.9	5	11	3.3	6	14	4.0
Northern Armenia												
Public: International	5	9	6.3	1	2	1.3	0	0	0.0	0	0	0.0
Public: National				16	18	20.2	16	15	20.2	0	0	0.0
Public: Provincial							9	9	11.4	3	6	3.8
Public: Municipal										2	4	2.5
Nonprofit: International												
Nonprofit: National												
Private: International												
Private: National												
TOTAL	5	9	6.3	17	20	21.5	25	24	31.6	5	10	6.3

K = Number of Interactions; N = Number of Actors; % = Percent of Total Interactions
Sources: *El Diario de Hoy*, San Salvador; *Hoy*, Quito, Ecuador; *Sovetakan Hayastan*, Yerevan, Armenia
[a]The term national is equivalent to union in Soviet Armenia; [b]The term provincial is equivalent to republic in

city and special districts — were combined in order to create a category for substate or sub-provincial government that was comparable to local government in other nations.

In the Loma Prieta Earthquake, a much larger and more complex system, the pattern shifts in a slightly different direction. More than one-third, or 35.1%, of the interactions occurred at the state and local level, but a sizeable number, 21.8%, involved nonprofit organizations and a small but noticeable proportion, 4.7%, involved private organizations. In the Marathwada, India Earthquake, the proportion of international and national agencies engaged in interactions for dis-

Chapter 9: Adaptation to Disaster 241

Source and Jurisdiction, Nonadaptive Systems

Nonprofit Organizations							Private Organizations							Total		
International			National			International			National					
K	N	%	K	N	%	K	N	%	K	N	%	K	N	%
2	4	1.3	5	13	3.2	0	0	0.0	2	4	1.3	30	60	19.2
6	12	3.8	28	43	17.9	0	0	0.0	7	16	4.5	91	161	58.3
0	0	0.0	0	0	0.0	0	0	0.0	0	0	0.0	0	0	0.0
0	0	0.0	5	18	3.2	0	0	0.0	0	0	0.0	7	22	4.5
1	2	0.6	3	8	1.9	1	3	0.6	0	0	0.0	5	13	3.2
			12	35	7.7	0	0	0.0	9	40	5.8	21	75	13.5
						1	2	0.6	1	2	0.6	2	4	1.3
									0	0	0.0	0	0	0.0
9	18	5.7	53	117	33.9	2	5	1.2	19	62	12.2	156	335	100.0
0	0	0.0	1	6	0.7	0	0	0.0	0	0	0.0	34	57	22.5
8	21	5.3	5	12	3.3	7	19	4.6	9	17	6.0	97	148	64.2
0	0	0.0	0	0	0.0	0	0	0.0	0	0	0.0	1	4	0.7
0	0	0.0	1	3	0.7	2	6	1.3	0	0	0.0	3	9	2.0
2	5	1.3	1	2	0.7	4	8	2.6	1	2	0.7	8	17	5.3
			4	7	2.6	1	2	0.7	0	0	0.0	5	9	3.3
						1	2	0.7	0	0	0.0	1	2	0.7
									2	5	1.3	2	5	1.3
10	26	6.6	12	30	8.0	15	37	9.9	12	24	8.0	151	251	100.0
7	14	8.8	1	2	1.3	1	2	1.3	0	0	0.0	15	29	19.0
2	3	2.5	3	6	3.8	0	0	0.0	0	0	0.0	37	42	46.8
0	0	0.0	3	8	3.8	0	0	0.0	0	0	0.0	15	43	19.0
0	0	0.0	0	0	0.0	0	0	0.0	0	0	0.0	2	4	2.5
1	2	1.3	3	6	3.8	1	2	1.3	0	0	0.0	5	10	6.3
			4	7	5.1	0	0	0.0	0	0	0.0	4	7	5.1
						1	3	1.3	0	0	0.0	1	3	1.3
									0	0	0.0	0	0	0.0
10	19	12.6	14	29	17.8	3	7	3.9	0	0	0.0	79	118	100.0

Soviet Armenia

aster response was substantial at 50%, but the combined state and municipal proportion was 22.2%. Nonprofit organizations, both international and national, engaged in disaster operations in the same proportion, 22.2%, of the total interactions. Clearly the burden of response for the direction and implementation of disaster operations was shifting closer to the community level in all three response systems.

In the Auto Adaptive set, Table 69, the Northridge and Hanshin response systems show different patterns of evolving response. Northridge demonstrated a dense pattern of interactions among organizations, with a total of 245, 47.3% of

Table 67. Summary, Frequency Distribution: Types of Interactions in Disaster Response by Funding

Type of Interaction	Public Organizations											
	International			National			Provincial			Municipal		
	K	N	%	K	N	%	K	N	%	K	N	%
Mexico City												
Public: International	4	7	1.8	13	26	5.9	2	4	0.9	0	0	0.0
Public: National				68	69	31.1	37	42	16.9	5	9	2.3
Public: Provincial							7	12	3.2	3	5	1.4
Public: Municipal										1	2	0.5
Nonprofit: International												
Nonprofit: National												
Private: International												
Private: National												
TOTAL	4	7	1.8	81	95	37.0	46	58	21.0	9	16	4.1
Costa Rica												
Public: International	7	9	4.2	38	35	22.7	1	2	0.6	0	0	0.0
Public: National				41	33	24.6	3	5	1.8	8	10	4.8
Public: Provincial							0	0	0.0	0	0	0.0
Public: Municipal										1	2	0.6
Nonprofit: International												
Nonprofit: National												
Private: International												
Private: National												
TOTAL	7	9	4.2	79	38	47.3	4	7	2.4	9	12	5.4
Erzincan, Turkey												
Public: International	5	15	7.6	9	10	13.6	0	0	0.0	0	0	0.0
Public: National				16	18	24.2	15	17	22.7	3	6	4.5
Public: Provincial							0	0	0.0	6	10	9.1
Public: Municipal										0	0	0.0
Nonprofit: International												
Nonprofit: National												
Private: International												
Private: National												
TOTAL	5	15	7.6	25	28	37.9	15	17	22.7	9	16	13.6

K = Number of Interactions; N = Number of Actors; % = Percent of Total Interactions
Sources: *Excelsior*, Mexico City; *La Nacion*, Costa Rica; *Cumhuriyet*, Erzincan, Turkey; and professional

which involved public organizations at the state and municipal levels. It also showed a sizeable involvement by national agencies, 41.6%, but no interactions with international agencies. National nonprofit organizations were engaged in 10.2% of the interactions. Data for the Hanshin response system, with a small set of reported interactions during the first three weeks, show that over half, 56.6%, of those interactions occurred among organizations at the national and international levels. The three-week response period immediately following the earthquake excludes the substantial pattern of interaction among the other prefectures of Japan with Hyogo Prefecture, in which the stricken cities were located. All forty-seven prefectures participated in organizing teams of trained personnel and volunteers to assist with water distribution and voluntary care to people in

Source and Jurisdiction, Emergent Adaptive Systems

| Nonprofit Organizations ||| ||| Private Organizations ||| ||| Total |||
| International ||| National ||| International ||| National ||| |||
K	N	%	K	N	%	K	N	%	K	N	%	K	N	%
4	9	1.8	2	4	0.9	0	0	0.0	0	0	0.0	25	50	11.4
4	15	1.8	22	40	10.0	4	14	1.8	7	20	3.2	147	209	67.1
0	0	0.0	6	10	2.7	0	0	0.0	3	5	1.4	19	32	8.7
0	0	0.0	2	5	0.9	0	0	0.0	0	0	0.0	3	7	1.4
3	10	1.4	5	6	2.3	5	9	2.3	1	2	0.5	14	18	6.4
			3	4	1.4	3	4	1.4	3	4	1.4	9	21	4.1
						0	0	0.0	2	4	0.9	2	4	0.9
									0	0	0.0	0	0	0.0
11	34	5.0	40	69	18.3	12	27	5.5	16	35	7.3	219	341	100.0
5	9	3.0	0	0	0.0	1	2	0.6	0	0	0.0	52	57	31.1
12	14	7.2	25	37	15.0	2	4	1.2	6	12	3.6	97	115	58.1
0	0	0.0	0	0	0.0	0	0	0.0	0	0	0.0	0	0	0.0
0	0	0.0	1	2	0.6	0	0	0.0	1	2	0.6	3	6	1.8
4	7	2.4	3	4	1.8	0	0	0.0	3	6	1.8	10	17	6.0
			2	5	1.2	0	0	0.0	1	2	0.6	3	7	1.8
						0	0	0.0	1	2	0.6	1	2	0.6
									1	2	0.6	1	2	0.6
21	30	12.6	31	48	18.6	3	6	1.8	13	26	7.8	167	206	100.0
8	14	12.1	0	0	0.0	0	0	0.0	0	0	0.0	22	39	33.3
0	0	0.0	2	3	3.0	0	0	0.0	0	0	0.0	36	44	54.5
0	0	0.0	0	0	0.0	0	0	0.0	0	0	0.0	6	10	9.1
0	0	0.0	0	0	0.0	0	0	0.0	0	0	0.0	0	0	0.0
1	2	1.5	0	0	0.0	1	3	1.5	0	0	0.0	2	5	3.0
			0	0	0.0	0	0	0.0	0	0	0.0	0	0	0.0
						0	0	0.0	0	0	0.0	0	0	0.0
									0	0	0.0	0	0	0.0
9	16	13.6	2	3	3.0	1	3	1.5	0	0	0.0	66	98	100.0

reports

shelters, but this organization came largely after the first three weeks which are documented in this analysis. Interactions among organizations are a product of information exchange. These interactions, under optimum circumstances, lead to the next phase of dynamic disaster operations, organizational learning.

Organizational Learning

Organizational learning is the most difficult of the dynamic processes leading to adaptive behavior to document. It needs to occur both within organizations in order for a single organization to change its perceptions and behavior, and between organizations in order to support coordinated response for the whole

Table 68. Summary, Frequency Distribution: Types of Interactions in Disaster Response by Funding

Type of Interaction	Public Organizations											
	International			National			Provincial			Municipal		
	K	N	%	K	N	%	K	N	%	K	N	%
Whittier Narrows, CA												
Public: International	0	0	0.0	0	0	0.0	0	0	0.0	0	0	0.0
Public: National				1	2	0.6	4	6	2.5	1	2	0.6
Public: State							5	7	3.2	20	23	12.7
Public: Local/Municipal										46	56	29.1
Nonprofit: International												
Nonprofit: National												
Private: International												
Private: National												
TOTAL	0	0	0.0	1	2	0.6	9	13	5.7	67	81	42.4
Loma Prieta, CA												
Public: International	0	0	0.0	1	2	0.5	0	0	0.0	0	0	0.0
Public: National				15	14	7.1	11	11	5.2	17	17	8.1
Public: State							20	14	9.5	20	28	9.5
Public: Municipal										17	27	8.0
Nonprofit: International												
Nonprofit: National												
Private: International												
Private: National												
TOTAL	0	0	0.0	16	16	7.6	31	25	14.7	54	72	25.6
Maharashtra, India												
Public: International	1	2	1.9	1	3	1.9	5	4	9.3	1	2	1.9
Public: National				4	7	7.4	8	16	14.8	2	6	3.7
Public: State							5	7	9.3	0	0	0.0
Public: Municipal										3	6	5.5
Nonprofit: International												
Nonprofit: National												
Private: International												
Private: National												
TOTAL	1	2	1.9	5	10	9.3	18	27	33.3	6	14	11.1

K = Number of Interactions; N = Number of Actors; % = Percent of Total Interactions
Source: *Los Angeles Times, San Francisco Chronicle, San Francisco Examiner, San Jose Mercury,* professional *Statesman,* Calcutta

community. These processes are exceptionally difficult under the stress of an actual event, and often occur, if at all, in periods of review and reflection after the event. Nonetheless, one measure from this analysis offers insight into the process of organizational learning. It is the relationship between communication/coordination and emergency response, vital to effective disaster operations, that is traced through the daily reports of activities in disaster operations for each of the eleven cases. While the correlation is not perfect, the evidence shows that, in most cases, reports of change in emergency response activities appeared to follow reports of change in communication/coordination activities. Figures 50–60 pre-

Source and Jurisdiction, Operative Adaptive Systems

| Nonprofit Organizations ||| ||| Private Organizations ||| ||| Total |||
| International ||| National ||| International ||| National ||| |||
K	N	%	K	N	%	K	N	%	K	N	%	K	N	%
0	0	0.0	0	0	0.0	0	0	0.0	0	0	0.0	0	0	0.0
0	0	0.0	8	8	5.1	0	0	0.0	2	3	1.3	16	21	10.1
0	0	0.0	2	3	1.3	0	0	0.0	1	2	0.6	28	35	17.7
0	0	0.0	44	33	27.8	0	0	0.0	15	20	9.5	105	109	66.5
0	0	0.0	0	0	0.0	0	0	0.0	0	0	0.0	0	0	0.0
			8	7	5.1	0	0	0.0	1	2	0.6	9	9	5.7
						0	0	0.0	0	0	0.0	0	0	0.0
									0	0	0.0	0	0	0.0
0	0	0.0	62	51	39.3	0	0	0.0	19	27	12.0	158	174	100.0
0	0	0.0	0	0	0.0	0	0	0.0	0	0	0.0	1	2	0.5
0	0	0.0	4	10	1.9	0	0	0.0	2	2	0.9	49	54	23.2
0	0	0.0	5	8	2.4	0	0	0.0	5	7	2.4	50	57	9.5
0	0	0.0	29	32	13.7	0	0	0.0	8	17	3.8	54	76	25.6
0	0	0.0	0	0	0.0	0	0	0.0	0	0	0.0	0	0	0.0
			3	6	1.4	1	2	0.5	42	44	19.9	46	52	21.8
						0	0	0.0	1	2	0.5	1	2	0.5
									10	14	4.7	10	14	4.7
0	0	0.0	41	56	19.4	1	2	0.5	68	86	32.2	211	257	100.0
1	2	1.9	1	5	1.9	0	0	0.0	1	8	1.9	11	26	20.4
0	0	0.0	0	0	0.0	1	4	1.9	1	2	1.9	16	35	29.6
1	2	1.9	1	2	1.9	0	0	0.0	1	3	1.8	8	14	14.8
1	2	1.9	0	0	0.0	0	0	0.0	0	0	0.0	4	8	7.4
0	0	0.0	3	4	5.5	1	2	1.9	1	4	1.8	5	10	9.2
			7	14	12.9	0	0	0.0	0	0	0.0	7	14	13.0
						0	0	0.0	0	0	0.0	0	0	0.0
									3	8	5.5	3	8	5.6
3	6	5.6	12	25	22.2	2	6	3.7	7	25	12.9	54	115	100.0

reports; *Times of India*, Bombay, *Times of India*, New Delhi, *The Hindustan Times*, New Delhi, and *The*

sent the marginal history charts of this relationship for the four sets of response systems.

Figures 50–52 show the records for the Nonadaptive Systems: San Salvador, Ecuador, and Armenia. These records show the daily change in news reports of activities in emergency response in correlation with the daily change in news reports of activities in communication/coordination. In San Salvador, reports of emergency response activities drop from Day 1 to Day 2, gradually climbing through Day 3 to a slight peak on Day 4, following a sharp increase in communication/coordination activities on Day 4. As communication/coordination reports decline on Day 5, reports of emergency response activities also decline, climbing

Table 69. Summary, Frequency Distribution: Types of Interactions in Disaster Response by Funding

Type of Interaction	Public Organizations											
	International			National			State			Local[a]		
	K	N	%	K	N	%	K	N	%	K	N	%
Northridge, CA												
Public: International	0	0	0.0	0	0	0.0	0	0	0.0	0	0	0.0
Public: National				29	15	11.8	20	18	8.2	37	36	15.1
Public: State							12	10	4.9	20	24	8.2
Public: Local[a]										43	40	17.6
Nonprofit: International												
Nonprofit: National												
Private: International												
Private: National												
TOTAL	0	0	0.0	29	15	11.8	32	28	13.1	100	100	40.8
Hanshin, Japan												
Public: International	0	0	0.0	4	5	7.5	1	2	1.9	0	0	0.0
Public: National				4	6	7.6	5	5	9.4	11	29	20.8
Public: State							0	0	0.0	2	4	3.8
Public: Municipal										1	2	1.9
Nonprofit: International												
Nonprofit: National												
Private: International												
Private: National												
TOTAL	0	0	0.0	8	11	15.1	6	7	11.3	14	35	26.4

K = Number of Interactions; N = Number of Actors; % = Percent of Total Interactions
[a]Three types of local government — regional, county, and city — were actively involved in response opera-ernment, to make the data more comparable to data for the Hanshin Earthquake response system
Sources: *The Los Angeles Times* and the *Daily News*, Los Angeles; *The Japan Times*, Tokyo, Japan

again on Day 8 after an increase in reports of communication/coordination. As the events of the disaster continue, reports of both functions tend to level off.

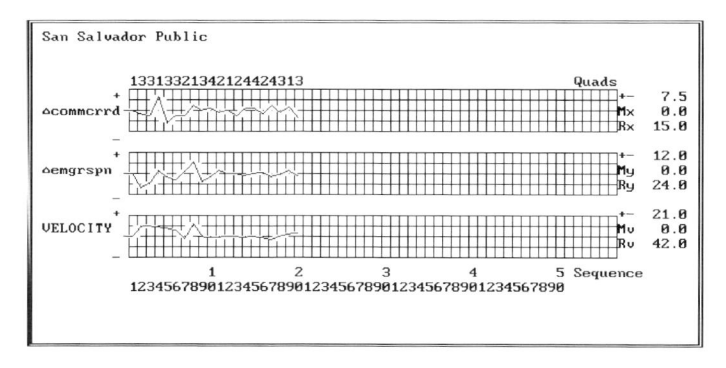

Figure 50. 1987 San Salvador Earthquake: Marginal History, Public Organizations, Emergency Response by Communication/Coordination

Chapter 9: Adaptation to Disaster

Source and Jurisdiction, Auto-Adaptive Systems

Nonprofit Organizations						Private Organizations						Total		
International			National			International			National					
K	N	%	K	N	%	K	N	%	K	N	%	K	N	%
0	0	0.0	0	0	0.0	0	0	0.0	0	0	0.0	0	0	0.0
0	0	0.0	11	13	4.5	0	0	0.0	5	3	2.0	102	85	41.6
0	0	0.0	4	8	1.6	0	0	0.0	2	4	0.8	38	46	15.5
0	0	0.0	23	30	9.4	0	0	0.0	12	19	0.4	78	89	31.8
0	0	0.0	0	0	0.0	0	0	0.0	0	0	0.0	0	0	0.0
			18	25	7.3	0	0	0.0	7	8	3.3	25	33	10.2
						0	0	0.0	0	0	0.0	0	0	0.0
									2	4	0.8	2	4	0.8
0	0	0.0	56	76	22.9	0	0	0.0	28	38	11.4	245	257	100.0
0	0	0.0	1	2	1.9	0	0	0.0	0	0	0.0	6	9	11.3
0	0	0.0	0	0	0.0	0	0	0.0	4	12	7.5	24	52	45.3
0	0	0.0	0	0	0.0	0	0	0.0	2	4	3.8	4	8	7.6
0	0	0.0	3	4	5.7	0	0	0.0	2	6	3.8	6	12	11.3
0	0	0.0	2	4	3.8	1	3	1.9	2	9	3.8	5	16	9.4
			4	10	7.5	2	9	3.8	2	4	3.8	8	23	15.1
						0	0	0.0	0	0	0.0	0	0	0.0
									0	0	0.0	0	0	0.0
0	0	0.0	10	20	18.9	3	12	5.7	12	35	22.6	53	120	100.0

tions for the Northridge Earthquake. These types have been combined to create a single category, local gov-

In Ecuador, the pattern is more dramatic. Reports of emergency response activities drop deeply on Day 2, climbing on Day 3 following a sharp increase in

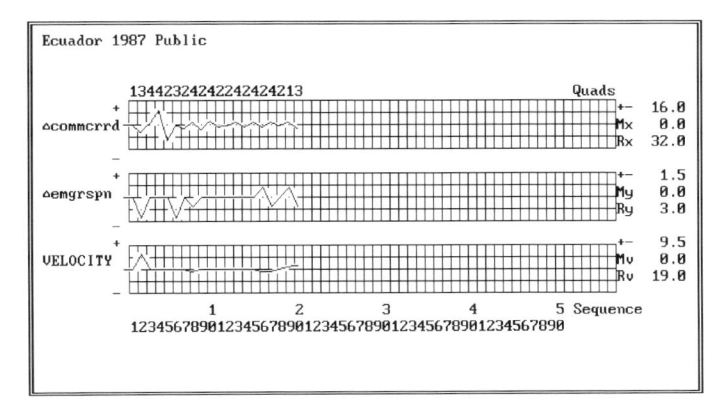

Figure 51. 1987 Ecuadorian Earthquakes: Marginal History, Public Organizations, Emergency Response by Communication/Coordination

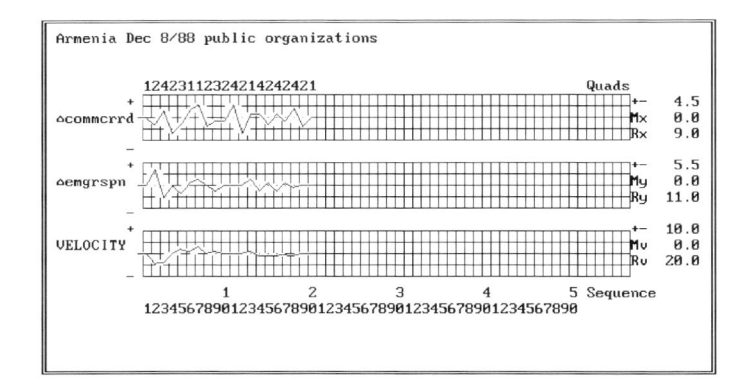

Figure 52. 1988 Armenia Earthquake: Marginal History, Public Organizations, Emergency Response by Communication/Coordination

reports of communication/coordination activities, dropping again on Day 6, following a sharp decline in reports of communication/coordination activities, but increasing again on Day 8 after reports of communication/coordination increase. The two functions then level off, with two late peaks in emergency response activities that likely represent actions taken in outlying villages affected by the earthquake. In Armenia, interestingly, the pattern appears to reverse the relationship, with reports of emergency response leading reports of communication/coordination. This may reflect the initial reluctance to report activities within Soviet Armenia, followed by Premier Gorbachev's decision on Day 2 to use the disaster as an example of the USSR's "Glasnost" policy of openness to the international community. In keeping with this policy, Premier Gorbachev agreed both to accept international aid and to allow the international press to cover the events of disaster operations.

The effect of this decision is shown by the increase in reported communication/coordination activities on Day 3. As reports of emergency response activities level off, reports of communication/coordination activities continue with marked increases and decreases over the three-week response period.

Figures 53–55 presents the marginal history charts of change in reported emergency response activities by change in reported communication/coordination activities for the Emergent Adaptive Systems: Mexico City, Costa Rica, and Erzincan, Turkey. In Mexico City, change in reported activity for both emergency response and communication/coordination functions drop from Day 1 to Day 2, but change in reported activity in communication/coordination increased sharply on Day 3, followed by an increase in reported emergency response activities. Reported emergency response activities continue, but drop again sharply after reported activities in communication/coordination fall on Day 7. Change in emergency response climbs again on Day 8, as reported activity in communication/coordination increases. Again, reports of emergency response activities

Chapter 9: Adaptation to Disaster   249

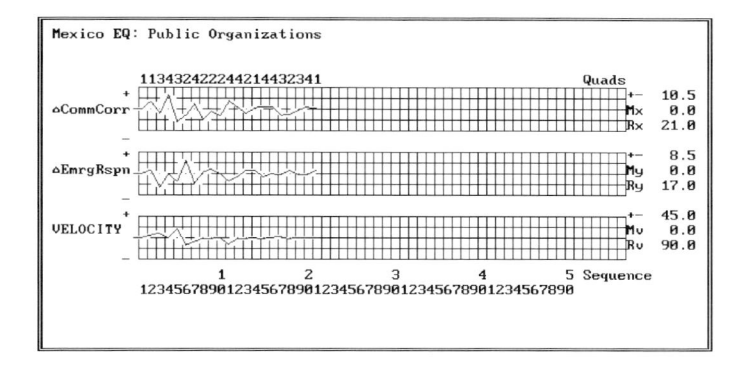

Figure 53. 1985 Mexico City Earthquake: Marginal History, Public Organizations, Emergency Response by Communication/Coordination

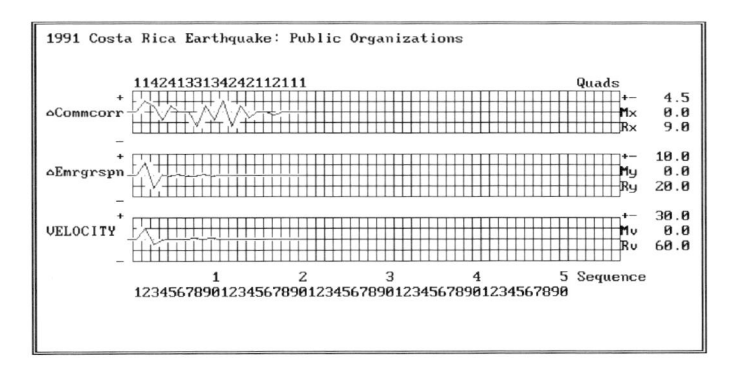

Figure 54. 1991 Costa Rica Earthquake: Marginal History, Public Organizations, Emergency Response by Communication/Coordination

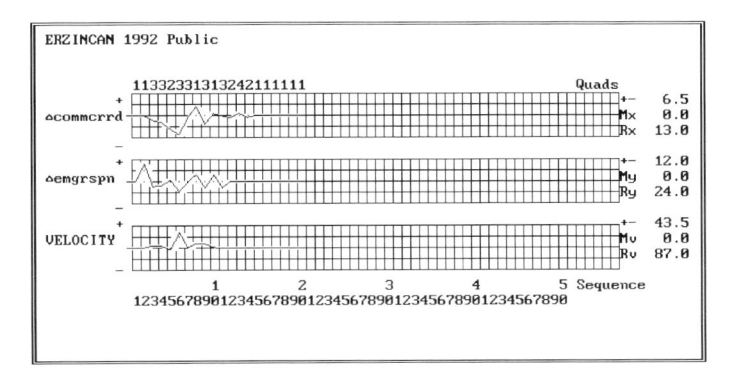

Figure 55. 1992 Erzincan, Turkey Earthquake: Marginal History, Public Organizations, Emergency Response by Communication/Coordination

leveled off after Day 9, while reports of communication/coordination activities continued to fluctuate.

In Costa Rica, change in reported emergency response activity increases on Day 2 with change in communication/coordination, but drops again sharply on Day 3 as reported activity in communication/coordination declines. Reported emergency response activity increases again on Day 4 and then levels off, as emergency response ended early in Costa Rica. Reported activity on communication/coordination continued to fluctuate with sharp increases and decreases, reflecting the tension in the relationships among the governmental departments engaged in disaster response.

The case of Erzincan, Turkey offers a different perspective. Reports of emergency response activities peaked on Day 2, while reports of communication/coordination actions declined steadily through Day 6, increasing strongly on Day 8, dropping again on Day 9 and then leveling off. Reported activities in emergency response appeared to be relatively independent of reported activities in communication/coordination until Day 6, when change in emergency response activities appeared to follow the increase in communication/coordination activities.

Figures 56–58 presents the pattern of relationships between the functions of emergency response and communication/coordination for the Operative Adaptive cases: Whittier Narrows, CA; Loma Prieta, CA, and Marathwada, India. In Whittier Narrows, change in reported emergency response activity increased sharply on Day 2, simultaneously with reported activity in communication/coordination. These reports drop sharply for emergency response on Day 3, but rise again with reports of communication/coordination on Day 4. The increases in reported activities for both functions on Day 4 likely reflects the occurrence of a sharp aftershock on Sunday, October 4, 1987 that reactivated the emergency response system. After that date, reported activities for emergency response level off, as this was a moderate earthquake, while reported activity in communication/coordination continue to fluctuate during the three-week response period.

In the Loma Prieta response system, change in reported activities for both emergency response and communication/coordination functions appeared relatively late, with a sharp increase in both only on Day 4, followed by a drop again on Day 5. Reported activities for both functions increase on Day 6, followed by a leveling off. Change in reported activities for both functions occurred nearly simultaneously, with emergency response following slightly behind communication/coordination. In Marathwada, India, change in emergency response reports appeared to lead reports on communication/coordination on Day 2; change in reported activities for both functions dropped on Day 3, increasing again on Days 4 and 5. Reported activities on communication/coordination continued to increase, dipping again on Day 6, but continued with minor fluctuations over the three-week period. Reports on emergency response leveled off after Day 10.

Figure 59–60 presents the marginal history of rates of change in reported response functions of communication/coordination and emergency response for

Chapter 9: Adaptation to Disaster 251

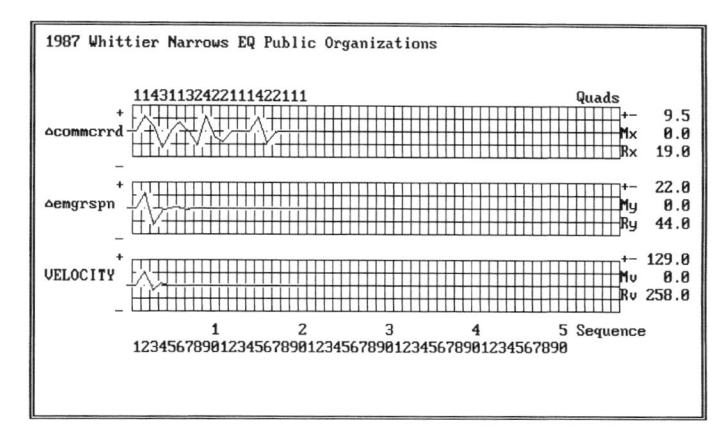

Figure 56. 1987 Whittier Narrows, CA Earthquake: Marginal History, Public Organizations, Emergency Response by Communication/Coordination

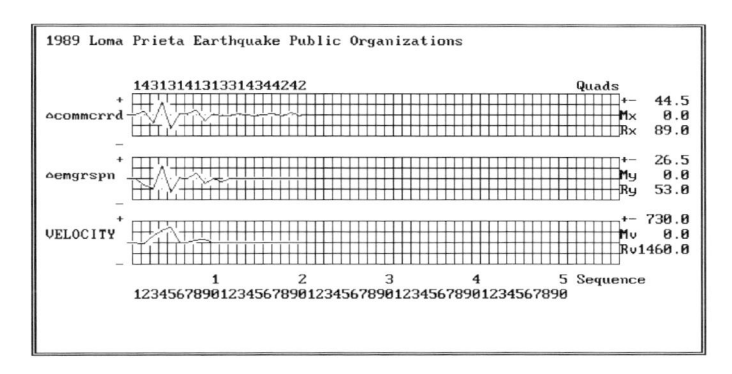

Figure 57. 1989 Loma Prieta, CA Earthquake: Marginal History, Public Organizations, Emergency Response by Communication/Coordination

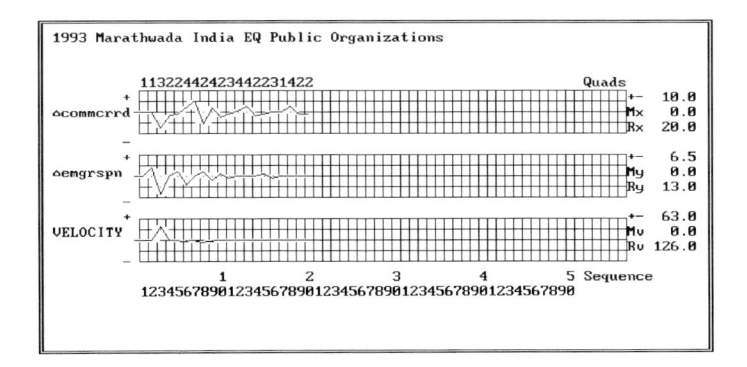

Figure 58. 1993 Marathwada, India Earthquake: Marginal History, Public Organizations, Emergency Response by Communication/Coordination

252   *Part III: Future Strategies*

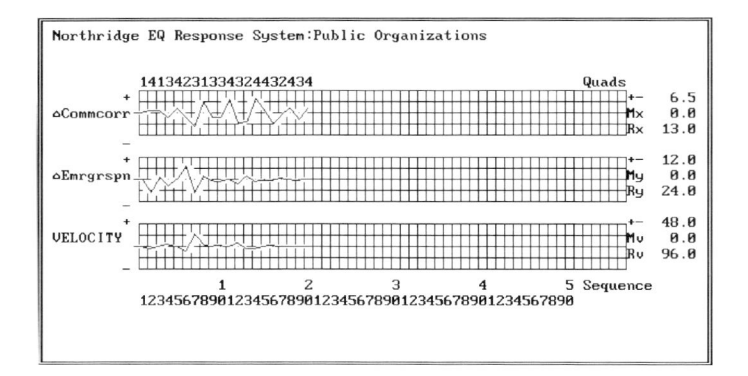

```
Northridge EQ Response System:Public Organizations
         14134231334324432434                    Quads
ΔCommcorr  ⩘⩗⩘⩗⩘⩗⩘                              +- 6.5
                                                 Mx 0.0
                                                 Rx 13.0

ΔEmrgrspn                                       +- 12.0
                                                 My 0.0
                                                 Ry 24.0

VELOCITY                                        +- 48.0
                                                 Mv 0.0
                                                 Rv 96.0

           1       2       3       4       5 Sequence
           12345678901234567890123456789012345678901234567890
```

Figure 59. 1994 Northridge, CA Earthquake: Marginal History, Public Organizations, Emergency Response by Communication/Coordination

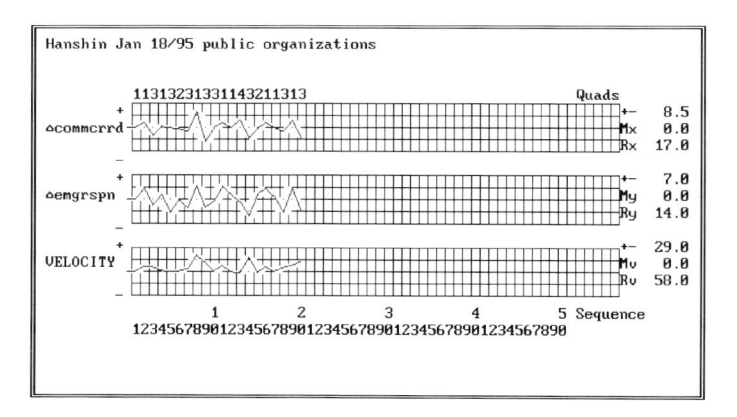

```
Hanshin Jan 18/95 public organizations
         11313231331143211313                    Quads
Δcommcrrd                                       +- 8.5
                                                 Mx 0.0
                                                 Rx 17.0

Δemgrspn                                        +- 7.0
                                                 My 0.0
                                                 Ry 14.0

VELOCITY                                        +- 29.0
                                                 Mv 0.0
                                                 Rv 58.0

           1       2       3       4       5 Sequence
           12345678901234567890123456789012345678901234567890
```

Figure 60. 1995 Hanshin, Japan Earthquake: Marginal History, Public Organizations, Emergency Response by Communication/Coordination

the Auto Adaptive Systems: Northridge, CA and Hanshin, Japan. These findings are interesting, as change in reported emergency response activities appears relatively independent of change in reported communication/coordination activities for the Northridge response system. Change in reported activities on communication/coordination is relatively even for the first three days, dipping slightly on Day 4, but rising again on Day 5, and then fluctuating with relatively sharp peaks and valleys for the next 14 days. Change in emergency response drops on Day 2, increases on Day 3, dips slightly on Day 4, then increases on Days 5 and 6, dropping on Day 7, and then leveling off as the response period wanes. This pattern reflects the field observation that experienced local response organizations in the Los Angeles region successfully followed their own procedures. Fluctuations in communication/coordination activities appeared to reflect continuing efforts to engage the wider community in response operations.

In Hanshin, Japan, change in reported activities in emergency response appeared to follow more closely the change in reported activities in communication/coordination. Fluctuations in both emergency response and communication/coordination reports continued over the full three-week response period. These findings indicate both the overwhelming impact of the event upon the local community, and the difficulty in mobilizing response effectively to meet the massive needs of the community.

While this analysis of the relationship between change in reported activities of emergency response by change in reported activities of communication/coordination is not conclusive, it offers evidence that the two functions are, in most cases, related. Further, the dominant pattern shows that change in emergency response activities follows change in communication/coordination activities. This pattern suggests that organizational learning, both within organizations and between organizations, occurs to some degree over the three-week period of disaster response operations. This finding suggests that a promising means for increasing effectiveness in emergency response is to increase the capacity for communication/coordination within communities vulnerable to seismic risk. Organizational learning creates the basis for adaptive behavior, the final phase of self-organization in communities vulnerable to seismic risk.

Adaptive Behavior

Adaptive behavior among organizations means the capacity to change not only actions, but also priorities for allocating resources and attention to meet new or immediate demands from the environment. In communities, organizational behavior is necessarily interdependent, and consequently adaptive behavior is more difficult, as it involves initiating change not only in one organization, but also in the set of organizations that are involved in executing the interdependent functions of the community. This change occurs most efficiently when communities are able to "self-organize", that is, to reorder their priorities and actions spontaneously, without imposition of external controls. In the urgent world of disaster response, this capacity is the most beneficial. It is also the most difficult to achieve, given different levels of understanding of risk, different requirements for action, and different responsibilities for mobilizing action among organizations participating in disaster response operations in an interdependent, interjurisdictional setting.

Tables 70–73 present the findings from a nonlinear logistic regression analysis of basic disaster response functions for the four sets of response systems. This analysis recognized the different levels of responsibility for action in disaster operations, and likely different levels of understanding of risk, by classifying the organizations participating in disaster operations according to their source of funding: public, private or nonprofit.

Public organizations have the legal responsibility for the protection of life and property in their respective communities. Nonprofit organizations have a mission of humanitarian assistance to their communities or constituents. Private organizations have a responsibility and interest in protecting the lives of their employees, the property of their respective organizations, and the economic functions of the community.

Self-organization operates most effectively when all component organizations of the community participate actively in response functions, at their respective levels of understanding and capacity to act. While this level of collective performance is difficult to achieve, the findings from the nonlinear logistic regression analysis summarized in Tables 70–73 present a measure of the extent to which the eleven response systems represent efforts by their respective communities to adapt its behavior during the three-week period immediately following the earthquake.

Table 70 summarizes the findings from this analysis for the Nonadaptive Systems: San Salvador, Ecuador and Armenia. For the San Salvador response system, public organizations report significant values for R^2 on two basic functions: Emergency Response ($R^2 = 0.79$, $F = 71.02$; $p = 0.000$) and Communication/Coordination, ($R^2 = 0.52$; $F = 20.97$; $p = 0.001$). This means that 79% of the variance in the reported emergency response activities at period t is explained by the function of emergency response at period t-1, while 52% of the variance in reported communication/coordination activities at period t are explained by the function of communication/coordination at period t-1. Disaster Relief produced no equation, and while the other three functions produced equations, they were not statistically significant. The k values range from 1 to 4, and are calculated as a measure of the stability of the function, as well as the subsystem and total system. With k values of 1.49 and 1.16, the two basic functions show some instability. Recovery/Reconstruction fell within the chaotic range (k = 3.83). Public organizations produced a significant subgroup response ($R^2 = 0.39$, $F = 12.1$; $p < 0.05$).

Nonprofit organizations report significant values of R^2 on two functions, although at a lesser degree: Emergency Response ($R^2 = 0.19$; $F = 4.42$; $p < 0.05$), and Recovery/Reconstruction ($R^2 = 0.34$; $F = 9.90$; $p < 0.01$). The k value for Emergency Response, however, is high at 3.99, indicating that this system was highly unstable. Values of k pass into the region of chaos at 3.66. Recovery/Reconstruction, with a k value of 1.00, is stable. Private organizations produced one significant equation in Communication/Coordination with $R^2 = 0.55$; $F = 23.6$; $p < 0.001$). The k value shows this function to be relatively stable at k = 1.21. Given the distribution of functions within the whole response system, the analysis produced a significant response that included public, private and nonprofit organizations: ($R^2 = 0.37$; $F = 11.24$; $p < 0.01$). With a k value of 1.00, the total system appears stable, but this apparent stability is likely fragile, given the weaknesses shown by the analysis. Four response functions report k values over the chaos threshold of 3.66; four additional functions report equations that are not significant, and nine functions report no equation at all.

Table 70. Summary, Logistic Regression Analysis, Nonadaptive Systems

Response Function	Public Organizations			Nonprofit Organizations			Private Organizations		
	R^2	F	p	k	R^2	F	p	k	R^2	F	p	k

San Salvador

Emergency Response	0.79	71.02	0.000	1.49	0.19	4.42	0.005	*3.99*	NE	NE	NE	NE
Damage Assessment	0.06	1.15	NS	1.00	NE	NE	NE	NE	0.11	2.39	NS	*3.95*
Communication/Coordination	0.52	20.97	0.001	1.16	NE	NE	NE	NE	0.55	23.58	0.001	1.21
Disaster Relief	NE	NE	NE	NE	0.07	1.43	NS	*3.83*	NE	NE	NE	NE
Recovery/Reconstruction	0.04	0.71	NS	*3.83*	0.34	9.90	0.01	1.00	NE	NE	NE	NE
Financial Assistance	0.12	2.69	NS	1.11	NE	NE	NE	NE	NE	NE	NE	NE
Subgroup Total	0.39	12.06	0.05	1.00	0.36	10.79	0.01	1.00	NE	NE	NE	NE
Grand Total	0.37	11.24	0.01	1.00	0.37	11.24	0.01	1.00	0.37	11.24	0.01	1.00

Ecuador

Emergency Response	0.85	107.39	0.000	1.25	NE	NE	NE	NE	NE	NE	NE	NE
Damage Assessment	0.79	72.02	0.000	1.77	0.18	4.23	NS	1.32	0.99	1981.99	0.000	1.90
Communication/Coordination	0.48	17.66	0.001	1.11	0.01	0.23	NS	2.77	NE	NE	NE	NE
Disaster Relief	NE	NE	NE	NE	0.04	0.71	NS	*3.99*	NE	NE	NE	NE
Recovery/Reconstruction	0.50	18.98	0.001	1.21	NE	NE	NE	NE	0.37	11.05	0.01	*4.00*
Financial Assistance	NE	NE	NE	NE	NE	NE	NE	NE	NE	NE	NE	NE
Subgroup Total	0.53	21.42	0.001	1.17	0.12	2.71	NS	*3.90*	0.07	1.48	NS	*3.92*
Grand Total	0.47	16.72	0.001	1.16	0.47	16.72	0.001	1.16	0.47	16.72	0.001	1.16

Armenia

Emergency Response	0.03	0.53	NS	1.00	0.43	14.98	0.01	1.20	NE	NE	NE	NE
Damage Assessment	NE	NE	NE	NE	NE	NE	NE	NE	NE	NE	NE	NE
Communication/Coordination	0.45	16.15	0.001	1.00	NE	NE	NE	NE	NE	NE	NE	NE
Disaster Relief	NE	NE	NE	NE	0.00	0.05	NS	*3.95*	NE	NE	NE	NE
Recovery/Reconstruction	0.04	0.75	NS	*3.94*	NE	NE	NE	NE	NE	NE	NE	NE
Financial Assistance	NE	NE	NE	*3.94*	NE	NE	NE	NE	NE	NE	NE	NE
Subgroup Total	0.25	6.75	0.05	1.00	NE	NE	NE	NE	NE	NE	NE	NE
Grand Total	NE	NE	NE	NE	NE	NE	NE	NE	NE	NE	NE	NE

NE = No equation; NS = Not significant
k values in italics indicate that the function is in chaos

The nonlinear logistic regression analysis for Ecuador identifies a logistic equation for the subsystem of public organizations that is statistically significant: $R^2 = 0.53$, $F = 21.42$, $p < 0.001$. The subgroups of nonprofit and private organizations produce equations, but neither is statistically significant and both produce k values that indicate the subsystems were operating in chaotic environments. The subgroup of public organizations produced significant values for R^2 on four of the six basic functions, which served to stabilize not only the public subset, but also the entire system. This equation for the Ecuadorian system ($R^2 = 0.47$; $F = 16.72$; $p < 0.001$; $k = 1.16$) appears fragile, given the findings in the nonprofit and private subsystems. There were no significant equations identified for functions in the nonprofit sector, although two — damage assessment and recovery/reconstruction — were identified for the private sector. Seven functions produced no equations, and two functions, in addition to the subgroup equations, produced k values in the chaotic range.

In Armenia, the subsystem findings for public organizations show a moderate response function for Communication/Coordination ($R^2 = 0.45$; $F = 16.15$; $p < 0.001$; $k = 1.00$), which stabilized the subsystem's performance ($R^2 = 0.25$; $F = 6.75$; $p < 0.05$; $k = 1.0$). Interestingly, the function of emergency response produced a significant equation for nonprofit organizations, but not for public organizations. In Soviet Armenia, there was no private sector, although 9 international private organizations contributed assistance. Three functions reported k values in chaos, and the system as a whole did not produce an equation. Given the difficult circumstances, it is remarkable that measurable coherence evolved among the subset of public organizations.

Table 71 presents findings from the Emergent Adaptive Systems: Mexico City, Costa Rica, and Erzincan, Turkey. The findings for Mexico City present a strong response system, which produced equations for all six functions in the subset of public organizations ($R^2 = 0.79$; $F = 74.4$; $p < 0.001$; $k = 1.05$). Only one of these functions, Financial Assistance, proved insignificant statistically. The function of disaster relief produced a significant equation for the nonprofit subgroup, $R^2 = 0.45$; $F = 16.2$; $p < 0.001$; $k = 1.0$. This moderately high value stabilized the subsystem of nonprofit organizations, resulting in a subgroup finding of $R^2 = 0.44$; $F = 15.8$; $p < 0.001$; $k = 1.00$. Nonetheless, signs of instability also exist. The functions of Emergency Response, Communication/Coordination, and Recovery/Reconstruction did not produce equations in the Nonprofit subgroup. The function of Financial Assistance produced an equation, but it was insignificant and also highly unstable. Only one function, Emergency Response, produced a significant equation for the Private subgroup, ($R^2 = 0.91$; $F = 203.2$; $p < 0.000$; $k = 1.76$), with four functions producing no equation, and one other, disaster relief, producing an insignificant equation. The cumulative findings for the three subsets produced a coherent system over the three-week period, with $R^2 = 0.82$; $F = 92.1$; $p < 0.001$; $k = 1.0$, based upon reported activities from *Excelsior*. This findings indicate that the Mexico City public organizations were able to adapt their performance to meet needs in the community over the three-week response

Chapter 9: Adaptation to Disaster

Table 71. Summary, Logistic Regression Analysis, Emergent Adaptive Systems

Response Function	Public Organizations				Nonprofit Organizations				Private Organizations			
	R^2	F	p	k	R^2	F	p	k	R^2	F	p	k
Mexico City												
Emergency Response	0.64	35.1	0.001	1.00	NE	NE	NE	NE	0.91	203.2	0.000	1.76
Damage Assessment	0.66	39.5	0.001	1.10	0.24	6.42	0.05	2.84	NE	NE	NE	NE
Communication/Coordination	0.49	19.5	0.00	1.11	NE	NE	NE	NE	NE	NE	NE	NE
Disaster Relief	0.28	7.96	0.05	1.00	0.45	16.24	0.001	1.00	0.11	2.59	NS	1.00
Recovery/Reconstruction	0.61	31.1	0.00	1.00	NE	NE	NE	NE	NE	NE	NE	NE
Financial Assistance	0.09	1.74	NS	1.05	0.13	2.99	NS	*3.99*	NE	NE	NE	NE
Subgroup Total	0.79	74.4	0.001	1.05	0.44	15.77	0.001	1.00	0.59	28.31	0.001	1.00
Grand Total	0.82	92.1	0.001	1.00	0.82	92.10	0.001	1.00	0.82	92.1	0.000	1.00
Costa Rica												
Emergency Response	0.23	6.04	0.05	1.11	NE	NE	NE	NE	NE	NE	NE	NE
Damage Assessment	0.22	5.62	0.05	1.00	NE	NE	NE	NE	NE	NE	NE	NE
Communication/Coordination	0.52	21.78	0.001	1.00	NE	NE	NE	NE	0.17	4.09	NS	*3.99*
Disaster Relief	0.25	6.61	0.05	1.00	0.57	26.53	0.001	1.00	NE	NE	NE	NE
Recovery/Reconstruction	0.42	14.54	0.01	1.00	NE	NE	NE	NE	NE	NE	NE	NE
Financial Assistance	0.07	1.48	NS	1.00	0.58	28.15	0.001	1.00	0.07	1.47	NS	1.00
Subgroup Total	0.55	24.82	0.001	1.00	0.57	26.48	0.001	1.00	0.57	26.48	0.001	1.00
Grand Total	0.57	26.48	0.001	1.00								
Erzincan, Turkey												
Emergency Response	0.47	16.91	0.001	1.11	NE	NE	NE	NE	NE	NE	NE	NE
Damage Assessment	0.40	12.45	0.01	1.22	NE	NE	NE	NE	NE	NE	NE	NE
Communication/Coordination	0.89	148.5	0.000	1.24	NE	NE	NE	NE	0.94	277.6	0.00	1.43
Disaster Relief	NE	NE	NE	NE	0.14	3.05	NS	1.00	NE	NE	NE	NE
Recovery/Reconstruction	0.24	6.13	0.05	1.25	NE	NE	NE	NE	NE	NE	NE	NE
Financial Assistance	NE	NE	NE	NE	0.09	1.77	NS	1.00	0.14	3.01	NS	*3.95*
Subgroup Total	0.63	32.43	0.001	1.11	0.60	28.97	0.001	1.11	0.60	28.97	0.001	1.11
Grand Total	0.60	28.97	0.001	1.11								

NE = No equation; NS = Not significant
k values in italics indicate that the function is in chaos

period, and did so with the assistance of nonprofit organizations in disaster relief and private organizations in emergency response. Gaps did exist, particularly in the nonprofit and private sectors, but the system achieved a distinctive coherence, as reported in daily news accounts, over the relatively brief response period.

The Costa Rica findings also indicate a coherent response system, stabilized by a moderately high value for R^2 for Communication/Coordination ($R^2 = 0.52$; $F = 21.8$; $p<0.001$; $k = 1.0$) in the subset of public organizations. While the values for R^2 were not as high as in the Mexico City case, the Costa Rica case did produce significant equations for five of the six basic functions of disaster response for the public subgroup. These findings were bolstered by a moderately high value for R^2 for Disaster Relief ($R^2 = 0.57$; $F = 24.8$; $p<0.001$; $k = 1.0$) in the nonprofit sector, which contributed to a significant value for the total response system, ($R^2 = 0.57$; $F = 26.5$; $p<0.001$; $k = 1.0$). Again, the analysis reveals substantial weakness in the total Costa Rica response system, with five of the six functions not producing equations in the nonprofit group, and only one equation produced for the private subgroup, which was not significant and highly unstable.

Findings for the Erzincan, Turkey case produced an interesting pattern, with strong equations for Communication/Coordination in the public and private subgroups, and only one, insignificant equation for Disaster Relief in the nonprofit subgroup. The functions of Emergency Response and Damage Assessment produced moderate values of R^2 for the public subgroup, which contributed to a measure of overall coherence ($R^2 = 0.63$; $F = 32.4$; $p<0.001$; $k = 1.11$). Strong measures of communication/coordination for both the public and private subgroups appeared to stabilize the entire system, contributing to the system equation ($R^2 = 0.60$; $F = 28.97$; $p<0.001$; $k = 1.11$). Again, the analysis revealed serious gaps in the response system, particularly in the nonprofit and private sectors.

Reviewing the logistic regression equations for the total set of basic functions for the Emergent Adaptive category, Mexico City, Costa Rica and Erzincan, Turkey, all show a significant advance in performance over the three cases in the Nonadaptive category. The data indicate greater adaptiveness in performance of basic functions of response under the urgent conditions of disaster.

Findings for the Operative Adaptive cases — Whittier Narrows, CA, Loma Prieta, CA and Marathwada, India are presented in Table 72. All three cases produced significant equations for the total response systems. Whittier Narrows presented significant equations for all three subsystems — public, private and nonprofit. The Loma Prieta case produced higher equations for the functions of Emergency Response, Damage Assessment, Communication/Coordination and Recovery/ Reconstruction than Whittier Narrows, resulting in a stronger equation for the subsystem of public organizations ($R^2 = 0.49$; $F = 18.2$; $p<0.001$; $k = 1.111$). Nonetheless, this case failed to produce significant equations for the nonprofit and private subgroups, resulting in a lower equation for the total response system ($R^2 = 0.35$; $F = 10.11$; $p<0.01$; $k = 1.0$) than Whittier Narrows. This finding indicates that while the public organizations demonstrated learning from the experience of the Whittier Narrows case in the same state, the nonprofit

Table 72. Summary, Logistic Regression Analysis, Operative Adaptive Systems

Response Function	Public Organizations				Nonprofit Organizations				Private Organizations			
	R^2	F	p	k	R^2	F	p	k	R^2	F	p	k
Whittier Narrows, CA												
Emergency Response	0.30	8.11	0.05	1.25	NE	NE	NE	NE	0.09	1.82	NS	1.25
Damage Assessment	0.46	16.35	0.00	1.12	NE	NE	NE	NE	0.23	5.55	0.05	1.11
Communication/Coordination	0.10	2.16	NS	1.00	NE	NE	NE	NE	0.40	12.82	0.010	1.25
Disaster Relief	NE	NE	NE	NE	0.11	2.46	NS	1.00	NE	NE	NE	NE
Recovery/Reconstruction	NE	NE	NE	NE	NE	NE	NE	NE	0.25	6.37	0.05	1.11
Financial Assistance	NE	NE	NE	NE	NE	NE	NE	NE	NE	NE	NE	NE
Subgroup Total	0.39	11.90	0.01	1.11	0.24	5.92	0.05	1.00	0.28	7.38	0.05	1.20
Grand Total	0.39	12.38	0.01	1.11	0.39	12.38	0.01	1.11	0.39	12.38	0.01	1.11
Loma Prieta, CA												
Emergency Response	0.76	61.79	0.000	1.00	0.40	12.41	0.01	1.20	NE	NE	NE	NE
Damage Assessment	0.60	28.72	0.001	1.11	0.20	4.88	0.05	1.11	NE	NE	NE	NE
Communication/Coordination	0.42	13.57	0.010	1.11	NE	NE	NE	NE	0.58	26.61	0.00	1.00
Disaster Relief	NE	NE	NE	NE	0.04	0.74	NS	1.00	NE	NE	NE	NE
Recovery/Reconstruction	0.17	3.98	NS	1.00	NE	NE	NE	NE	0.45	15.74	0.00	1.17
Financial Assistance	NE	NE	NE	NE	NE	NE	NE	NE	0.10	2.16	NS	2.84
Subgroup Total	0.49	18.22	0.001	1.11	0.18	4.11	NS	1.00	0.06	1.18	NS	1.00
Grand Total	0.35	10.11	0.01	1.00	0.35	10.11	0.01	1.00	0.35	10.11	0.01	1.00
Marathwada, India												
Emergency Response	0.69	42.52	0.001	1.25	0.22	5.21	0.05	1.11	NE	NE	NE	NE
Damage Assessment	0.85	105.98	0.000	1.81	NE	NE	NE	NE	NE	NE	NE	NE
Communication/Coordination	0.64	34.06	0.001	1.22	0.05	1.00	NS	*4.00*	NE	NE	NE	NE
Disaster Relief	0.62	31.50	0.001	1.22	0.44	15.09	0.001	1.11	0.12	2.55	NS	*3.87*
Recovery/Reconstruction	NE	NE	NE	NE	NE	NE	NE	NE	NE	NE	NE	NE
Financial Assistance	NE	NE	NE	NE	NE	NE	NE	NE	NE	NE	NE	NE
Subgroup Total	0.77	63.17	0.001	1.16	0.46	16.44	0.001	1.11	0.08	1.63	NS	*3.88*
Grand Total	0.77	62.82	0.001	1.00	0.77	62.82	0.001	1.00	0.77	62.82	0.001	1.00

NE = No equation; NS = Not significant
k values in italics indicate that the function is in chaos

and private organizations were not as fully engaged. Importantly, none of the functions in either of these two cases exhibited instability, although eighteen functions among the two sets produced no equations, indicating underdevelopment.

The Marathwada, India case produced very interesting findings. India, one of the poorest economies included in the set, nonetheless produced a remarkably high set of findings for the public subset, ($R^2 = 0.77$; $F = 63.17$; $p < 0.001$; $k = 1.16$), which stabilized performance for the total system ($R^2 = 0.77$; $F = 62.8$; $p < 0.001$; $k = 1.0$). The performance of the public sector was strengthened by moderate performance in Disaster Relief and less strong, but still significant performance in Emergency Response by the nonprofit sector. Three functions in the Marathwada case veered into chaos, Communication/Coordination in the nonprofit sector, and Disaster Relief in the private sector, which also contributed to an unstable equation for the private subgroup.

The findings for this set of Operative Adaptive cases show a different pattern of development than the Emergent Adaptive cases. While the two California cases show weaker equations for the overall response systems than the Emergent Adaptive cases, they also show greater engagement of the private and nonprofit sectors in basic response functions. This pattern is strengthened by the findings from the interactions analysis that shows local public organizations were interacting more directly with organizations in their own communities. While the pattern is not yet established in these two California systems, the direction is confirmed by these findings. Especially indicative is the relatively strong finding for Communication/Coordination in the private sector for the Loma Prieta case. The Marathwada, India case demonstrates this same pattern. Led by strong participation from the public sector, participation from the nonprofit sector clearly influenced the overall performance of the system, with lesser participation from the private sector.

Table 73 presents the results of the nonlinear logistic regression analysis for the Auto Adaptive systems: Northridge, CA and Hanshin, Japan. Both cases fall more accurately at the 'edge of chaos' than clearly within the category of self-organizing systems. The Northridge case shows stronger indications of adaptive performance, but there are weaknesses as well. The case produced a full set of equations for the six basic response functions, but the equations for two functions – Disaster Relief ($R^2 = 0.09$; $F = 1.9$; $k = 3.31$) and Financial Assistance ($R^2 = 0.12$; $F = 2.60$; $k = 3.92$) – were not significant. Further, both functions had high k values, indicating instability, and Financial Assistance as well as the function of Reconstruction/Recovery, revealed chaotic performance. These weaknesses were countered by strong performance in the functions of Emergency Response by the nonprofit sector and Damage Assessment and Communication/Coordination by the private sector. The functions of Disaster Relief and Recovery/Reconstruction in the nonprofit sector produced equations, but neither was statistically significant and both indicated chaotic performance. These findings indicate that the Northridge response system was performing at the outer edge of its capacity, stabilized by a strong Emergency Response function and a

Table 73. Summary, Logistic Regression Analysis, Auto-Adaptive Systems

Response Function	Public Organizations			Nonprofit Organizations			Private Organizations		
	R^2	F	p	k	R^2	F	p	k	R^2	F	p	k

Response Function	R^2	F	p	k	R^2	F	p	k	R^2	F	p	k
Northridge, CA												
Emergency Response	0.65	34.6	0.001	1.00	0.84	102.11	0.000	1.82	NE	NE	NE	NE
Damage Assessment	0.50	18.9	0.001	1.46	NE	NE	NE	NE	0.89	159.39	0.000	1.27
Communication/Coordination	0.33	9.7	0.01	1.00	NE	NE	NE	NE	0.57	25.61	0.001	2.04
Disaster Relief	0.09	1.9	NS	3.31	0.05	0.99	NS	3.99	NE	NE	NE	NE
Recovery/Reconstruction	0.20	4.7	0.05	*3.95*	0.08	1.63	NS	*3.82*	NE	NE	NE	NE
Financial Assistance	0.12	2.6	NS	*3.92*	NE	NE	NE	NE	NE	NE	NE	NE
Subgroup Total	0.57	25.65	0.00	1.00	0.02	0.37	NS	*3.93*	NE	NE	NE	NE
Grand Total	0.43	14.51	0.01	1.00	0.43	14.51	0.01	1.00	0.43	14.51	0.01	1.00
Hanshin, Japan												
Emergency Response	0.23	5.7	0.05	*3.95*	0.16	3.59	NS	*3.95*	NE	NE	NE	NE
Damage Assessment	0.55	23.6	0.001	2.09	NE	NE	NE	NE	0.51	19.61	0.001	2.21
Communication/Coordination	0.56	24.0	0.001	*3.95*	NE	NE	NE	NE	NE	NE	NE	NE
Disaster Relief	0.01	0.2	NS	*3.95*	0.05	1.07	NS	1.12	0.12	2.70	NS	*3.34*
Recovery/Reconstruction	NE	NE	NE	NE	NE	NE	NE	NE	NE	NE	NE	NE
Financial Assistance	0.53	21.2	0.001	1.42	NE	NE	NE	NE	NE	NE	NE	NE
Subgroup Total	0.33	9.3	0.01	*3.95*	0.06	1.31	NS	*3.95*	0.14	3.06	NS	*3.34*
Grand Total	0.11	2.25	NS	*3.95*	0.11	2.25	NS	*3.95*	0.11	2.25	NS	*3.95*

NE = No equation; NS = Not significant
k values in italics indicate that the function is in chaos

stable, if moderate, Communication/Coordination function in the public sector. But the overall findings for the system indicate likely problems in adaptation, if subjected to a more severe earthquake of longer duration.

The findings for the case of Hanshin, Japan reveal a response system under severe stress. The function of Emergency Response produced a modest equation ($R^2 = 0.23$; $F = 5.7$; $p < 0.05$; $k = 3.95$) which is statistically significant, but the high k value indicates chaotic performance. Further, two other basic functions, Communication/Coordination and Disaster Relief, show k values of high instability. The result is that both the public and nonprofit subsets show k values indicating chaotic performance, and the private sector, with a k value of 3.34, is just at the edge of chaos. The total system produced an equation ($R^2 = 0.11$; $F = 2.25$), but this finding was not statistically significant, and the k value at 3.95 indicates chaotic performance. Although a response system did emerge, it was unable to adapt successfully within the three-week period to the urgent demands of the environment. In the succeeding weeks, the response system did stabilize, and performance improved notably in the reconstruction/recovery period.

The Process of Transition

The findings presented in this chapter demonstrate that each of the eleven response systems underwent some form of transition and adaptation to the sudden, urgent threats posed by the earthquakes to their respective communities. Yet, the process of this transition was clearly more difficult in some cases than others. Consistent with this process is the central role of Communication/Coordination, and the clear focus on saving lives and protecting property in prompt emergency response.

The initial model of adaptation appears to hold. It identifies the set of Initial Conditions as the existing state in which the hazardous event occurs. This set of conditions influences the system's subsequent evolution through sequential processes of Information Search, Information Exchange, and Organizational Learning to reach the (temporary) outcome state of Adaptive Performance. Although the initial conditions in which the eleven earthquakes occurred differ, the same phases of transition can be observed across the set of eleven response systems. Transition is viewed as an evolving sequence of interactions that mark the exchange of information, resources, and action between the vulnerable communities and their seismic environments.

While improvement in performance is demonstrated over the set of eleven response systems, it is not chronological and it is not necessarily related to the magnitude of the earthquake event. Rather, performance appears to be influenced much more by the degree of complexity in interdependent systems at the local level, and by the effectiveness of the communication and coordination systems already in place. There is much work to be done in terms of developing resilient communities that are able to identify the risks to which they are vulnerable and allocate their resources appropriately to reduce those risks.

CHAPTER TEN

SOCIOTECHNICAL SYSTEMS AND THE REDUCTION OF GLOBAL RISK

Complexity in Disaster Environments

The responsibility for addressing shared risk remains a daily task in vulnerable communities, but reflection on the eleven cases and their implications for collective action to reduce seismic risk suggests new insights and fresh approaches. The field studies present compelling evidence of rapidly evolving complexity in disaster environments. This complexity arises from the interaction of technical with organizational systems, designed to support the ordinary transactions of human communities. Efforts to improve the performance of either technical systems or organizational systems in disaster environments separately are likely to prove ineffective, given the interdependent functions they perform. If one fails, the other is disabled, unless alternate strategies are anticipated and placed in reserve. The two types of systems need to be integrated carefully in order to increase a community's capacity to reduce risk and respond effectively to threats when they occur. Such integration requires a sociotechnical approach, linking organizations, computers, physical monitoring systems, and community residents in a coherent, adaptive process to reduce risk and reallocate resources and energy to meet changing needs.

A sociotechnical approach requires a shift in the conception of response systems as reactive, command-and-control driven systems to one of inquiring systems, activated by processes of inquiry, validation, and creative self-organization. Inquiring systems function best with an appropriate investment in information infrastructure and organizational training that enables the system to assess accurately the conditions in a community that precipitate risk and to act quickly to reduce threat or minimize the consequences when destructive events occur. Combining technical with organizational systems appropriately enables communities to face complex events more effectively by monitoring changing conditions and adapting its performance accordingly, increasing the efficiency of its use of limited resources. It links human capacity to learn with the technical means to support that capacity in complex, dynamic environments.

The 'Edge of Chaos'

Each community appears to have a threshold point at which it is either able to absorb the shocks and damage inflicted by an earthquake and, with sufficient external assistance, to form a new, more effective mode of operation, or it dissipates its energies in unproductive efforts to maintain its previous pattern of operations without adequate assistance. Under the latter conditions, public managers responsible for the protection of life and property in the community are unable to meet the social and economic needs of large segments of the population, or to forestall the inevitable criticism of its operations as the population bears the brunt of the cost and losses. In either case, each community included in this study formed a distinct disaster response system that focused specifically on meeting needs generated by the disaster event.

Under favorable conditions, the community's basic operations are reinforced by external assistance as the response system widens the scope of its actions and accesses a broader range of resources to achieve a new mode of operation required to meet the demands of the disaster. In unfavorable conditions, the community's ordinary operating functions are seriously disrupted by the earthquake. Destructive consequences multiply, and the whole community slides toward chaos. Eventually, a new order is restored, but at substantial cost to the community, its residents, and the wider society that sends resources and personnel to assist.

Dynamic Processes in the Evolution of Response Systems

Information processes drive the dynamic evolution of response systems in either constructive or destructive ways. The effectiveness of these processes — information search, exchange, and organizational learning — depends not only upon the initial conditions in which the event occurs, but also upon the cumulative effects of their performance. That is, if any one process is absent, disrupted or blocked, the evolution of the system to adaptive behavior or self-organization is distorted or stalled. The four subsets of field studies of disaster response operations offer insight into the dynamics of evolving strength and failure in these complex, dynamic systems. The reported experience of the field studies suggests a set of conditions that consistently strengthen and facilitate the rapid evolution of response systems, and equally important, a set of weaknesses that repeatedly stall or deter their evolution. These patterns, observed in the sequence of evolving states of response among the subsets of field studies, are summarized below.

Creating the Context for Self-Organization

What have we learned from this comparative study of disaster response systems in eleven very different economic, social, legal, and cultural settings? Differences and similarities can be identified between the types of disaster response systems on four critical issues that are instructive as thoughtful policy makers consider means of reducing risk from seismic events on a global scale. These differences and similarities are summarized briefly by type of adaptive system on the issues of timing, balance between structure and flexibility, self-organization, and sustainability.

Timing in the Evolution of the Response System

Response is driven by the threat to life and the brief period in which life can be sustained following injury or severe threat in disaster. The ability to interpret a threatening event accurately and to mobilize response accordingly is crucial to effective response. The varying capacities of the response systems to meet this criterion effectively served as a distinguishing measure among the four types of systems.

Nonadaptive Systems: San Salvador, Ecuador, Armenia

Interdependent emergency response organizations were unable to make rapid transition to an emergency response system vital to saving lives in the first hours following the earthquake. Critical factors inhibited this transition in each case. They include:
— The basic technical means needed to support the search for, and exchange of, information in the dynamic disaster environment either did not exist or were not functioning
— The existing organizational structure for emergency response was inadequate for the magnitude of the earthquake disaster and its impact upon the region
— The community organizations and population had little knowledge of, or preparation for, seismic risk

Given these conditions, the organizations involved in disaster response operations found it difficult to mobilize action at multiple organizational levels and sustain it effectively over time. Each of these communities succeeded in overcoming these difficulties to achieve a remarkable level of coherence in practice, but the cost in lives and property was very high.

Emergent Adaptive Systems: Mexico City, Costa Rica, Erzincan

After the initial shock of the event, these response systems mobilized quickly to respond to the needs of their respective communities. Yet, the actions were carried out primarily by the national governments supported by international

organizations, with relatively little participation by, or engagement with, community organizations. Factors inhibiting the adaptation of these systems include:
- Existing emergency plans proved inadequate or unworkable in the specific contexts of the disaster
- Inadequate organizational linkages existed between national, provincial and local levels of government, inhibiting the timely exchange of information among emergency personnel and an accurate assessment of needs for the population
- Inadequate knowledge of the structure, facilities, resources, and vulnerabilities of the stricken area resulted in inadequate action to reduce seismic risk to the population

In each case, the national public organizations demonstrated capacity to innovate and respond to the needs of the population, but the response was more a reaction to the specific earthquake event than a conscious implementation of a policy to reduce known seismic risk for the population.

Operative Adaptive Systems: Whittier Narrows, CA; Loma Prieta, CA; Marathwada, India

In each of these three cases, the timing of the response was rapid and effective. Action was facilitated both by technical means and organizational training. These factors included:
- Access by local personnel to advanced communications equipment and facilities
- Sufficient training among local personnel to use this equipment effectively in a disaster event, and to organize a collective response
- Immediate recognition by policy makers of the impact of the event upon the community and prompt action to alleviate the losses and continuing risks

In these three cases, the integration of technical and organizational support facilitated the activation of disaster response. The larger task of providing an informed, multijurisdictional knowledge base to facilitate coordinated decision-making, however, had not yet been undertaken.

Auto Adaptive Systems: Northridge, CA and Hanshin, Japan

Both of these communities had the technical capacity to build infrastructure to support decision-making in event of disaster, but each chose different means to develop organizational preparedness. The earthquakes in the two communities, both metropolitan areas with dense populations, generated unexpected tests of the existing emergency plans. The results offer sober insights into the limitations of both methods, and the potential for either to contribute to a genuine process of self-organization to reduce risk.
- Northridge, CA: Informed knowledge base regarding seismic risk in California; primary investment for risk reduction made in training and preparedness

for public organizations with legal responsibility for protection of lives and property in communities; some investment in public education for population regarding seismic risk
- Northridge, CA: Technical equipment to improve communication was brought in after the event to facilitate response and recovery
- Northridge, CA: Outreach to community organizations was impromptu and not always systematic or timely
- Hanshin, Japan: Informed professional knowledge base regarding seismic risk in Japan; primary investment for risk reduction made in technical design of buildings, transportation structures
- Hanshin, Japan: Organizational strategies to respond to needs of disaster environment developed slowly after the event, with insufficient prior planning or training of personnel; little investment in public education for population regarding seismic risk
- Hanshin, Japan: Outreach to community organizations came late in the response period; volunteer groups formed, exhibiting characteristics of self-organization

The two strategies showed very different results in practice. In Northridge, the prior organizational training and experience of emergency personnel, supported by an advanced communications network, resulted in rapid mobilization of an emergency response network. Technical designs for buildings and transportation structures failed in unexpected ways, disrupting the daily operations of the community. Practiced organizational procedures quickly brought external technical equipment, personnel, and financial assistance from state and federal sources. The response system escalated very quickly, but experienced observers indicated it was operating at the full limit of its ability to absorb new demands. This observation is confirmed by nonlinear logistic regression analysis, showing a number of response functions that were operating within the chaotic range.

In Hanshin, when the technical structures for transportation, buildings, communications and lifeline systems failed under the force of the severe earthquake, the existing organizational structures were unable to respond promptly to assess the extent of the damage and mobilize a response system. A response system was mobilized, but the delay was costly in terms of lives and property lost.

Balance between Structure and Flexibility

The balance between structure and flexibility is the criterion that defines the adaptiveness of a response system in meeting its goals in a dynamic environment. It depends upon the integration of technical and organizational components to enable human managers to adapt quickly and effectively to new demands under the escalating complexity of a disaster environment. The four types of systems exhibited different balances of structure and flexibility, and different degrees of ability to adjust this balance under the duress of an actual disaster.

Nonadaptive Systems: San Salvador, Ecuador, Armenia

Each of the three systems had difficulty in adjusting the balance of structure and flexibility in their response systems, the result of competing pressures and criticisms in their wider social and political environments as well as inadequate information infrastructures. Efforts were made toward flexibility in each case, but strong and hostile criticism was also leveled against the changes proposed by public organizations to meet community needs. The outcome in each case limited the needed flexibility of the public institutions to adapt their performance effectively to meet the needs of their respective communities. Obstacles to increased flexibility included:
— Efforts to create a unified structure for disaster response wavered under strong criticism and eventually collapsed
— Pre-existing tensions in the society were exacerbated by the disaster, creating a divisive rather than unifying force to address the difficult problems of response, recovery and reconstruction
— Actions taken to cope with the tensions reinforced prior conflicts of interest, reducing flexibility to make necessary adaptations to meet the needs of the afflicted communities

In each case — San Salvador, Ecuador, Armenia — the existing government was operating under strong criticism or hostile attack prior to the occurrence of the earthquake. The governments had scant reserves of trust or good will from their respective populations to enable them to alter their procedures in order to cope more effectively with the demands from the disaster. Efforts to increase flexibility were viewed with suspicion and rejected or limited by wary populations.

Emergent Adaptive Systems: Mexico City, Ecuador, Erzincan, Turkey

Each of the three systems took innovative steps in response to the disaster. Sustaining the innovations across organizations, jurisdictions and over time proved more difficult. Several factors inhibited the secure establishment of innovative measures in disaster response operations. They included:
— Lack of adequate, timely, two-way communication between field and headquarters in the conduct of disaster operations
— Lack of an adequate knowledge base of shared information to support interorganizational decision-making in the conduct of disaster operations, as well as to sustain innovations across all jurisdictional levels and over time
— Economic pressures seeking to minimize the cost of disaster, resume normal economic operations as quickly as possible, and alleviate the social and economic cost to the nation
— Underestimation of the long-term damage incurred from the disaster in terms of the continuing social and economic development of the country

In each case, resources allocated for immediate disaster relief proved inadequate for the long-term reconstruction and rehabilitation efforts necessary to sustain innovative efforts to reduce seismic risk for the community. The flexibility achieved during disaster operations receded to the previous levels of structure over time.

Operative Adaptive Systems: Whittier Narrows, CA; Loma Prieta, CA; Marathwada, India

These three response systems effectively adapted the balance of structure and flexibility in their existing systems of civil protection to achieve operational capability during the crucial first hours of response. As competing demands increased during the second and third weeks of response, these systems found it more difficult to calibrate an appropriate balance in the dynamic environment of response leading to recovery. The transition to response proved effective; the second transition to recovery proved more difficult. Factors contributing to this difficulty were:
- Lack of a cumulative knowledge base of disaster operations for the entire response system, each of which had multiple sites and operations offices
- Lack of an adequate community knowledge base prior to the disaster, which could inform long-term decisions regarding recovery operations
- Lack of a clear, continuing program for integrating the process of rebuilding community structures with the simultaneous reduction of future seismic risk
- Lack of recognition of the need to engage the wider community in a continuing process of seismic risk reduction

In each case, the primary focus of the response system stayed on response, even as the needs of the respective communities moved to recovery, reconstruction, and rehabilitation. Strategies which proved effective during response operations needed to be revised in working directly with community organizations to engage their participation and effort in the process of repair, reconstruction, and recovery. The response systems appeared to reach a limit to their flexibility which inhibited their transition to the next phase of recovery.

Auto Adaptive Systems: Northridge, CA and Hanshin, Japan

The balance between structure and flexibility reflected the prior choices made by each community in terms of acknowledging its vulnerability to seismic risk. In Northridge, emergency response personnel were well-trained and experienced, and were able to adapt effectively within the parameters of their training and experience for first response. This same capability for adaptation was not apparent in the response functions of disaster relief and financial assistance, which involved other organizations in the public sector. The balance was even less evident among the nonprofit and private organizations.

In Hanshin, emergency response personnel were overwhelmed by the immediate demands of the disaster and hindered by lack of investment in organizational training at the local level. Upper administrative levels of the system, following

the established procedures, found it difficult to assess the damage and the needs of the community without adequate input from the local level. The balance of structure and flexibility was distorted, given the extraordinary demands of the disaster, in the first hours and days. This balance was restored, but late into disaster operations. In both cases, finding the appropriate – and changing – balance between structure and flexibility over the dynamic course of disaster events proved difficult. Factors influencing this balance – in either direction – included:
- Presence (or absence) of advanced technical means for the exchange of information among operations personnel and policy makers through a satellite system and related radio and cell phone networks
- Level of organizational training, professional experience and planning for operations in a major, urban disaster
- Existence of predefined emergency plans which governed response operations, constraining adaptation by requiring specific procedures, even if unworkable
- Presence (or absence) of adequate feedback procedures among the participating organizations in the response system
- Informed support and cooperation (or lack of same) from organizations in the wider community – nonprofit, private, and residential populations

Finding the appropriate balance between structure and flexibility and adapting it to the different requirements and conditions of disaster response is an extremely demanding task, requiring not only shared knowledge, but also trust among response personnel and resonance between response personnel and the local communities. This task is far more difficult when it is addressed for the first time under the urgent stress of an actual disaster. In Northridge, response personnel had the advantage of considerable experience in disaster operations, but the balance must be defined appropriately for each specific disaster. In Hanshin, response personnel had limited experience with a disaster of the magnitude of the earthquake in a densely populated urban community. The limits and potential of each response system depend to a very large degree on the ability of responsible personnel to recognize and define the appropriate balance for this operational equation.

Self-Organization

Self-organization is spontaneous action based upon informed choice taken to achieve a collective goal. It exemplifies citizens acting together, voluntarily, to meet a common need. It represents the fullest type of adaptation in a complex system that engages participants in collective action to reduce risk. It is likely the most efficient means of risk reduction, as well as the most acceptable to residents of communities vulnerable to seismic risk. But self-organization is very difficult to achieve consistently in communities that confront actual seismic events, which occur infrequently in any one location, but with relative frequency across the world. To what extent was self-organization identified in the field studies of disaster operations? If present, can it be incorporated into disaster planning as a continuing objective of disaster response?

Elements of self-organization were observed in each response system, in terms of the spontaneous acts of individuals, organizations, or sets of organizations. Yet self-organization as a recognizable, sustained process for whole response systems appeared elusive or at best, limited to temporary adjustments that reverted back to their original patterns once the obvious threat was past. Individual acts of self-organization are the easiest to recognize. For example, they include the group of Guatemalan workers who heard news of the earthquake in San Salvador on their radio, tossed their equipment into a pick-up truck and drove to San Salvador, arriving as the first rescue team at the disaster site. Such acts also include the Protestant church worker who used his ham radio to send news of the earthquake disaster from his village in Napo Province, Ecuador to his Church's offices in Quito when the telephone communications were disabled. These acts also include the spontaneous organization of volunteers, for example, the students at Yerevan State University in Armenia who dropped their books, picked up shovels, and climbed into University buses to travel to Leninakan and dig through the rubble to rescue victims.

Self-organization achieves more influence in a response system when links are established to other organizations. For example, in the Mexico City disaster, the US Search and Rescue Team arrived with the support of the US Office of Foreign Disaster Assistance, but at the spontaneous suggestion of a group of trained search dog owners who had learned of the earthquake while attending a National Association of Search and Rescue (NASAR) conference and volunteered to go immediately. That experience served as the basis for organizing a national roster of volunteers for search and rescue teams within the United States, with a set of national standards for performance for acceptance on the team. In Marathwada, India, people of nearby cities and villages volunteered professional skills and days of work to assist people who had lost their homes and belongings in the Latur and Osmanabad districts. They did so most successfully when they worked through charitable organizations which provided structure for their efforts and direction to their tasks.

Sustaining the process of self-organization in a continuing way requires access to communication for all of the participants to support the exchange of information, stored memory for actions taken that allow reflection and redesign, and evaluation and feedback from the other participants in the group. In no response system was this capacity fully or consistently available, nor was it recognized by many disaster managers as essential for the conduct of wider disaster operations. It is not surprising that processes of self-organization were not observed for the whole community in any response system. The more difficult question is whether the concept of self-organization at the community level represents a feasible approach to the reduction of seismic risk.

Given the findings from this analysis, self-organization appears not only feasible, but it is now currently within reach of virtually any community in the world that confronts seismic risk. It represents a voluntary response to the reduction of risk, based upon informed choice. It depends most heavily on a widely available resource, human capacity for learning. But it does require investment in both technical and organizational infrastructure to provide the structure and flexibility necessary to support processes of information search and exchange that lead to learning.

Self-organization, supported by an adequate information infrastructure, provides content to the concept of coordination. No longer would coordination be a hollow guise for external control. With access to the means for easy exchange and validation of information, managers will be able to propose new actions, track their performance, invite rigorous review, redesign projects on the basis of valid information, offer timely feedback on actions taken, and act cooperatively with others in more informed, responsible ways to achieve collective goals. Coordination, facilitated by appropriate technical and organizational functions, is transformed into interorganizational learning.

Sustainability of Changes in Community Performance

The transition from response to recovery is as difficult for organizations as the transition from routine operations to disaster response. It requires different skills of negotiation and readjustment, relinquishment of emergency powers, and acceptance of new responsibilities, frequently with fewer resources. It means continuing previous operations, often in a physically diminished setting, while coping with loss and the disruption of reconstruction and reorganization of work and community life.

Since the period of analysis for this study covered only the period of three weeks immediately following the disaster, this transition from response to recovery is noted, but not fully not addressed. In all eleven cases, this three-week period covered the response phase, and included at lease a partial transition to recovery. But also in all eleven cases, the recovery phase continued for months beyond the first three weeks, and in most cases, for years beyond the actual event.

Only in the instance of the three California earthquakes, occurring within seven years under the same state administrative structure and emergency response plan, is it possible to get a glimpse of the dynamic processes involved in securing sustainability. Changes initiated in community performance following one earthquake were tested in action in subsequent earthquakes. The risk is that, without systematic evaluation and broad participation in this review, partial conclusions may be drawn and false inferences made which may place the community at greater risk in the next event. Experience from the earlier earthquakes reinforced the need for training prior to the Northridge event, but the intensive focus on training for public personnel may have diminished the importance of training for nonprofit and private organizations as well. Without broad community support, the public organizations lose resiliency — the ability to bounce back — from a major disaster. The intent is to design a community-wide system that reduces vulnerability to earthquakes before they occur, yet responds effectively when they do.

This study notes the interdependence of response with recovery, and the importance of linking the two sets of processes in a coherent effort to increase the sustainability of communities vulnerable to seismic risk. This linkage can be supported by a disaster-specific knowledge base which monitors the continuing actions of response personnel and the incoming information regarding the two processes for legitimate participants in the response system. Elements that contributed to an effective response system were often improvisational and rudimentary. These elements could be greatly strengthened through systematic development, training, information exchange, and systematic monitoring or feedback among participating organizations to increase their effectiveness in continuing efforts to reduce seismic risk. In this respect, recovery is not an event that is completed with finality. For communities vulnerable to seismic risk, it is a continuing process of limiting risk in the ordinary transactions of community life.

Noting differences and similarities among response systems, this analysis confirms important characteristics of a nonlinear model of disaster response. First, disaster creates a symmetry-shattering event that both disrupts established patterns of thought and action and creates the opportunity to redesign an emergency response system that 'fits' the environment more effectively.

Second, the critical function of aggregating units from different levels of an intergovernmental disaster response system easily into a wider system of response underscores both the difficulty of this task under linear models of organization and the interdependence of these units in a massive, large-scale disaster. This function and its capacity to mobilize resources — personnel, equipment, supplies, skills, and knowledge — requires a mechanism of information exchange to achieve a shared system-wide goal: protection of life and property. This function is performed more effectively in rapidly changing disaster environments by a nonlinear, dynamic system that is able to coordinate diverse resources, materials, and personnel across previously established organizational and jurisdictional boundaries through means of information exchange guided by a clear 'internal model' or goal for action and prompt feedback. Such a system uses processes of self-organization in which informed participants initiate action, but adjust their action to that of others operating toward the same goal to achieve a timely, efficient response. It is essentially an organizational system operating in parallel, supported by a strong, distributed information system.

Third, the goal of the disaster response system serves as an 'internal model' for self-organizing processes. This goal allows participants from diverse perspectives, experience, and resources to adjust their actions and contributions to that of other participants in the system. Finally, an 'epistemic community' (Haas 1990) of knowledgeable people from diverse backgrounds, experience, and organizations that focuses on the shared problem is vital to the articulation of a common goal for the formulation of strategies for risk reduction and action that can be communicated to a wider set of responsible actors. This step is essential to the development of 'resonance' or willingness to support shared action when necessary to sustain the goal of a responsible, civil society. There are some indications that the formation of such a self-organizing community of knowledgeable people is already taking place in, for example, the international effort to establish a Global Disaster Information Network, initiated at a conference sponsored by the US Department of State in Washington, DC in July, 1998. The challenge is to build upon this spontaneous base of interest and experience to foster a continuing exchange of information, knowledge, and skills in the mitigation of seismic risk in vulnerable nations.

Conclusions

Returning to the research questions posed in Chapter 3, I can now indicate the conclusions that have been reached throughout this study. The questions, briefly restated, are:
1. How do the content and exchange of information affect the decision-making capacity of public/private/nonprofit managers engaged in disaster operations?
2. What kinds of information are required by public/private/nonprofit managers to coordinate actions of their respective organizations appropriately in a disaster environment?
3. In what ways can information content and exchange be structured to maximize adaptive performance within and between organizations in a disaster environment?

As demonstrated by the evidence presented in the analyses of the field studies, the content and exchange of information are critical to the ability of practicing policy makers and response personnel to adapt the balance between structure and flexibility in their operations to meet the changing needs of a dynamic disaster environment. Improving their capacity to do so is likely the most effective investment that any community can make in terms of the continuing reduction of seismic risk. Such an investment, however, may be beyond the resources available for any single community. It is an investment that likely requires national and global support. The returns on such investment, however, will also be national and global, as reduction in risk at the local level creates a fundamental reduction in risk at national and global levels. This finding is clearly supported by the evidence that increased capacity at local levels decreases dependence upon national and international organizations for disaster relief.

The evidence presented by the field studies also indicates clearly that two kinds of information are essential to enable managers of public, nonprofit and private organizations to coordinate their actions effectively in disaster response. First, these managers need an accurate assessment of both vulnerabilities and resources in their existing communities prior to the earthquake. Such information is at once interdisciplinary and interjurisdictional. Building the knowledge base for effective risk reduction is a collective process. Recognized as such, with the appropriate investments made in both technical and organizational development, such a knowledge base becomes the focus both for continuing organizational learning and the capacity of the community to monitor its own risk and adjust its performance accordingly.

Second, responsible managers need real-time information disseminated simultaneously to relevant participants in the policy and response process. The technical means to create such systems are now available. Linking these means to organizational processes effectively is the next task for effective risk reduction.

Structuring the content and exchange of information to maximize adaptive performance within and between organizations in disaster environments, as shown by the evidence in this study, can only be accomplished effectively by the thoughtful design of sociotechnical systems. Advanced information technology currently makes it possible to achieve a level of timeliness and accuracy under the urgent stress of a disaster environment not previous considered. The organizational design and procedures to incorporate this technical capacity fully into vulnerability assessment and response operations has not yet been fully recognized and accepted. This is the major task that needs to be undertaken in a global effort to reduce seismic risk, and to enable processes of self-organization at local, national, and global levels of jurisdiction.

In conclusion, the eleven cases of the disaster response systems indicate that processes of self-organization in disaster response are dependent upon a sociotechnical infrastructure that supports the timely, accurate exchange of information in a rapidly changing environment. The N-K model offers a means of assessing disaster response systems in different contexts in the effort to gain insight into the dynamics of the rapidly evolving process of disaster response and recovery. This approach offers a significant opportunity to design information strategies for disaster environments that would facilitate the constructive emergence of self-organizing systems guided by the system-wide goal of protection of life and property. Toward this objective, I offer the following recommendations:

1. Evaluate the model of a rapidly evolving disaster response system and the methodology for assessing such systems in practice in the context of actual organizations. Such an evaluation could be most appropriately carried out in the context of a computer-based simulated disaster operations exercise, which would facilitate and track the processes of information search, exchange, and adaptation in action
2. Increase investment to build information infrastructures that include both technical and organizational components in communities vulnerable to seismic risk. Such investment may involve a range of resources: international, national, state/provincial and local. The design is likely most effectively carried out through a consortium of public, nonprofit and private organizations, but the goal must be clearly defined as serving the interests of the community as a whole. That is, the information infrastructure should be designed for

wide access to responsible community members, providing them with the means to learn and incentives for cooperative performance
3. Design the knowledge base for the information infrastructure at the local level, with full participation of personnel who would be responsible for disaster response operations, in a manner consistent with global standards to facilitate the exchange of information quickly if an event escalates to national or international scales of response
4. Define standards of responsible performance for access to the community knowledge base, and organize programs of training and development for participants with different roles in community response and different levels of responsibility for community functions; these programs would also be consistent with global standards for disaster response
5. Implement the design for a Global Disaster Information Network that includes physical international, national, and local networks of computers, organizations, people and monitoring systems to provide a current source of valid, timely information regarding risk reduction to people living in communities vulnerable to seismic risk

Such strategies, further, would contribute to our broader theoretical understanding of complex adaptive systems and how they facilitate or fail to support proposed actions in the international arena. Seismic risk is shared, not only among individuals and organizations within specific communities vulnerable to earthquakes, but also among the thirty-six nations of the world that are exposed to seismicity from their physical environments. Shared risk can best be confronted by increasing the capacity of local communities to assess, monitor, and reduce their own risk, supported by a wider network of resources from state, national, and global jurisdictions.

NOTES

1. Linda Wallace, Member, National Association of Search and Rescue, and Member, United States Search Dog Team to Mexico City Earthquakes, September 19–20, 1985. Interview, Pittsburgh, PA, November 22, 1985.
2. Several initiatives are already under way to initiate such a network. These include a Task Force established by Vice President Albert Gore to explore the development of a Global Information Network. (Disaster Information Task Force Report 1997) and two initiatives by the United Nations: the Risk Assessment Tools for Diagnosis of Urban Areas Against Seismic Disasters (RADIUS) Programme, directed by Kenji Okazaki, United Nations Department of Humanitarian Affairs, and The Global Programme, directed by J.M. Col, United Nations Department for Development Services and Management Support, (Col and Chu 1995).

BIBLIOGRAPHY

Aguirre, B.E. 1991. "Social Aspects of the Costa Rica Earthquake of April 22, 1991." College Station, TX: Department of Sociology, Texas A & M University, Technical Report.

Almond, G. and S. Verba. 1962. *The Civic Culture*. Princeton, NJ: Princeton University Press.

American Psychiatric Association. 1987. *Diagnostic and Statistical Manual of Mental Disorders*, 3rd edn. (revised). Washington D.C.

Argyris, Chris. 1980. *Inner Contradictions of Rigorous Research*. New York, NY: Academic Press, Inc.

Argyris, Chris. 1982. *Reasoning, Learning and Action: Individual and Organizational*. San Francisco, CA: Jossey-Bass.

Argyris, Chris. 1985. *Strategy, Change and Defensive Routines*. Boston, MA: Pitman.

Argyris, Chris. 1990. *Overcoming Organizational Defenses*. Boston, MA: Allyn & Bacon.

Argyris, Chris. 1991. "Teaching Smart People how to Learn." *Harvard Business Review*, 69(3):99–109.

Argyris, Chris. 1993. *Knowledge for Action: A Guide to Overcoming Barriers to Organizational Change*. San Francisco, CA: Jossey-Bass.

Argyris, Chris and Donald A. Schon. 1974. *Theory in Practice: Increasing Professional Effectiveness*. San Francisco, CA: Jossey-Bass.

Argyris, Chris and Donald A. Schon. 1978. *Organizational Learning: A Theory of Action Perspective.* Reading, MA: Addison-Wesley Publishing Co.

Axelrod, R.M. 1984. *The Evolution of Cooperation.* New York, NY: Basic Books.

Bak, P. and K. Chen. 1991. "Self-Organized Criticality." *Scientific American*, January:46–53.

Bardach, Eugene. 1977. *The Implementation Game: What Happens After a Bill Becomes a Law.* Cambridge, MA: MIT Press.

Bartels, Larry M. and Henry E. Brady. 1993. "The State of Quantitative Political Methodology." In: Ada W. Finifter (ed.) *Political Science: The State of the Discipline.* Washington D.C.: American Political Science Association, 121–159.

Barton, Allen H. 1969. *Communities in Disaster: A Sociological Analysis of Collective Stress Situations.* Garden City, NY: Doubleday.

Bateson, Gregory. 1980. "Men are Grass: Metaphor and the World of Mental Process." *The Lindesfarne Letter.* West Stockbridge, MA: Lindesfarne Press.

Benuska, L. (ed.) 1990. "Loma Prieta Earthquake Reconnaissance Report." *Earthquake Spectra: Socioeconomic Impacts and Emergency Response*, 6(Suppl.): 393–451.

Bermudez. M.Ch. 1993. "The 1991 Telire-Limon, Costa Rica Earthquake: Management and its Implications." Department of Sociology, Universidad de Costa Rica, San Jose, Costa Rica. A revised version of this paper was published in United Nations. 1996. *Guidelines for Disaster Management.* New York, NY: United Nations Department for Development Support and Management Services, Division of Public Administration and Development Management, 67–82.

Blalock, H.M. 1972. *Social Statistics.* New York, NY: McGraw-Hill.

Bruzewicz, Andrew J. and Harlan L. McKim. 1995. "Remote Sensing and Geographic Information Systems (GIS) for Emergency Management." Proceedings, International Symposium on Spectral Sensing Research 1995. Melbourne, Australia, November 26 – December 1.

Burt, R.S. 1982. *Toward a Structural Theory of Action: Networks of Social Structure, Perception and Action.* New York, NY: Academic Press.

Caiden, Naomi and Aaron Wildavsky. 1974. *Planning and Budgeting in Poor Countries.* New York, NY: Wiley.

California Office of Emergency Services. 1989. *Introduction to the Incident Command System*. Sacramento, CA: State Board of Fire Services.

California Seismic Safety Commission. 1992. *California at Risk: Reducing Earthquake Hazards, 1992-1996*. Sacramento, CA: Seismic Safety Commission: Report SSC 91-08.

Census of India. 1991. New Delhi, India: Government of India Printing Office.

Chambers, Robert. 1974. *Managing Rural Development: Ideas and Experience from East Africa*. Uppsala, Sweden: Scandinavian Institute of African Studies.

Chengappa, R. with A.K. Menon. 1993. "Earthquakes: Where Next?" *India Today*, October 31:54–55.

Churchman, C.W. 1971. *The Design of Inquiring Systems: Basic Concepts of Systems and Organizations*. New York, NY: Basic Books.

Civil Defense, Armenia SSR. 1989. Briefing. Civil Defense Headquarters, Yerevan, Armenia, March 21.

Cohen, Michael D. 1984. "Conflict and Complexity: Goal Diversity and Organizational Effectiveness." *American Political Science Review*, 78(2):435–451.

Cohen, Michael D. 1986. "Artificial Intelligence and the Dynamic Performance of Organizational Designs." In: J.G. March and R. Weissinger-Baylon (eds.) *Ambiguity and Command: Organizational Perspectives on Military Decision Making*. Marshfield, WI: Pitman, 53–71.

Cohen, Michael D., James G. March and Johan P. Olsen. 1972. "A Garbage Can Model of Organizational Choice." *Administrative Science Quarterly*, 17:1–25.

Cohen, W.M. and Levinthal, D.A. 1990. "Absorptive Capacity: A New Perspective on Learning and Innovation." *Administrative Science Quarterly*, 35(1):128–152.

Col, J.M. and Chu, Jean. 1996. "Executive Summary, Global Programme for the Integration of Public Administration and the Science of Disasters." In: United Nations *Guidelines for Disaster Management*. New York, NY: Department for Development Support and Management Services, Division of Public Administration and Development Management, 43–51.

Comfort, L.K. 1985. "Action Research: A Model for Organizational Learning." *Journal of Policy Analysis and Management*, 5(1):100–118.

Comfort, L.K. 1986. "International Disaster Assistance in the Mexico City Earthquake." *New World*, 1(2):12–43.

Comfort, L.K. and A.G. Cahill. 1988. "Increasing Problem Solving Capacity between Organizations: The Role of Information in Managing the 31 May, 1995 Tornado Disaster in Western Pennsylvania." In: L.K. Comfort (ed.) *Managing Disaster: Strategies and Policy Perspectives.* Durham, NC: Duke University Press, 280–314.

Comfort, L.K. 1989a. "The San Salvador Earthquake." In: Uriel Rosenthal, Michael T. Charles, and Paul t'Hart (eds.) *Coping with the Crisis: The Management of Disasters, Riots, and Terrorism.* Springfield, IL: Charles C. Thomas, 323–339.

Comfort, L.K. 1989b. "Field Report, 1988 Armenia Earthquake." Pittsburgh, PA: University of Pittsburgh, Interdisciplinary Study Team, Moscow-Pittsburgh-Yerevan.

Comfort, L.K. 1989c. "Interorganizational Coordination in Disaster Management: A Model for an Interactive Information System." National Science Foundation Grant CES 88-04285.

Comfort, L.K. with T.M. Woods and J.E. Nesbitt. 1990a. "Designing an Emergency Information System: The Pittsburgh Experience." In: *Advances in Telecommunications Management, vol. 3, Information Technology and Crisis Management.* Greenwich, CT: JAI Press, 13–33.

Comfort, L.K. 1990b. "Turning Conflict into Cooperation: Organizational Designs for Community Response in Disaster." *International Journal of Mental Health*, 19(1):89–108.

Comfort, L.K. 1991a. "Organizational Interaction in Response and Recovery." In: Robert L. Schuster (ed.) *The March 5, 1987 Ecuadorian Earthquakes: Mass Wasting and Socioeconomic Effects.* Washington D.C.: National Research Council, Committee on International Disasters, 5:122–163.

Comfort, L.K. 1991b. "Designing an Interactive, Intelligent, Spatial Information System for International Disaster Assistance." *International Journal of Mass Emergencies and Disasters*, 9(3):339–353.

Comfort, L.K. 1993. "Integrating Information Technology into International Crisis Management and Policy." *Journal of Contingencies and Crisis Management*, 1(1):17–29.

Comfort, L.K. 1994. "Self-Organization in Complex Systems." *Journal of Public Administration Research and Theory*, 4(3):393–410.

Comfort, L.K. 1995. "Self-Organization in Disaster Response: Global Strategies to Support Local Action." Presented at the Workshop on "The United Nations, Multilateralism and Catastrophes." Institute of International Studies, University of California, Berkeley, November 9–10.

Comfort, L.K. 1996. "Self-Organization in Disaster Response: The Great Hanshin, Japan Earthquake of January 17, 1995." Boulder, CO: University of Colorado at Boulder, Natural Hazards Center, QR78. 51pp.

Comfort, L.K. 1997. "Shared Risk: A Dynamic Model of Interorganizational Learning and Change." In: James L. Garnett and Alexander Kouzmin (eds.) *Handbook of Administrative Communication*. New York, NY: Marcel Dekker Inc. 395–411.

Comfort, L., A. Tekin, E. Pretto, B. Kirimli, D. Angus and Other Members of the International, Interdisciplinary Disaster Research Group. 1998. "Time, Knowledge, and Action: The Effect of Trauma upon Community Capacity for Action." *International Journal of Mass Emergencies and Disasters*, 16(1):73–91.

Coveney, Peter and Roger Highfield. 1995. *Frontiers of Complexity: The Search for Order in a Chaotic World*. New York, NY: Fawcett Columbine.

Crecine, J.P. 1986. "Defense Resource Allocation: Gabage Can Analysis of C3 Procurement." In: J.G. March and R. Weissinger-Baylon. *Ambiguity and Command: Organizational Perspectives on Military Decision Making*. Marshfield, WI: Pitman, 72–119.

Cumhuriyet. 1992. Istanbul, Turkey. March 13 – June 2.

The Daily News. 1994. Los Angeles, CA. January 18 – February 6.

Deming, W. Edwards. 1986. *Out of the Crisis*. Cambridge, MA: Massachusetts Institute of Technology, Center for Advanced Engineering Study.

Deutsch, Karl W. 1963. *The Nerves of Government*. New York, NY: The Free Press.

Drabek, Thomas E. 1990. *Emergency Management: Strategies for Maintaining the Organizational Integrity*. New York, NY: Springer-Verlag.

Dryzek, J. 1987. *Rational Ecology: Environment and the Political Economy*. Oxford, UK: Basil Blackwell.

Dynes, Russell R. 1969. *Organized Behavior in Disaster: Analysis and Conceptualization*. Columbus, OH: Disaster Research Center.

Dynes, Russell R. and Kathleen J. Tierney (eds.) 1994. *Disasters, Collective Behavior, and Social Organization*. Newark, DE: University of Delaware Press.

Earthquake Engineering Research Institute Newsletter. 1992. Special Earthquake Report, "Erzincan, Turkey Earthquake of March 13, 1992," 26(5):1–4.

Earthquake Engineering Research Institute. 1994. *Northridge Earthquake, January 17, 1994, Preliminary Reconnaisance Report*, Chapter 9, "Social Impacts and Emergency Response," 86–89.

EQE Engineering. 1987. *Summary of the October 1, 1987 Whittier, California Earthquake*. San Francisco, CA: EQE Engineering.

EQE International. 1991. *The April 22, 1991 Valle de la Estrella Costa Rica Earthquake: A Quick Look Report*. (May). San Francisco, CA: EQE International Inc.

El Diario de Hoy. 1986. San Salvador, El Salvador. October 14 – November 3.

El Comercio. 1987. Quito, Ecuador. June 14 – July 15.

Excelsior. 1985. Mexico City, DF. September 20 – October 11.

Federal Emergency Management Agency. 1994. "Situation Reports, January 18 – February 6, 1994." Pasadena, CA: Federal/State Disaster Field Office.

Federal Response Plan for Public Law 93-288, as amended. Washington D.C.: U.S. Government Printing Office, 1994-514-748/80726.

Forrester, Jay W. 1987. "Nonlinearity in High-order Models of Social Systems." *European Journal of Operational Research*, 30:104–109.

Frankl, Victor E. 1970. *Man's Search for Meaning: An Introduction to Logotherapy.* New York, NY: Simon & Schuster Trade.

French, S. and G. Rudholm. 1990. "Damage to Public Property in the Whittier Narrows Earthquake: Implications for Earthquake Insurance." *Earthquake Spectra*, 6(1): 105–123.

Gilbreth, Frank B. and L.M. Gilbreth. 1917. *Applied Motion Study: A Collection of Papers on the Efficient Method to Industrial Preparedness*. New York, NY: Sturgis & Walton Company.

Geertz, Clifford. 1973. *The Interpretation of Cultures: Selected Essays*. New York, NY: Basic Books.

Gell-Mann, M. 1994. *The Quark and the Jaguar: Adventures in the Simple and the Complex*. New York, NY: W.H. Freeman & Co.

Goodman, Paul S, Sproull, Lee S. and Associates. 1990. *Technology and Organizations*. San Francisco, CA: Jossey-Bass.

Government of Maharashtra. 1993. *A Preliminary Report on the September 30, 1993 Earthquake*. Bombay, India.

Gupta, T.N. et al. 1993. "Action Plan for Reconstruction in Earthquake Affected Regions of Maharashtra." New Delhi, India: Ministry of Urban Development, Government of India, IV:1–7.

Gulkan, Polat. 1992. "A Preliminary Field Reconnaissance Report on the Erzincan Earthquake of 13 March, 1992." Ankara, Turkey.

Gulkan, Polat, and Oktay Ergunay. 1992. "The Turkish Disaster Management Seminar of the UNDRO under the UNDP, 2–5 June, 1992." Ankara, Turkey: General Directorate of Disaster Affairs, Ministry of Public Works and Settlement.

Haas, E.B. 1990. *When Knowledge is Power: Three Models of Change in International Organizations*. Berkeley, CA: University of California Press.

Habermas, Jurgen. 1979. *Communication and the Evolution of Society*. Boston, MA: Beacon Press.

Harlow, David H., Michael J. Rymer, and Randy A. White. 1986. "The San Salvador Earthquake of October 10, 1986 and Implications of the Regional Earthquake History." Menlo Park, CA: U.S. Geological Survey, Preliminary Report.

Hayes-Roth, F., D. Waterman and D. Lenat. 1983. *Building Expert Systems*. Reading, MA: Addison-Wesley Publishing Company.

Hindustan Times. 1993. New Delhi, India. October 1–2.

Holland, John. 1975. *Adaptation in Natural and Artificial Systems*. Ann Arbor, MI: University of Michigan Press.

Holland, John. 1995. *Hidden Order: How Adaptation Builds Complexity*. Reading, MA: Addison-Wesley.

Hoy. 1987. Quito, Ecuador. March 7–31.

India Today. 1993. October 11, 54

Japan Times. 1995. Tokyo, Japan. January 18 – February 6.

Kauffman, S.A. 1993. *The Origins of Order: Self-Organization and Selection in Evolution.* New York, NY: Oxford University Press.

Kauffman, S.A. and A.S. Perelson. 1991. *Molecular Evolution on Rugged Landscapes: Proteins, RNA, and the Immune System: Proceedings of the Workshop on Applied Molecular Evolution and the Maturation of the Immune Response, March, 1989, Santa Fe, New Mexico.* Redwood City, CA: Addison-Wesley Publishing Co.

Kiel, L.D. 1994. *Managing Chaos and Complexity in Government.* San Francisco, CA: Jossey-Bass.

Kerlinger, F.N. 1986. *Foundations of Behavioral Research.* Chicago, IL: Holt, Rinehart, and Winston, Inc.

King, G., R.O. Keohane, and S. Verba. 1994. *Designing Social Inquiry: Scientific Inference in Qualitative Research.* Princeton, NJ: Princeton University Press.

Knoke, David and James H. Kuklinski. 1982. *Network Analysis.* Beverly Hills, CA: Sage Publications.

Kobe City Fire Department. 1995. "Hanshin-Awaji Daishinsai (Kobe Shiiki) ni okeru Shobokatsudo no Kiroku." Kobe, Japan.

Krackhardt, David. 1992. "The Strength of Strong Ties: The Importance of Philos in Organizations." In: N. Nohria & R. Eccles (eds.) *Networks and Organizations: Structure, Form, and Action.* Boston, MA: Harvard Business School Press, 216–239.

Kurian, George Thomas. 1981. *The New Book of World Rankings.* New York, NY: Facts on File Inc. 11–12.

La Nacion. 1991. San Jose, Costa Rica. April 23 – May 15.

La Republica. 1991. San Jose, Costa Rica. April 23 – May 15.

Landau, M. 1991. "Multiorganizational Systems in Public Administration." *Journal of Public Administration Research and Theory,* 1(1):5–18.

Laporte, Todd R. and Paula M. Consolini. 1991. "Working in Practice, but not in Theory: Theoretical Challenges of 'High-Reliability Organizations.'" *Journal of Public Administration and Theory*, 1(1):19–48.

Lavell, A. 1991. "Prevention and Mitigation of Disasters in Central America: Social and Political Vulnerability to Disasters at the Local Level." Paper presented to the Developing Areas Research Group, British Geographers and the Royal Geographical Society on Disasters Vulnerability and Response, London, 3–4 May.

Lavell, A. 1993. "Estudio de Caso de los Desastres de Limon, Costa Rica de 1991." In: A. Maskrey and A. Lavell (eds.) *Manejo de Desastres y Mecanismos de Respuesta: Un Analysis Comparativo del Alto Mayo, Peru y Limon, Costa Rica*. Lima, Peru: La Red.

Leyendecker, E.V., L.M. Highland, M. Hopper, E.P. Arnold, P. Thenhaus, and P. Powers. 1988. "Early Results of Isoseismal Studies and Damage Surveys." *Earthquake Spectra: The Whittier Narrows, California Earthquake of October 1, 1987*, 4(1).

Lima, B.R. 1989. "Disaster Severity and Emotional Disturbance: Implications for Primary Mental Health Care in Developing Countries." *Acta Psychiatrica Scandinavica*, 79:74.

Lima, B.R. 1989. *Psicosociales consecuencias de desastre: La esperiencia Latinoamericana*. Chicago, IL: Hispanic American Family Center.

Lindblom, Charles E. 1992. *Inquiry and Change: The Troubled Attempt to Understand and Shape Society*, 2nd edn. New Haven, CT: Yale University Press.

Lindblom, Charles E. and David K. Cohen. 1979. *Usable Knowledge: Social Science and Social Problem Solving*. New Haven, CT: Yale University Press.

Linstone, Harold A. 1984. *Multiple Perspectives for Decision Making: Bridging the Gap between Analysis and Action*. New York, NY/North Holland: Elsevier Science Ltd.

Los Angeles Times. 1987. Los Angeles, CA. October 1–28.

Los Angeles Times. 1994. Los Angeles, CA. January 18 – February 6.

Los Angeles County. 1994. Emergency Operations Plan. Los Angeles, CA: Emergency Operations Center.

Los Angeles County. 1994. Briefing. Emergency Operations Center, Los Angeles, CA. February 4.

Luhmann, Niklas. 1989. *Ecological Communication*. Chicago, IL: University of Chicago Press.

March, J.G. 1988. *Decisions and Organizations*. Oxford, UK: Basil Blackwell.

Maskrey, A. and A. Lavell. 1993. *Manejo de Desastres y Mecanismos de Respuesta: Un Analisis Comparativo del Alto Mayo, Peru y Limon*, Costa Rica. Lima, Peru: La Red.

Meltsner, Arnold and Christopher Bellavita. 1983. *The Policy Organization*. Beverly Hills, CA: Sage Publications.

Miller, G. 1967. "The Magical Number Seven, Plus or Minus Two: Some Limits on Our Capacity for Processing Information." In: *Psychology of Communication*. New York, NY: Basic Books, 14–44.

Mitchell, William. 1993. "Social Impacts and Emergency Response: Turkish Earthquake Renaissance Report." June. This preliminary report served as the basis for Chapter 7, "Social Impacts and Emergency Response." In: *Earthquake Spectra: Erzincan, Turkey Earthquake of March 13, 1992*, 9(Suppl):101–111.

Mohr, Lawrence. 1982. *Explaining Organizational Behavior*. San Francisco, CA: Jossey-Bass.

National Emergency Committee. 1991. *National Emergency Plan*. San Jose, Costa Rica.

National Land Agency. 1995. *Summary of Reports from Ministries regarding Status of Hanshin-Awaji Disaster Operations*. Tokyo, Japan, March 29 – April 23.

Nechaev, E.A. and M.I. Reznik. 1990. "Methodological Approaches to the Problems of Medicine in Extreme Situations." *Voenno-medicinskiji Journal*, 4:5–10 (In Russian).

Nelson, Richard R. and Sidney G. Winter. 1982. *An Evolutionary Theory of Economic Change*. Cambridge, MA: Belknap Press of Harvard University Press.

Newell, Allen and Herbert A. Simon. 1972. *Human Problem Solving*. Englewood Cliffs, NJ: Prentice-Hall.

Nicolis, G. and I. Prigogine. 1989. *Exploring Complexity: An Introduction.* New York, NY: W.H. Freeman and Co.

Oakland Tribune. 1989. Oakland, CA. October 18 – November 14.

Olson, Mancur. 1971. *The Logic of Collective Action: Public Goods and the Theory of Groups*, 2nd edn. Cambridge, MA: Harvard University Press.

Parsons, Talcott. 1951. *Toward a General Theory of Action.* Cambridge, MA: Harvard University Press.

Peitgen, H., H. Jurgens, and D. Saupe. 1992. *Chaos and Fractals: New Frontiers of Science.* New York, NY: Springer-Verlag.

Perrow, C. 1972. *Complex Organizations.* Glenview, IL: Scott Foresman.

Perrow, C. 1984. *Normal Accidents: Living with High Risk Technologies.* New York, NY: Basic Books Inc.

Petak, William J. and Arthur Atkisson. 1982. *Natural Hazard Risk Assessment and Public Policy: Anticipating the Unexpected.* New York, NY: Springer Verlag.

Piaget, Jean. 1980. *Adaptation and Intelligence.* Chicago, IL: University of Chicago Press.

Priesmeyer, H. Richard. 1992. *Organizations and Chaos: Defining the Methods of Nonlinear Management.* Westport, CT: Quorum Books.

Priesmeyer, H. Richard. 1994a. *The Chaos System Software.* Fair Oaks Ranch, TX: Management Concepts, Inc.

Priesmeyer, H. Richard. 1994b. "A Logistic Regression: A Method for Describing, Interpreting and Forecasting Social Phenomena with Nonlinear Equations." Presented at the Chaos and Society Workshop. Université du Québec á Hull, Hull, Québec, CA.

Priesmeyer, H. Richard and W.T. Andrews. 1994. "Logistic Regression: Forecasting with Chaos Theory." San Antonio, TX: School of Business and Administration, St. Mary's University.

Priesmeyer, H. Richard and Lawrence F. Sharp. 1995. "Practical Uses of Chaos Theory." San Antonio, TX: School of Business and Administration, St. Mary's University.

Prigogine, I. 1987. "Exploring Complexity." *European Journal of Operational Research*, 30(2):97–103.

Prigogine, I. and I. Stengers. 1984. *Order Out of Chaos: Man's New Dialogue with Nature*. New York, NY: Bantam Books Inc.

Quarantelli, E.L. 1978. *Disasters: Theory and Research*. London, UK: Sage Publications.

RADIO Magazine. Moscow, April, 1990.

Rivlin, Alice. 1971. *Systematic Thinking for Social Action*. Washington D.C.: Brookings Institution.

Rivlin, Alice M. 1992. *Reviving the American Dream: The Economy, the States, & the Federal Government*. Washington D.C.: Brookings Institution.

Roberts, Karlene H. 1993. *New Challenges to Understanding Organizations*. New York, NY: Macmillan.

Rochlin, Gene I., Todd R. LaPorte, and Karlene H. Roberts. 1987. "The Self-Designing High-Reliability Organization: Aircraft Carrier Flight Operations at Sea." *Naval War College Review*, 40(4):76–90.

Rochlin, Gene I. 1989. "Organizational Self-Design as a Crisis-Avoidance Strategy: U.S. Naval Flight Operations as a Case Study." *Industrial Crisis Quarterly*, 3159–3176

Rochlin, Gene I. 1993. "Defining High Reliability Organizations in Practice: A Taxonomic Prolegomenon." In: Karlene H. Roberts (ed.) *New Challenges to Understanding Organizations*. New York, NY: Macmillan, 11–32.

Rochlin, Gene I. (ed.) 1996. "Special Issue: New Directions in Reliable Organizational Research." *Journal of Contingencies and Crisis Management*, 4(2).

Rossi, Peter, James D. Wright and Eleanor Weber-Burdin. 1982. *Natural Hazards and Public Choice: The State and Local Politics of Hazard Mitigation*. New York, NY: Academic Press.

Rueschemeyer, D. 1991. "Different Methods – Contradictory Results? Research on Development and Democracy." In: C.C. Ragin (ed.) *Issues and Alternatives in Comparative Social Research*. Leiden: E.J. Brill, 9–38.

Ruelle, D. 1991. *Chance and Chaos*. Princeton, NJ: Princeton University Press.

San Francisco Chronicle. 1989. San Francisco, CA. October 18 – November 14.

San Francisco Examiner. 1989. San Francisco, CA. October 18 – November 14.

San Jose Mercury. 1989. San Jose, CA. October 18 – November 14.

Sarkis, T., M.D. 1993. "Limon Earthquake: April 22, 1991." Paper prepared for the United Nations Interregional Seminar on Disaster Management, Jakarta, Indonesia, December 13–18.

Schneider, Walter. 1992. "Skill Acquisition, Transfer and Retention for High Workload Performance." Pittsburgh, PA: University of Pittsburgh, Learning Research and Development Center.

Schon, Donald. 1983. *The Reflective Practitioner: How Professionals Think in Action.* New York, NY: Basic Books.

Schon, Donald. 1987. *Educating the Reflective Practitioner: Toward a New Design for Teaching and Learning in the Professions.* San Francisco, CA: Jossey-Bass.

Schuster, Robert L. (ed.) 1991. *The March 5, 1987 Ecuadorian Earthquakes: Mass Wasting and Socioeconomic Effects.* Washington D.C.: National Research Council, Committee on International Disasters, vol. 5.

Sharan, V., O. Gupta, and S. Sethi. 1993. *A Comprehensive Note on the Latur Earthquake.* Latur, India: District Collector's Office. Prepared under the Guidance of P. Pardeshi.

Shrivastava, Paul. 1992. *Bhopal: The Anatomy of Disaster*, 2nd edn. London: P. Chapman Publishers.

Simon, H.A. 1983. *Reason in Human Affairs.* Stanford, CA: Stanford University Press.

Simon, H.A. 1981. *The Sciences of the Artificial*, 2nd edn. Cambridge, MA: MIT Press.

Smart, C. and I. Vertinsky. 1977. "Designs for Crisis Decision Units." *Administrative Science Quarterly,* 22(4):640–657.

Sovetakan Hayastan. 1988. Yerevan, Armenia. December 8–27.

The Statesman. 1993. Calcutta, India. October 1–7, 10.

Staw, Barry. 1991. "Rationality and Justification in Organizational Life." In: L.L. Cummings and Barry M. Staw (eds.) *Information and Cognition in Organizations.* London, UK: Jai Press, Inc.

Sutphen, Sandra and Virginia Bott. 1990. "Issue Salience and Preparedness as Perceived by City Managers." In: R.T. Sylves and W.L. Waugh Jr. (eds.) *Cities and Disaster: North American Studies in Emergency Management.* Springfield, IL: Charles C. Thomas, Publisher, 133–153.

Taghavi, M. 1987. "Summary of October, 1987 Earthquake Damage Reports." Structural Engineering Division, City of Los Angeles, October 19.

Tata Institute of Social Sciences. 1994. *Survey of People Affected by the Earthquake in the Latur and Osmanabad Districts, 1993.* Bombay, India: Joint Action Group of Institutions for Social Work Education. Final Report, February.

Taylor, F.W. 1967. *The Principles of Scientific Management*, 2nd edn. New York, NY: W.W. Norton.

Tarun Bharat. Solapur, India. October 1–31.

t'Hart, Paul and Bert Pijenburg. 1989. "The Heizel Stadium Tragedy." In: Uriel Rosenthal, Michael T. Charles and Paul t'Hart (eds.) *Coping with Crisis: The Management of Disasters, Riots, and Terrorism.* Springfield, IL: Charles C. Thomas, 197–224.

Thompson, James D. 1967. *Organizations in Action: Social Science Bases of Administrative Theory.* New York, NY: McGraw-Hill.

Tierney, K. 1988. "The Whittier Narrows, California Earthquake of 1987 – Social Aspects." *Earthquake Spectra*, 4(1):11–23.

Times of India. 1993. Bombay, India. October 1–21.

Times of India, New Delhi, India. October 1–21.

Times of India, Calcutta, India. October 1–21.

Train, H.D. 1986. "Decision Making and Managing Ambiguity in Politico-Military Crisis." In: J.G. March and R. Weissinger-Baylon (eds.) *Ambiguity and Command: Organizational Perspectives on Military Decision Making.* Marshfield, MA: Pitman, 298–308.

United Nations Economic Commission for Latin America and the Caribbean. 1987. "The Natural Disaster of March, 1987 in Ecuador and its Impact on Social and Economic Development." Santiago, Chile: Report LC/G. 1465, May 6.

United States Agency for International Development/El Salvador. 1986. "Assessment of Damages Resulting from the San Salvador Earthquake of October 10, 1986." San Salvador, El Salvador: USAID Mission.

United States Embassy. 1986. "The Question of Private Relief to the Archdiocese of San Salvador." Bulletins One and Two, San Salvador, El Salvador, October 16.

United States Office of Foreign Disaster Assistance. 1991. "Situation Reports, Costa Rica Earthquake of April 22, 1991." No. 4, April 30.

United States Office of Foreign Disaster Assistance. 1986. "Situation Reports, San Salvador Earthquake of October 10, 1986." Washington D.C. October – November.

United States Office of Foreign Disaster Assistance. 1985. "Situation Reports, Mexico City Earthquakes of September 19–20, 1985." Washington D.C. September – October.

United States Geological Survey. 1997. *Significant Earthquakes*. National Earthquake Information Center. Denver, Colorado.

Web, F.H. 1987. "Whittier Narrows Earthquake: Los Angeles County." *California Geology*, 40(12):275–281.

Weick, K.E. 1990. "Technology as Equivoque: Sensemaking in New Technologies." In: P.S. Goodman, L.S. Sproull, and Associates (eds.) *Technology and Organizations*. San Francisco, CA: Jossey-Bass, 1–44.

Weick, K.E. 1993. "The Collapse of Sensemaking in Organizations: The Mann Gulch Disaster." *Administrative Science Quarterly*, 22(3):606–639.

Weick, K.E. and K.H. Roberts. 1993. "Collective Mind in Organizations: Heedful Interrelating on Flight Decks." *Administrative Science Quarterly*, 38(3):357–381.

Weiner, S.S. 1976. "Participation, Deadlines and Choice." In: J.G. March and J.P. Olsen (eds.) *Ambiguity and Choice in Organizations*. Bergen, Norway: Universitetsforlaget, 225–250.

Weiss, Carol H. 1977. *Using Social Research in Public Policy Making*. Lexington, MA: Lexington Books.

Weiss, Carol H. and Michael J. Bucuvalas. 1984. *Social Science Research and Decision-Making*. New York, NY: Columbia University Press.

Weiss, Carol H. 1998. *Evaluation: Methods for Studying Programs and Policies*, 2nd edn. Upper Saddle River, NJ: Prentice Hall.

Weissinger-Baylon, R. 1986. "A Garbage Can Decision Processes in Naval Warfare." In: J.G. March and R. Weissinger-Baylon (eds.) *Ambiguity and Command: Organizational Perspectives on Military Decision Making*. Marshfield, MA: Pitman, 36–52.

Wildavsky, A. 1988. *Searching for Safety*. New Brunswick, NJ: Transaction Books.

Wilson, James Q. 1989. *Bureaucracy: What Government Agencies Do and Why They Do It*. New York, NY: Basic Books, 268–274.

Yin, Robert K. 1993. *Applications of Case Study Research*. Newbury Park, CA: Sage.

Yin, Robert K. 1994. *Case Study Research: Design and Methods*, 2nd edn. Thousand Oaks, CA: Sage.

APPENDICES

Appendix A: Estimation Procedures for Nonlinear Logistic Regression   294

Appendix B: Map of Disaster Zones, Ecuador, 1987   296

Appendix C: Sample Characteristics
  Table 1 — 1985 Mexico City Earthquake Study   297
  Table 2 — 1986 San Salvador Earthquake Study   298
  Table 3 — 1987 Ecuadorian Earthquakes Study   299
  Table 4 — 1987 Whittier Narrows, CA Earthquake Study   300
  Table 5 — 1988 Armenia Earthquake Study   301
  Table 6 — 1989 Loma Prieta, CA Earthquake Study   302
  Table 7 — 1991 Costa Rica Earthquake Study   303
  Table 8 — 1992 Erzincan, Turkey Earthquake Study   304
  Table 9 — 1993 Marathwada, India Earthquake Study   305
  Table 10 — 1994 Northridge, CA Earthquake Study   306
  Table 11 — 1995 Hanshin, Japan Earthquake Study   307

Appendix D: Characteristics of the Latur and Osmanabad Districts, Marathwada, India
  Table 1 — Initial Conditions, Characteristics of the Latur District   308
  Table 2 — Initial Conditions, Characteristics of the Osmanabad District   309
  Table 3 — A Comparative Assessment of Damage by District, 1993 Marathwada Earthquake   310

APPENDIX A

Estimation Procedures for Nonlinear Logistic Regression

The following procedures were developed by H. Richard Priesmeyer, and are implemented in his Chaos! software. This software was used to analyze the data from the content analysis of newspapers for the eleven cases of rapidly evolving response systems included in this study.

Following are the steps taken to estimate k and X in any data set:

Step 1: Standardize Target Data

Compute the mean and standard deviation of the target data (T_o) and compute Z values for each observation. Label this vector T_s for "standardized target values".
The equation is:

$$Z_n = \frac{X_t - \mu}{\partial_t}$$

where:

$$\partial_t = \frac{\sum (X-\mu)^2}{n}$$

Step 2: Compute Logistic Values

Compute repeated series of logistic values with n observations using the formula:

$$X_{n+1} = k * X_n * (1 - X_n)$$

Increment X with increments of k; k ranges from 0 to 4 while X ranges from 0 to 1. Step the incrementing of k and X by 0.01 or smaller.

Step 3: Standardize the Logistic Values

Standardize each series of the logistic values computed in Step 2 in the same way as the target values were standardized. Label this vector L_s for "standardized logistic values".

Step 4: Compute SSE and SSR

For each series of logistic values, compute the sum of squares error (SSE) between the standardized target values and the standardized logistic values. Specifically, compute:

$$SSE = \sum(T_s - L_s)^2$$

Compute the sum of squares regression (SSR) and R^2 as measures of the quality of fit between the logistic values and the target values. Compute SSR as SSTO-SSE where SSTO is the sum of squares total and is computed as SSTO = ΣT_s^2. Compute R^2 as SSR/SSTO (an F ratio can also be computed).

Step 5: Compare and Test

Compare SSR to the computed SSR on the previous trial. Retain values of k and X if SSR exceeds its previously computed highest values. Return to Step 2 until all step increments of k and X are completed.

Step 6: Compute Fitted Values and Forecasts

Generate the logistic values associated with the k and X parameters that had the highest SSR. To generate forecasted values, continue incrementing beyond the number of observed values. Standardize these data, then multiply each standardized value by the standard deviation of the target values and add the mean of the target data to each one. The resulting values are those that best fit the observed data. Those generated beyond the current number of observations are forecasted measures based on the underlying logistic equation that best fits the data. The equation for computing the fitted measures (F) from an iterative logistic expression is:

$$F_i = u_t + ((kx_{i+1}(1 - x_i) - u_1) * (\partial_t / \partial_1))$$

Where:
- F_i = fitted estimate for each observation
- u_t = mean of the target values
- k = the logistic constant
- x_i = the logistic variable at time i
- x_{i+1} = the logistic variable at time $i + 1$
- u_1 = mean of the logistic values
- ∂_t = standard deviation of target values
- ∂_1 = standard deviation of logistic values

APPENDIX B

Map of Disaster Zones, 1987 Ecuadorian Earthquakes

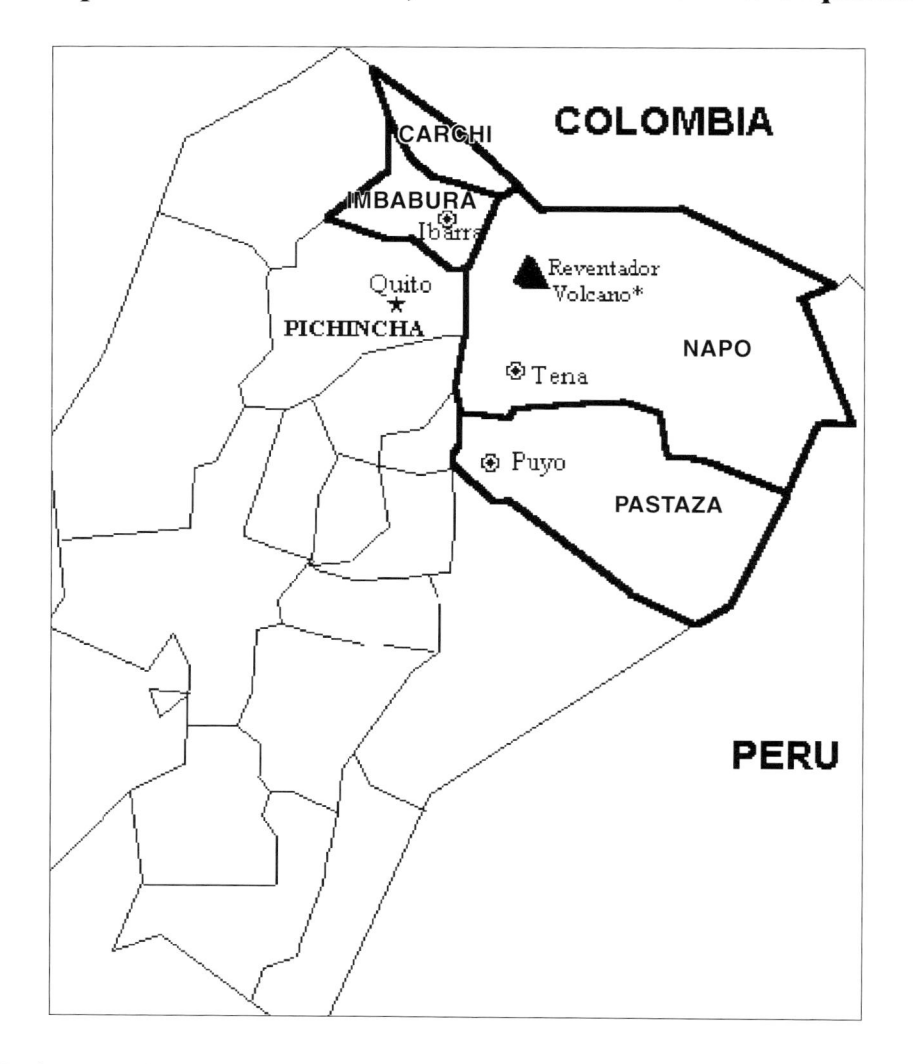

*Epicenter of March 5–6, 1987 Ecuadorian Earthquakes.
The four provinces outlined in bold – Carchi, Napo, Imbabura and Pastaza – were declared disaster zones by the President of Ecuador, Leon Febres Cordero.

APPENDIX C

Table 1. Sample Characteristics, 1985 Mexico City Earthquake Study

Field Interviews, Managers of Public, Private and Nonprofit Organizations

Source of Funding	N	%	Jurisdiction	N	%
Public	13	35.1	International	20	54.1
Private	5	13.5	National	14	37.8
Nonprofit	19	51.4	Municipal	3	8.1
Total	37	100.0	Total	37	100.0

Emergency Response Function	N	%	Gender	N	%
Emergency Response	3	8.1	Male	30	81.1
Damage Assistance	0	0.0	Female	7	18.9
Communications/Coord.	5	13.5	Total	37	100.0
Disaster Relief	23	62.2			
Recovery/Reconstruction	5	13.5			
Financial Assistance	1	2.7			
Total	37	100.0			

Survey, Residents of Disaster-Affected Areas of Mexico City

Occupation: Interviewee	N	%	Occupation: Head of Household	N	%	Age	N	%
Professionals	78	10.7	Professionals	141	19.4	17–23 yrs	169	23.2
Small businessmen	66	9.1	Small businessmen	112	15.4	24–30 yrs	177	24.3
White collar workers	176	24.2	White collar workers	249	34.2	31–40 yrs	149	20.5
Blue collar workers	103	14.1	Blue collar workers	152	20.9	41–50 yrs	105	14.4
Housewives	170	23.4	Housewives	36	4.9	50+ yrs	128	17.6
Students	113	15.5	Students	3	0.4	Total	728	100.0
Retired	19	2.6	Retired	29	4.0			
Farmers	0	0.0	Farmers	4	0.5			
Unemployed	2	0.3	Unemployed	1	0.1			
No Answer	1	0.1	No Answer	1	0.1			
Total	728	100.0	Total	728	100.0			

Education	N	%	Income(Pesos)*	N	%	Gender	N	%
3/4ths of elementary	48	6.6	<40,000	110	15.1	Male	339	46.6
Completed Elementary	137	18.8	40,001–80,000	197	27.1	Female	387	53.2
Completed High School	139	19.1	80,001–200,000	276	37.9	NR	2	0.3
College	209	28.7	200,001–400,000	66	9.1	Total	728	100.0
University	179	24.6	Ps. 400,001	39	5.4			
Post Graduate University	16	2.2	No Response	40	5.5			
Total	728	100.0		728	100.0			

*US$1 = 350 pesos (1985)

Table 2. Sample Characteristics, 1986 San Salvador Earthquake Study

Field Interviews, Managers of Public, Private and Nonprofit Organizations

Source of Funding	N	%	Jurisdiction	N	%	Emergency Function	N	%
Public	29	87.9	International	30	90.9	Emergency Response	13	39.4
Private	1	3.0	National	2	6.1	Damage Assessment	3	9.1
Nonprofit	3	9.1	Municipal	1	3.0	Communication/Coordination	3	9.1
Total	33	100.0	Total	33	100.0	Disaster Relief	11	33.3
						Recovery/Reconstruction	2	6.1
Sex						Financial Assistance	1	3.0
						Total	33	100.0
Female	1	3.0						
Male	32	97.0						
	33	100.0						

All field interviews were conducted in the City of San Salvador

Survey, Residents of Disaster-Affected Zones, San Salvador

Age	N	%	Income*	N	%	Occupation	N	%
20–24 yrs	68	20.3	c600 or less	229	68.4	Professional	3	0.9
25–29 yrs	49	14.8	c601–800	32	9.6	Technical	24	7.2
30–39 yrs	84	25.1	c801–1000	14	4.2	Community & Service		
40–49 yrs	52	15.5	c1001–1500	12	3.6	Employees	91	27.2
50 + yrs	80	23.9	c1501–2500	13	3.9	Construction Workers	15	4.5
Missing	2	0.6	c2501 +	9	2.7	Mechanics & rel.	6	1.8
	335	100.2	Don't Know	26	7.6	Agricultural Workers	1	0.3
				335	100.0	Industrial Workers	20	6.0
						Others	21	6.3
						Unemployed	154	46.0
							335	100.0

*US$1 = 5 colones (1986).

Gender	N	%	Education	N	%
Male	83	24.78	3 yrs or less	163	48.7
Female	252	75.22	Primary	69	20.6
	335	100.0	Some secondary	42	12.5
			Secondary	46	13.7
			Some university	12	3.6
			University	3	0.9
				335	100.0

Tabl3 3. Sample Characteristics, 1987 Ecuadorian Earthquakes Study

Field Interviews, Managers of Public, Private and Non-Profit Organizations

Location	N	%	Source of Funding	N	%	Jurisdiction	N	%
Pichincha Province (Quito)	30	60.0	Public	32	64.0	International	11	34.4
Pichincha Province (Outside Quito)	5	10.0	Private	2	4.0	National	11	34.4
Imbabura Province	4	8.0	Nonprofit	16	32.0	Provincial	4	12.5
Napo Province	7	14.0	Total	50	100.0	Municipal	6	18.8
Napo Province (Lago Agrio)	4	8.0				Total	32	100.0
	50	100.0						

Emergency Function	N	%	Gender	N	%	Age	N	%
Emergency Response	7	14.0	Male	45	90.0	31–40	8	16.0
Damage Assessment	4	8.0	Female	5	10.0	41–50	21	42.0
Communication/Coordination	7	14.0	Total	50	100.0	51–60	20	40.0
Disaster Relief	20	40.0				61+	1	2.0
Recovery/Reconstruction	12	24.0				Total	50	100.0
Financial Assistance	0	0.0						
Total	50	100.0						

Survey, Residents of Disaster-Affected Areas

Education	N	%	Monthly Income*	N	%	Urban/Rural	N	%
Illiterate	54	17.8	Peasant	53	17.5	Rural	167	55.1
3rd grade	80	26.4	$26 or less	61	20.1	Urban	129	42.6
Primary	117	38.6	$27–$52	55	18.2	No Response	7	2.3
Some secondary	29	9.6	$53–$105	78	25.7		303	100.0
Secondary	17	5.6	$106–$158	17	5.6			
Some university	4	1.3	$159–$211	4	1.3			
University	1	0.3	$211+	4	1.3			
NR	1	0.3	NR	31	10.2			
Total	303	100.0	Total	303	100.0			

*USD$1=190 sucres (1987)

Age	N	%	Gender	N	%
16–19	17	5.6	Male	139	45.9
20–24	33	10.9	Female	163	53.8
25–29	39	12.9	No Response	1	0.3
30–39	69	22.8	Total	303	100.0
40–49	56	18.5			
50+	87	28.7			
NR	2	0.7			
Total	303	100.0			

Table 4. Sample Characteristics, 1987 Whittier Narrows, CA Earthquake

Field Interviews: Operations Service Chiefs (Public Organizations)

Location	N	%	Emergency Function	N	%	Jurisdiction	N	%
Alhambra	8	12.1	Emergency Response	22	33.3	City	46	69.7
Los Angeles	24	36.4	Damage Assessment	16	24.2	County	18	27.3
Monterey Park	7	10.6	Communication/Coord.	25	37.9	State	2	3.0
Pasadena	7	10.6	Disaster Relief	2	3.0	Federal	0	0.0
Rosemead	6	9.1	Recovery/Reconstruction	1	1.5		66	100.0
Santa Fe Springs	7	10.6	FA	0	0.0			
Whittier	7	10.6	Total	66	100.0			
Total	66	100.0						

Gender	N	%	Education	N	%	Age	N	%
Male	63	95.5	High School	1	1.5	21–30	2	3.0
Female	3	4.5	AA/Some College	10	15.2	31–40	16	69.6
	66	100.0	BA/BS	21	31.8	41–50	24	36.4
			MPA/MBA/MS	31	47.0	51–60	23	34.8
			NR	3	4.5	61+	1	1.5
				66	100.0		66	100.0

Field Interviews. Managers of Auxiliary Organizations (Nonprofit, Private and Public)

Location	N	%	Emergency Function	N	%	Source of Funding	N	%	Years of Service	N	%
Alhambra	3	6.5	Emer. Response	3	23.1	Nonprofit	14	30.4	1–5 yrs	8	17.4
Burbank	1	2.2	Emer. Preparedness	4	8.7	Private	21	45.7	6–10 yrs	5	10.9
Covina	1	2.2	Damage Assessment	4	8.7	Public	11	23.9	11–15 yrs	5	10.9
Hollywood	1	2.2	Communications/			Total	46	100.0	16–20 yrs	10	21.7
Los Angeles	9	19.6	Coordination	15	32.6				21–25 yrs	7	15.2
Monterey Park	1	2.2	Disaster Relief	8	3.8				26–36 yrs	4	8.7
Pasadena	3	6.5	Recovery/Reconstruction	12	26.1				Missing	7	15.2
Pico Rivera	1	2.2	Total	46	100.0	Public			Total	46	100.0
Pomona	1	2.2				Jurisdictions:			Gender		
Rosemead	5	10.9	Education			City	5	45.5	Female	12	26.1
San Gabriel	1	2.2	High School	11	23.9	County	1	9.1	Male	34	73.9
San Marino	1	2.2	A.A.	2	4.3	State	3	27.3	Total	46	100.0
Santa Fe Springs	4	8.7	B.A./B.S.	17	37.0	Federal	2	18.2			
South Gate	1	2.2	MA/MBA/MSW	8	17.4	Total	11	100.0			
Whittier	13	28.3	J.D.	1	2.2						
Total	46	100.0	Missing	7	15.2						
			Total	46	100.0						

Field Interviews: Survey, Field Personnel

Municipality	N	%	Agency	N	%	Jurisdiction	N	%
Alhambra	33	17.9	Building & Safety	14	17.6	County	37	20.1
Los Angeles	24	13.0	Engineering/Public Works	25	13.6	City	146	79.6
Monterey Park	21	11.4	Fire	51	27.7	Missing	1	0.5
Pasadena	22	12.0	Police	65	35.3	Total	184	100.0
Rosemead	10	5.4	Social Services	29	15.8			
Santa Fe Springs	14	7.6	Total	184	100.0	Years of		
Whittier	22	12.0				Service		
Los Angeles			Education			1–5 yrs	35	19.0
County	37	20.1	High School	27	14.7	6–10 yrs	35	19.0
Missing	1	0.1	A.A.	60	32.6	11–15 yrs	29	15.8
Total	184	100.0	B.A.	66	35.9	16–20 yrs	25	13.6
			M.P.A	12	6.5	21–30 yrs	20	10.9
			MBA/JD/other	6	3.2	31 yrs +	32	17.3
			Total	184	100.0	Missing	8	4.3
						Total	184	100.0

Table 5. Sample Characteristics, 1988 Armenia Earthquake Study

Field Interviews, Managers of Public and Nonprofit Organizations

Location	N	%	Source of Funding	N	%	Jurisdiction	N	%
Moscow	10	24.4	Nonprofit	14	34.1	International	18	43.9
Yerevan	20	48.8	Public	27	65.9	All-union	7	17.1
Spitak	4	9.8		41	100.0	Republic	8	19.5
Leninakan	4	9.8				Municipal	8	19.5
Maralick	1	2.4					41	100.0
Washington, DC	2	4.9						
	41	100.0						

Emergency Function	N	%	Gender	N	%
Disaster Relief	19	46.3	Female	3	7.3
Emergency Response	7	17.1	Male	38	92.7
Emergency Medicine	11	26.8		41	100.0
Other Medical Services	3	7.3			
Recovery/Reconstruction	1	2.4			
	41	100.0			

Table 6. Sample Characteristics, 1989 Loma Prieta, CA Earthquake Study

Field Interviews: Operations Service Chiefs

Location	N	%	Emergency Function	N	%	Source of Funding	N	%
Oakland	18	58.1	Emergency Response	16	51.6	Private	3	9.7
Mountain View	6	19.4	Damage Assessment	2	6.5	Public	28	90.3
Pleasant Hill	2	6.5	Communic./Coordination	9	29.0		31	100.0
Richmond	1	3.2	Disaster Relief	4	12.9			
Sacramento	2	6.5	Recovery/Reconstruction	0	0.0			
San Francisco	2	6.5	Financial Assistance	0	0.0			
	31	100.0		31	100.0			

Gender	N	%	Public Jurisdictions:*	N	%
Female	3	9.7	Federal	2	7.1
Male	28	90.3	Federal, Region IX	4	14.3
	31	100.0	State	11	39.3
			State, Region II	2	7.1
			Municipal	9	32.1
				28	100.0

*Federal = Washington, DC
Federal, Region IX = San Francisco
State = Sacramento
State, Region II = Pleasant Hill

Table 7. Sample Characteristics, 1991 Costa Rica Earthquake Study

Field Interviews, Managers of Public, Private and Nonprofit Organizations

Location	N	%	Source of Funding	N	%	Jurisdiction	N	%
San Jose	25	61.0	Public	30	73.2	International	9	22.0
Limon	14	34.2	Private	6	14.6	National	25	61.0
Mattina	1	2.4	Nonprofit	5	12.2	Provincial	1	2.4
Valle d'Estrella	1	2.4		41	100.0	Municipal	6	14.6
	41	100.0					41	100.0

Emergency Function	N	%	Gender	N	%
Emergency Response	7	17.1	Male	34	82.9
Damage Assessment	8	19.5	Female	7	17.1
Communication/Coordination	5	12.2		41	100.0
Disaster Relief	19	46.3			
Recovery/Reconstruction	2	4.9			
Financial Assistance	41	100.0			

Table 8. Sample Characteristics, 1992 Erzincan, Turkey Earthquake Study

Field Interviews: Managers of Public and Nonprofit Organizations

Location	N	%	Source of Funding	N	%
Erzincan	9	47.4	Nonprofit	1	5.3
Erzurum	1	5.2	Public	18	94.7
Ankara	9	47.4		19	100.0
	19	100.0			

Emergency Function	N	%	Jurisdiction	N	%
Emergency Response	6	31.6	Municipal	9	47.4
Damage Assessment	3	15.8	National	9	47.4
Communication/Coordina.	5	26.3	International	1	5.2
Disaster Relief	3	15.8		19	100.0
Recovery/Reconstruction	2	10.5			
Financial Assistance	0	0.0			
	19	100.0			

All respondents were men.
All interviews were conducted between July 2 and July 10, 1992, a briefer stay than for the other field visits. Consequently, the number of interviews obtained is smaller.

Frequency Distribution: Sample Characteristics

Age	N	%	Role in Disaster Response	N	%
61 and over	5	2.9	National coordinator	2	1.3
51–60	10	5.9	Local coordinator	18	11.4
41–50	41	24.1	Search and rescue	7	4.4
31–40	50	29.4	Doctor/nurse	27	17.1
21–30	47	27.7	Citizen/lay person	104	65.8
Under 20	17	10.0	Total	158	100.0
Total	170	100.0	(Missing cases=27)		
(Missing cases=15)					

Gender	N	%
Male	121	71.2
Female	49	28.8
Total	170	100.0
(Missing cases=15)		

Table 9. Sample Characteristics, 1993 Marathwada, India Earthquake Study

Location	N	%	Source of Funding	N	%	Public Organizations: Jurisdiction	N	%
Solapur	8	16.7	Public	24	50.0	International	0	0.0
Latur	8	16.7	Private	7	14.6	National	2	8.3
Latur District	4	8.3	Nonprofit	17	35.4	State	4	16.7
Killari	2	4.2		48	100.0	Region	0	0.0
Sastur	3	4.2				District	5	20.8
Koral	3	6.3				Municipal	13	54.2
Osmanabad District	3	6.3					24	100.0
Bombay	9	18.8						
Omerga District	2	4.2						
Pardhewadi	2	4.2						
Nadihattaraga	1	2.1						
Pune	1	2.1						
Ambogojai	1	2.1						
Aurangabad	1	2.1						
	48	100.0						

Emergency Function	N	%	Age	N	%	Gender	N	%
Emergency Response	7	14.6	20–29	6	12.5	Male	38	79.2
Damage Assessment	0	0.0	30–39	15	31.3	Female	10	20.8
Communication/Coordination	13	27.1	40–49	12	25.0		48	100.0
Disaster Relief	21	43.8	50–59	8	16.7			
Recovery/Reconstruction	7	14.6	60–69	5	10.4			
Financial Assistance	0	0.0	70–79	2	4.2			
	48	100.0		48	100.0			

Table 10. Sample Characteristics, 1994 Northridge, CA Earthquake Study

Field Interviews, Managers of Public, Private and Nonprofit Organizations

Location	N	%	Source of Funding	N	%	Emergency Support Function*	N	%
Los Angeles	15	32.6	Public	37	80.4	Transportation	2	6.5
Northridge	1	2.2	Nonprofit	9	19.6	Communication	2	6.5
Pasadena	28	60.9	Total	46	100.0	Public Works	1	3.2
Van Nuys	2	4.3				Fire	2	6.5
	46	100.0	**Public Organizations:**			Intelligence	3	9.7
						Mass Care	6	19.4
Gender	N	%	**Jurisdiction**	N	%	Health/Medical	1	3.2
Men	38	82.6	Federal	18	48.6	Resource	1	3.2
Women	8	17.4	State	8	21.6	Urban Search and Rescue	12	38.7
	46	100.0	County	5	13.5	Hazardous Materials	1	3.2
			City	6	16.2	Food	0	0.0
			Total	37	100.0	Energy	0	0.0
						Total	31	100.0

*These are the twelve Emergency Support Functions defined under the Federal Response Plan, U.S. Federal Emergency Management Agency. They represent a more detailed classification of response functions for federal agencies.

Emergency Function	N	%
Emergency Response	12	26.1
Damage Assessment	3	6.5
Communication/Coordination	20	43.5
Disaster Relief	8	17.4
Recovery/Reconstruction	2	4.3
Financial Assistance	1	2.2
	46	100.0

Table 11. Sample Characteristics, 1995 Hanshin, Japan Earthquake Study

Field Interviews, Managers of Public, Private and Nonprofit Organizations

Location	N	%	Source of Funding	N	%
Itami City	4	8.9	Public	28	62.2
Kobe	16	35.6	Private	12	26.7
Nishinomiya	6	13.3	Nonprofit	5	11.1
Osaka	14	31.1	Total	45	100.0
Tokyo	5	11.1			
	45	100.0			

Gender	N	%	Jurisdiction: Public Organizations	N	%
Male	44	97.8	National	8	17.8
Female	1	2.2	Prefectural	6	13.3
Total	45	100.0	Municipal	14	31.1
			Total	28	62.2

Age	N	%	Emergency Function	N	%
21–30	1	2.2	Emergency Response	21	46.7
31–40	10	22.2	Damage Assessment	13	28.9
41–50	20	44.4	Communication/Coordination	5	11.1
51–60	11	24.4	Disaster Relief	6	13.3
61+	3	6.7	Recovery/Reconstruction	0	0.0
Total	45	100.0	Financial Assistance	0	0.0
			Total	45	100.0

APPENDIX D

Table 1. Initial Conditions: Characteristics of Latur District

Physical and Administrative:
Latur District was separated from Osmanabad District in August, 1981.
1. Total area of District   7,157 square kilometers
2. Total number of talukas (subdistricts):   7
   Latur, Ahmedpur, Chakur,
   Renapur, Udgir, Nilanga, Ausa
3. Total number of villages   936

Social:	N	%
1. Total population	1,677,000	100.0
2. Urban population	342,000	20.0
3. Rural population	1,335,000	80.0
4. Scheduled castes*, number and percent of total population	228,600	18.0
5. Scheduled tribes*, number and percent of total population	31,750	2.5
6. Literacy rate, total population	968,690	58.0
7. Proportion male literates	629,590	65.0
8. Proportion female literates	339,100	35.0
9. Sex ratio (number of females per 1,000 males)	944	

Occupational:	N	%
1. Agricultural:		
a. Landholders	258,428	39.4
b. Landless laborers	256,672	39.1
c. Livestock, forestry	5,057	0.8
2. Non-agricultural		
a. Trade & commerce	36,207	5.5
b. Other manufacturing	23,145	3.5
c. Other trades & services	76,734	11.7
Total number of workers	656,243	100.0

Source: *Government of India, Census of India, 1991.* Centre for Monitoring Indian Economy, 1985.

*Scheduled caste people are members of the formerly 'untouchable' caste, who are now regarded as equal members of the Indian society, but who are still seriously disadvantaged by their low socioeconomic status. Scheduled tribes are indigenous peoples who are also disadvantaged by low socioeconomic status.

Table 2. Initial Conditions: Characteristics of Osmanabad District

Physical and Administrative:
1. Total area   7,567 square kilometers
2. Total number of talukas (subdistricts):   6
   Tuljapur, Kalamb, Omerga, Bhum, Paranda
   Osmanabad
3. Total number of villages   704

Social:	N	%
1. Total population	1,275,000	100.0
2. Urban population	193,000	15.0
3. Rural population	1,082,000	85.0
4. Scheduled caste* population (Number and percent of total)	190,500	15.0
5. Scheduled tribe* population (Number and percent of total)	12,700	1.0
6. Literacy rate in district	561,000	44.0
7. Percent male literacy	364,650	65.0
8. Percent female literacy	196,350	35.0
9. Sex ratio (number of females per 1,000 males)	943	

Occupational:	N	%
1. Agricultural:		
a. Landholders/cultivators	214,496	40.5
b. Landless laborers	217,527	41.1
c. Livestock, forestry	6,069	1.1
2. Non-agricultural:		
a. Trade & commerce	18,029	3.4
b. Other manufacturing	14,022	2.6
c. Other trades & services	59,282	11.3
Total number of workers	529,425	100.0

Source: *Government of India, Census of India, 1991.* Centre for Monitoring Indian Economy, 1985.

*Scheduled caste people are members of the formerly 'untouchable' caste, who are now regarded as equal members of the Indian society, but who are still seriously disadvantaged by their low socioeconomic status. Scheduled tribes are indigenous peoples who are also disadvantaged by low socioeconomic status.

Table 3. A Comparative Assessment of Damage by District, Marathwada Earthquake, September 30, 1993

Type of Damage	Latur N (%)	Osmanabad N (%)	Total N (%)
Total number of villages	936 (57.1)	704 (42.9)	1,640 (100.0)
Number of villages severely damaged	817 (68.6)	374 (31.4)	1,191 (100.0)
Number of homes severely damaged	85,000 (58.6)	60,000 (41.4)	145,000 (100.0)
Number of dead	3,726 (49.1)	3,856 (50.9)	7,582 (100.0)
Number of injured	6,283 (40.4)	9,283 (59.6)	15,566 (100.0)
Number of cattle dead	1,083 (51.6)	1,017 (48.4)	2,100 (100.0)
Number of cattle injured	8,345 (64.0)	4,699 (36.0)	13,044 (100.0)

Source: *A Preliminary Report of the September 30, 1993 Earthquake.* Government of Maharashtra, Bombay, India, 1993.

INDEX

accountability, public managers 4
action science 50
adaptation
   adaptive performance 34, 35, 53, 62, 262
   evolving patterns of response 231–76
   mechanisms 32
adaptive systems 75–76, 160
aggregate models 14–15
Aguirre, B.E. 1991 135
Alameda County, CA
   Emergency Communications System 171
   fire and ambulance services 171
   Highland Hospital 171
Alaska 15
Alhambra, CA 162
   Chamber of Commerce 164
Amagasaki, Japan 212
anarchy 25, 26, 30
Andean oil pipeline 68
Andrews, W.T. 1994 52
archival records 47, 49–50
Argyris, C. 1982, 1985 50
Armenia 75, 107
   Civil Defense 105
Armenia earthquake 1988 13, 16, 45, 46
   adaptive behavior 108–114
   communication and coordination 113
   damage assessment 113
   damage and losses 104
   data collection 48
   disaster relief 113
   economic loss 66
   financial assistance 113
   funding sources 107, 110
   information exchange 106–7
   information search 104–5
   loss of life 66
   nonlinear logistic regression analysis 114
   organizational learning 107–8
   phase planes 113
   recovery/reconstruction 113
   republic organizations 113

   response system evaluation 68–70, 75, 114–5, 231–76
   Richter measurement 104
   sample characteristics 301
   triage system 108
   union organizations 111
"arrow of time" 8
Ashiya, Japan 212
assessment indicators 65–66
Atkisson, Arthur 1982 8
authority 25
auto adaptive systems 73, 197, 197–227
   characteristics 73–5, 197–98
   comparisons 76, 224–26
   evolution of response systems 265–67
   funding sources 223, 246
   information exchange 239–43
   information search 236–38
   initial conditions 234
   nonlinear logistic regression analysis 260–62
   organizational learning 252–53
   organizational response 232–34
   structure and flexibility 267–70
   technical structure 73, 197
   see also Hanshin earthquake, Japan 1995; Northridge CA earthquake 1994
autopoiesis 29, 59, 60, 63
Awaji island, Japan 212
Axelrod, R.M. 27

Bak, P. 1991 57
Bartels, Larry M. 1993 14
Barton, Allen H. 1969 20
behavior 8
   collective 4
Bellavita, Christopher 1983 28, 29
Bermudez, M. 1993 135
Bihar state 18
Blalock, H.M. 1972 38
Bombay *see* Mumbai
Brady, Henry E. 1993 14

311

Bruzewicz, Andrew J. 1995 59
Bucuvalas, Michael J. 1984 27
building construction and seismic risk 8, 69, 71, 159, 181
Burt, R.S. 1982 28

Caiden, Naomi 1974 19
Calderon, *President* Rafael 136, 137
California State 8, 161, 199
  Department of Transportation 173, 174
  Emergency Plan 172
  Highway Patrol 173, 174
  Joint Federal/State Coordinating Office 174
  Joint Federal/State Disaster Field Office, Pasadena 201
  Office of Emergency Services (OES) 22, 74, 164, 174, 201
  Office of Emergency Services, Regional Operations 172, 174
  Operational Area Satellite Information System (OASIS) 201
  Training Institute 162
Caltech-USGS Broadcast of Earthquakes/ Rapid Earthquake Data Integration (CUBE/REDI) system 201
Carchi province, Ecuador 94, 96
case profiles 54
case selection criteria 45
case study *see* response systems assessment
catastrophic trauma 149
Catholic Relief Services 97
causality 7, 42
  in nonlinear systems 38, 39
change 58, 272
  mechanisms 8, 57
chaos 57
Chaos System Software 49, 55, 294
Chen, K. 1991 57
co-evolution 47
coding of organizations 49
cognitive capacity 6, 11
Cohen, David K. 1979 27
Cohen, W.M. 1990 24
collective action 18–20, 32, 35
  Mexico City 56
collective learning 12, 22, 23
collective response 3–16, 20, 41
Colorado, University of, Natural Hazards Research and Applications Information Center 54–5

Comfort, L.K. 1985 50
Comfort, L.K. 1986 27
Comfort, L.K. 1991a 59, 93
Comfort, L.K. 1991b 59
command and control 24, 25, 26, 30, 263
common knowledge 19, 31, 32
common knowledge base 22, 23, 73
communications
  infrastructure 73, 159
  nonadaptive systems 81
  patterns of 8, 12–13, 20, 30
  processes 36
  technology 6, 10, 76, 160, 183
  and transition 22, 262
communicative acts 20
community performance 272
Community Recovery Centers 209
  see also Disaster Assistance Centers
complex adaptive systems 42, 160
complex systems 7, 57
complexity 263–64
  in non linear systems 38, 39, 40
  and response to risk 7, 18
Consolini, Paula M. 1991 32
constructs of time 10
continuum of performance 58
cooperation 20
coordination of response 8, 13, 19–20, 262
Cordero, *President* Leon Febres 93
Cordillera de Talamanca, Costa Rica 134
Cordillera Real, Ecuador 93
cost 12, 13
  see also damage; damage and losses, under each earthquake entry
Costa Rica 13, 16
  damage assesment 144
  Comision Nacional Emergencia 135
  Guardia Civil 136
  National Emergency Plan 136, 138
  Refineria Costarricense de Petroleo (RECOPE) 135
Costa Rica earthquake 1991 13, 16, 45, 46
  adaptive behaviour 144–45
  communication and coordination 139–142, 144
  damage assessment 144
  damage and losses 134–35
  disaster relief 144–45
  evaluation 70–2, 75
  financial assistance 144
  funding sources 139–42

information exchange 137–38
information search 136–37
nonlinear logistic regression analysis 144–45
organizational learning 138–44
recovery/reconstruction 145
response system evaluation 75, 156–7, 231–76
Richter measurement 134
sample characteristics 303
Coveney, Peter 1995 7
creative response 197
creative self-expression 29
Crecine, J.P. 1986 24, 26
CUBE/REDI system (Caltech-USGS Broadcast of Earthquakes/Rapid Earthquake Data Integration system) 201
cultural openness
  autoadaptive systems 73
  emergent adaptive systems 70
  importance of 64, 197
  nonadaptive systems 68, 75, 81
  operative adaptive systems 72
cultural values 66

DACS (Disaster Assistance Centers) 207
damage 12, 60
  see also damage and losses, under each earthquake entry
damage, potential 18
data analysis 49–50, 64
data collection 47–49
data interpretation 50–4
deaths 12–13
  see also damages and losses, under each earthquake
Deccan plateau, India 76, 181
decision-making
  capacity of managers 44
  interdependent 7, 18, 19, 21
  organizational 25
dependency 8
destruction, potential 18
Deutsch, Karl W. 1963 28
DFO (Disaster Field Office) 209
Diario de Hoy, San Salvador 85
Disaster Assistance Centers (DACS) 207
disaster declarations 200
Disaster Field Office (DFO) 209
disaster management 34

disaster response 3–16, 19–22
disaster response systems *see* response systems
divergent goals 30
Drabek, Thomas E. 1990 20
Dryzek, J. 1987 57
Duarte, *President* José Napoleón 84
dynamic disaster response system 20–22, 33, 34
dynamic environment 198
dynamic processes 264
Dynes, Russell R. 1969 20
dysfunction 197–227

"earthquake for the poor", Ecuador 94
earthquakes
  disaster response *see* response systems
  preparedness 3, 8
  see also under each earthquake entry
Ecuador 13, 16
  Civil Defense 95, 97
  municipal councils 96
  National Emergency Committee 96
Ecuador earthquake, 1987 13, 16, 45, 46
  adaptive behavior 98–102
  chaos 103
  communication and coordination 96–7, 99, 100–2
  damage assessment 102–3
  damage and losses 92–3
  data collection 48
  disaster relief 98
  funding sources 98–100
  information exchange 95–7
  information infrastructure 104
  information search 93–5
  map of disaster zones 296
  nonlinear logistic regression analysis 99, 102–3
  organizational learning 97–8
  recovery/reconstruction 103
  response system evaluation 68–70, 75, 114–5, 231–76
  Richter measurement 92
  sample characteristics 299
Ecuadorian State Petroleum Corporation 93
"edge of chaos" 14, 22, 38, 264
  auto adaptive systems 197, 198
  measurement of response systems 15, 46, 56–77

EDIS (Emergency Digital Information Services) 201
El Salvador 13, 16
Emergency Digital Information Service (EDIS) 201
emergent adaptive systems 119–58
  characteristics 70–72, 119
  comparisons 75, 156–7
  evolution of response systems 265–66
  funding sources 157, 242–43
  information exchange 239–43
  information search 237–38
  initial conditions 235
  nonlinear logistic regression analysis 256–58
  organizational learning 248–50
  organizational response 232–34
  structure and flexibility 268–69
  technical structure 70, 119
  see also Costa Rica earthquake 1991; Erzincan earthquake, Turkey 1992; Mexico City earthquake 1985
environments of shared risk 18–37
epidemic avoidance, Mexico City 56
epistemic community 225
Erzincan earthquake, Turkey 1992 13, 16, 45, 46
  adaptive behavior 154–6
  communication and coordination 155–6
  damage and losses 146–7
  data collection 48
  disaster relief 155
  financial assistance 155
  funding sources 151–53
  information exchange 150
  information search 147–50
  nonlinear logistic regression analysis 156
  organizational learning 151–54
  phase planes 154–5
  recovery/reconstruction 155
  response system evaluation 70–2, 75, 156–7, 231–76
  Richter measurement 146
  sample characteristics 304
Erzincan, Turkey, seismic risk 12
evaluation, standard methods 28
evolution 58
evolution of response systems 53–4, 81, 264–67
evolving response systems, recommendations 275–76
Excelsior (newspaper) Mexico City 126, 127
existing state 41

Federal Emergency Management Agency 74, 200–01, 205–7
feedback
  auto adaptive systems 197
  response coordination 9, 10, 19
  transition 23, 27, 31
feedback loops 8, 32, 62
FEMA see Federal Emergency Management Agency
field studies 47–9
  comparisons of results 231–76
  see also auto adaptive systems; emergent adaptive systems; nonadaptive systems; operative adaptive systems; and "data collection" under each earthquake entry
flexibility
  Mexican government 71
  response systems 22–25, 64, 81, 267–70
Florida 21
fractal organizations 9
Frankl, Victor E. 1970 29
frequency data 49
functional differentiation 59
functional operating system 231
funding sources 62, 233, 237, 253
  auto adaptive systems 223, 246–7
  emergent adaptive systems 157, 242–3
  nonadaptive systems 115, 240–1
  operative adaptive systems 194, 244–5
  see also under each earthquake entry

Geertz, Clifford 63
Gell-Mann, Murray 1994 8, 57
Geographic Information System (GIS) 201
Gilbreth, Frank B. 1917 24
Gilbreth, L.M. 1917 24
"Glasnost" policy, USSR 248
Global Disaster Information Network 274, 276
global solutions 11, 12
Goodman, Paul S. 1990 31
Gorbachev, *Premier* Nikolai, Soviet Union 106, 248
Gram Panchayats, India 183

Haas, E.B. 1990 27, 225

Habermas, Jurgen 1979 28
Hanshin earthquake, Japan 1995 13, 16, 21, 45–6, 212–26
  adaptive behavior 220
  communications capability 214
  communication and coordination 221–2
  damage assessment 223
  damage and losses 12, 212–14
  disaster relief 223
  economic loss 66
  financial assistance 222–23
  funding sources 218, 220
  information exchange 215
  loss of life 66, 75
  nonlinear logistic regression analysis 222–23
  organizational learning 217
  phase planes 221
  recovery/reconstruction 222
  response system evaluation 74–75, 198, 224–6, 231–76
  Richter measurement 212, 213
  sample characteristics 307
Hanshin region, Japan 213
Harlow, David H. 1986 82
health threat, Mexico City 56
"heedful interrelating" 32
Highfield, Roger 1995 7
Holland, John 1995 27
Honshu island, Japan 213
"hotline" 206
housing, disaster response 18
"Housing Network" 209
Hurricane Andrew 21, 199
Hyogo Prefecture, Japan 212–13
  satellite communication system 215

Imbabura province, Ecuador 94, 96
Incident Command System, California 22, 24
India 13, 15, 16
  Administrative Service 182
  Gram Panchayats *see* village councils
  Meteorological Department 181
  National Informatics Centre (NIC) 181
  seismic risk 181
  village councils 183, 193
  voluntary associations 182
Indian National Satellite System 73, 159, 181, 183
indicators 64

individual response 6
Indonesia 15
information
  base 21
  communication of 61
  and discretionary choices 21
  dissemination 5, 6, 14
  incoming 22
  processing and transmission 5, 6, 14, 28
  quality of 19
  technology 7, 10–12, 11, 14
  transmission 28
information exchange 19, 20
  data interpretation 53
  decision making 44
  evolution of response system 231–32
  facilitating adaptation 76
  measurement 47
  producing change 58
  requirement for transition 31, 35, 62, 81
  see also under each earthquake entry
information flow 197
information infrastructure 76, 160, 263
information management 28–9
information processes 14, 36, 44, 160, 264
information search 34, 53, 61, 160, 198
  evolution of response system 231–32
  see also under each earthquake entry
information technology 29, 30, 31, 36, 197
  future strategy 275
information–rich environment 5
informed choice 58
infrastructure 14, 60, 72
initial conditions
  data analysis to determine 49
  definition 34, 41, 53
  measures to determine 53
  sensitive dependency upon 8
  to initiate transition 61, 64, 160, 231–32
inquiring systems 197, 263
  definition 28
inquiry models 27, 28, 30
Instituto Geofísico de la Escuela Politécnia Nacional Quito 93
integration 10
inter-organizational learning, evolution of response system 231–32
interaction 20, 23, 56
  organizational 62
interdependency 6, 14
interjurisdictional systems 20

internal model 225
international aid, Mexico City earthquake 1985 123–26
international disaster response 18
international organizations, distribution 232–34
International Red Cross 83, 97, 135, 138
interorganizational
  communication 41
  decision processes 81
  learning 62
  response 42, 47
  transition 60, 61
interviews 48
  data 49–50
intra-organizational learning 35, 36, 62
  evolution of response system 231–32
Itami City, Japan 212

Japan 12, 13, 15, 16, 21
  National Land Agency 13, 214
Japan Times 217, 218, 220
Jurgens, H. 1992 9, 38
jurisdictional systems 20–21

Kansai Electric Co., Japan 213
Kansai region 12, 21
Karabakh, Azerbaijan 106
Kauffman, Stuart A. 1993 21, 22, 25, 32, 47
Keohane, R.O. 1994 41
Kerlinger, F.N. 1986 38
Killari, Latur District, India 180, 182, 183
King, G. 1994 41
Kirovakan, Armenia 104
Knoke, David 1982 47
knowledge, *see also* information
knowledge base, future strategy 274–75
knowledge, common 19
Kobe, Japan 212–14, 220
  Fire department 214, 215, 220
Krackhardt, David 1992 47
Kuklinski, James H. 1982 47

Lago Agrio, Ecuador 94
LaPorte, Todd R. 1987 32
Latin America 75
Latur, Marathwada region, India 73, 180–84, 308, 310
Lavell, A. 1991 143
Lavell, A. 1993 135, 143
League of International Red Cross Societies *see* International Red Cross
Leninakan, Armenia 104
  seismic risk 12
Levinthal, D.A. 1990 24
life, loss of 12
Limon (city) Costa Rica 135
Limon earthquake *see* Costa Rica earthquake
Limon (province) Costa Rica 136
Lindblom, Charles E. 1979 27
linear models 38
linearity 5–7
"lock-out" 24
logistic regression equation 52
Loma Prieta, CA earthquake 1989 8, 13, 16, 74, 170–80
  adaptive behavior 178
  communications 73, 172–73
  communication and coordination 178–80
  damage assessment 179
  damage and losses 170–1
  economic loss 66
  financial assistance 180
  funding sources 175–76
  information exchange 172–74
  information search 171–72
  information technology 172
  nonlinear logistic regression analysis 179, 180
  organizational learning 174
  phase planes 178
  recovery/reconstruction 179, 180
  response system evaluation 72, 75–6, 194, 231–76
  Richter measurement 170
  sample characteristics 302
Long Beach, CA earthquake 1933 8
Los Angeles city 21, 161, 162, 199
  Earthquake Damage Evaluation Teams 162
  Emergency Operations Bureau 162, 164
  Engineering Department 162
  helicopter team 200
  response teams 74
  seismic risk 12
Los Angeles county 21, 161
  Coordinator of Emergency Services 172
  Fire Department 201
  fire services 163
  response teams 74
  Sheriff's Office 163
Los Angeles Times 165, 204

loss estimation model 200
losses
  life and property 12–13
  see also damage and losses, under each earthquake entry
Luhmann, Niklas 1989 19–21, 29, 33

McKim, Harlan L., 1995 59
macro levels 28, 46
Madrid, *President* Miguel de la 121–2
Maharashtra earthquake Marathwada earthquake, *see* India 1993,
Maharashtra state, India 159
  communications 182
  Ministry of Public Works 185
  role in disaster response 187
Marathwada earthquake, India 1993 16
  adaptive behavior 187
  case selection 45, 46
  communication and coordination 186, 188–9, 191
  damage assessment by district 310
  damage and losses 13, 182–3
  data collection 48
  disaster relief 189–90
  funding sources 187–89
  information exchange 185
  information search 183
  initial conditions
    Latur district 308
    Osmanabad district 309
  nonlinear logistic regression analysis 189, 192
  organizational learning 186
  phase planes 187, 190
  response system evaluation 72–3, 75–6, 180–94, 231–76
  Richter measurement 180, 182
  sample characteristics 305
Marathwada region, India
  Regional Disaster Coordinator 186
  seismic risk 181
March, James G. 1972 25
March, J.G. 1988 25
Maskrey, A. 1993 143
measurement
  nonlinear systems 38–55
  qualitative 41, 47
  quantitative 41, 47
Medecins sans Frontiers, Netherlands 185
medical care, disaster response 18

Meltsner, Arnold 1983 28, 29
MERS (Mobile Emergency Response System) 201
Meseta Central, Costa Rica 135
methodology 15, 47–54
Mexico, National and Metropolitan Emergency Commission 122
Mexico City 56
  seismic risk 12
Mexico City earthquake 1985 13, 16, 45, 46
  adaptive behavior 128–34
  communication and coordination 128–29, 130–31, 133
  damage assessment 125, 135
  damage and losses 120
  data collection 48
  disaster relief 131–33
  financial assistance 133
  funding sources 128–30
  information exchange 122–26
  information search 120
  international aid 123–26
  loss of life 120
  nonlinear logistic regression analysis 132–4
  organizational learning 126–32
  phase planes 130
  response system evaluation 70–2, 75, 156–7, 231–76
  Richter measurement 120
  sample charteristics 297
micro levels 28
Miller, G. 1967 6
mitigation, mechanisms of 23
Mobile Emergency Response System (MERS) 201
mobilization, disaster response 4, 19, 41
Mohr, Lawrence 1982 5
Monterey Park, CA 162, 163
  Police Department 164
Moral, *General* Antonio Moral 97
Mountain View, CA 174
multiway communication 64
Mumbai (formerly Bombay), India 182, 185

N-K methodology 54, 85, 126–27, 204, 217
N-K model 275
N-K system 15, 47, 186
Nacion, La, Costa Rica 138, 139
Napo province earthquake *see* Ecuador earthquake

Napo province, Ecuador 96
NASA (National Aeronautics and Space Administration) 200
NASAR 271
National Association of Search and Rescue (NASAR) 271
national disaster response 15
National Science Foundation, Washington 55
National Teleregistration Center (NTC) 207
Natural Hazard Research and Applications Center 54–5
natural selection 58
needs assessment 18–19
networked communications technology 76
networks 29
Newell, Allen 27
news accounts 47, 49–50
Nishinomiya, Japan 212
nonadaptive systems 75, 81–118, 194
 characteristics 68–70
 comparisons 75, 114–15
 evolution of response systems 265
 funding sources 115, 240–1
 information exchange 239–41
 information search 237–38
 initial conditions 235
 nonlinear logistic regression analysis 254, 255
 organizational learning 245–48
 organizational response 232–34
 structure and flexibility 268
 technical structure 68, 81
 see also Armenia earthquake 1988; Ecuador earthquake 1987; San Salvador earthquake 1986
nonlinear analysis 53
 disaster response 50–4
nonlinear dynamic social systems 7–10, 14
nonlinear dynamic systems 225
nonlinear dynamics 5
nonlinear logistic regression 63
 analysis 253–62
  see also under each earthquake entry
 estimation procedures 294–95
nonlinear methods of problem solving 5–7
nonlinear software program 49
nonlinear systems 32, 45
 methodology 54
 performance 38
Northern Armenian earthquake see Armenian earthquake

Northridge, CA earthquake 1994 3, 4, 16, 21
 adaptive behavior 205
 damage and losses 8, 13, 199
 economic loss 66
 funding sources 204, 206–8
 information exchange 200–3
 information search 199–200
 loss of life 66
 nonlinear logistic regression analysis 210–12
 organizational learning 204
 phase planes 209, 210
 response system evaluation 74, 198, 199–212, 224–6, 231–76
 Richter measurement 199
 sample characteristics 306
NTC (National Teleregistration Center) 207

Oakland, CA 170–2
 Allied Ambulance company 173, 174
 Cypress Street viaduct collapse 170, 172, 173
 Cypress Structure Command Post 174
 Emergency Operating Center 171
 Fire Department 173, 174
 Police Department 173, 174
Oakland Tribune 174
OASIS (Operational Area Satellite Information System) 201
Olsen, Johan P. 1972 25
Olson, Mancur 1965 3
Omerga, Marathwada region, India 182
ONUCA (Observatores de Naciones Unidas, Central America) 138
operative adaptive systems 159–96
 characteristics 72–3, 159–60
 comparisons 75–6, 194
 evolution of response systems 266
 funding sources 194, 244–5
 information exchange 239–43
 information search 236–38
 initial conditions 235
 nonlinear logistic regression analysis 258–60
 organizational learning 250
 organizational response 232–34
 structure and flexibility 269
 technical structure 72, 159
 see also Loma Prieta, CA earthquake 1989; Marathwada earthquake, India 1993, Whittier Narrows, CA

earthquake 1987
Orange County, CA 161, 163
organizational action 28
organizational behavior 27
organizational decision making 25, 26
organizational disaster response systems, distribution 232—34
organizational flexibility 64, 65, 72, 73, 197
organizational information structure 235
organizational infrastructure 64
organizational learning 30, 243—53
    future strategy 274—5
    and initial conditions 53
    measurement indicators 44
    nonadaptive systems 82
    and transition 59
organizational memory 35
organizational networks 29
organizational performance 36, 61
    data analysis 50—4
organizational systems 263
organizational training 263
organizational transition 59, 60
organizations, data analysis of 49
Osaka Bay, Japan 213, 220
Osaka Gas Co., Japan 213
Osmanabad, Marathwada District, India 73, 181, 184, 185, 309—10

Pacific Telephone 173
PAHO (Pan American Health Organisation) 135, 138
Pan American Health Organisation 135, 138
Pan Pacific Forum 226
parallel-processing systems 58
parróquias, Ecuadorian Sierra 98
Pasadena, CA 162, 164
Pastaza province, Ecuador 96, 296
Peitgen, H. 1992 9, 38
perestroika, Soviet Union 105
Perrow, C. 1972 24
Petak, William J. 1982 8
phase planes 51—3
    see also under each earthquake entry
Philippines 15
Pichincha province, Ecuador 94
Pittsburgh, University of
    Center for International Studies 55
    Center for Latin American Studies 55
    Center for Russian and East European Studies 55

Center for Urban and Social Research 55
    Graduate School of Public and International Affairs 55
    Office of Research 55
planners 19
policy analysis, standard methods 28
policy makers 19
post traumatic stress 147, 149, 150
preparedness 12, 19, 23
Priesmeyer, H. Richard 1992 5, 50
Priesmeyer, H. Richard 1994 52, 55
Priesmeyer, H. Richard 1995 51
Priesmeyer, H. Richard, Chaos! software 294
Prigogine, Ilya 1987 7, 8
problem solving
    nonlinear vs linear methods 5—7
    organizational 25, 27—8, 59
    related to seismic risk 77
    requirements for shared risk 10—12
professional social inquiry 27, 28, 30
profitability, and risk 5
property, loss of 12, 13
psychological effects 12
public managers, accountability 4
public organizations, distribution 232—34
public policies 47
public practice 69
public risk 3—17

qualitative measures 41, 47
quantitative measures 41, 47
quantitative methods 14
Quarantelli, E.L. 1978 20
Quito, Ecuador 93—4, 96
    seismic risk 12

random events 39
rankings 64
rapidly evolving response systems *see* response systems
RECOPE refinery 135
recovery 23, 272
Recovery Channel 201
Red Cross *see* American Red Cross; International Red Cross
redundancy 26, 27, 30
research premises 13
resilience 21
resonance 60
resources
    allocation 60, 274

knowledge base 23, 76
  organization of 8–9
  use of 61
response mechanisms of 23
response systems 45
  assessment 42–54, 67
    auto-adaptive systems 68, 73–5
    classification scheme 67, 68
    data analysis 49–50
    data collection 47–9
    data interpretation 50–4
    emergent adaptive systems 68, 70–2
    indicators 65–6
    measurement 38–55, 41, 47
    methodology 47
    non-adaptive systems 68–70
    nonlinear analysis 50–54
    operative adaptive systems 68, 72–3
    preliminary rankings 67
    selection of cases 45–7, 66
  characteristics 46
  community response systems 23
  comparisons of field study results 231–76
  coordination 8, 12–13, 19–20, 262
  evolution 53–4, 81, 264–67
  flexibility 46, 267–70
  performance 46
  rapidly evolving 275–76
  requirements 39
  structure 46, 267–70
  see also information exchange; information search; initial conditions; intra-organizational learning; and under each earthquake entry
Richter magnitudes 13
  see also under each earthquake entry
Ring of Fire 15
risk
  reduction 4
    sociotechnical systems 263–76
  seismic 8
  shared 3–17
Rivlin, Alice M. 1993 57
Roberts, Karlene H. 1993 27
Roberts, Karlene H. (with Rochlin and LaPorte) 1987 32
Roberts, K.H. (with K.E. Weick) 1993 29, 31, 32
Rochlin, Gene I. 1987 32
Rochlin, Gene I. 1989 27, 29, 32
Rokko Mountains, Japan 213

Roman Catholic Church 84
  see also Catholic Relief Services
Rosemead, CA 162, 163, 164
Ruelle, D. 1991 21
Rueschemeyer, D. 1991 43
Ruiz, *General* Germán 97
Rymer, Michael J. 1986 82

Sacramento, CA 172
safety, collective 3
Salvador, El
  Comité de Emergencia Nacional 84
  Frente Marti Farabundo para la Liberación Nacional 84
San Andreas fault 170
San Francisco CA
  Bay area 170
  Bay Bridge 172
  Emergency Operating Center 171
  Marina district 170, 172
  seismic risk 12
San Francisco Chronicle 174
San Francisco Examiner 174
San Jose, CA, Emergency Operating Center 171
San Jose Mercury 174
San Salvador, Archbishop of 84
San Salvador earthquake 1986 13, 16, 45, 46
  civil war 82–92 *passim*
  communications 83, 88
  coordination 88
  Cuerpos de Bomberos 83
  damage assessment 91
  damage and losses 82
  data collection 48
  disaster relief 84
  funding sources 85–7
  information exchange 84–5
  information search 82–3
  instability of response system 91
  international private organizations 88
  local public organizations 89
  loss of life 72
  non-profit organizations 91–2
  nonlinear logistic regression analysis 91–2
  organizational learning 85–92
  phase planes 90
  response system evaluation 68–70, 75, 91, 115, 231–76
  Richter measurement 82
  Ruben Dario Building 83

sample characteristics 298
Santa Cruz, CA 170, 172
   Emergency Operating Center 171
   Pacific Garden Mall 170
Santa Cruz mountains, CA 170
Santa Fe Springs, CA 162, 163, 164
   Fire Department 163
   police services 163
Sarkis, T. 1993 135
satellite communications
   Hyogo prefecture, Japan 215
   India 73, 76, 159, 181, 183
Saupe, D. 1992 9, 38
Schneider, Walter 1992 54
Schon, Donald A. 1974 29
seismic awareness 47
seismic intensities 46
seismic policy 3, 8
seismic risk 69
   reduction 81
self-expression, creative 29
self-organization 4, 40
   and auto adaptive systems 197—227
   creating the context for 265, 270—72
   measurement of processes 53
   preliminary model 20, 33—6
self-organizing systems *see* auto adaptive systems
self-protection 3
"sense-making" 27, 29, 30, 35
"Severity of psychological stresses scale" 147
sewage systems, Mexico City 56
shared goals 81
shared knowledge *see* common knowledge
shared risk 3—17
Sharp, Lawrence F. 1995 51
Shrivastava, Paul 20
Sierra, Ecuador 94
Simon, H.A. 1972 27
Simon, H.A. 1981 6, 10—11, 21, 26, 28—9, 31
Simon, H.A. 1983 28
small-n comparative study 42, 43
Smart, C. 1977 20
social inquiry *see* inquiry models
social systems, failure 21
sociotechnical infrastructure 76
sociotechnical systems 5
   future strategy 275
   interdisciplinary 197, 198
   primary function 41

and risk reduction 263—76
and shared risk 10, 14
use of information technology 197
software program 49, 55, 294
Solapur, Marathwada region, India 182
Southern Hyogo Prefecture, Japan 212—13
Soviet All—Union administration 106
spatiality in nonlinear systems 38, 39, 40
Spitak, Armenia 104
Sproull, Lee S. 1990 31
"state of the system" 51—2
Stepanavan, Armenia 104
stochasticity 8
strange attractors 9, 32
strategy development 18
structure and flexibility 21—2, 23, 64, 267—70
systemic change 59

Takarazuka, Japan 212
Tangshan, China, seismic risk 12
Taylor, F.W. 1967 24
"team approach" 198
technical design 64
technical information structure 235
technical structure
   auto adaptive systems 73, 197
   data analysis 64, 65
   emergent adaptive systems 70, 119
   nonadaptive systems 68, 81
   operative adaptive systems 72, 159
technical systems 263
tectonic plates 15, 71, 213
Telefonos de Mexico (TELMEX) 120
"thick description" 63
time
   construct of 11
   irreversibility of 39
   and response to risk 7, 10, 18, 19
"time, arrow of" 8
time in nonlinear systems 38, 39, 40
timing, evolution of response systems 265—67
Tokyo Region, Japan 213
Tokyo, seismic risk 12
Train, H.D. 1986 24, 26
Trans-Ecuadorian Pipeline 93—4
transition 10, 18—37, 31, 53, 262
   from response to recovery 272
   Mexico City 57
trauma 147

traumatic stress 149, 150
triage system 108
"trial and error" 27, 30
Turkey 13, 16
  National Earthquake Research Centre 148
  National Emergency Plan 150
Turkey earthquake *see* Erzincan earthquake
Turrialba, Costa Rica 136

uncertain conditions
  action in 24, 26, 27
  decision making in 59
  "edge of chaos" 197–98
  reducing uncertainty 30
  and shared risk 18–19
unit of analysis 44
unit of observation 44
United Nations Observer Group in Central America (ONUCA) 138
United States
  Agency for International Development, El Salvador 82
  Federal Emergency Management Agency (FEMA) 200–1, 205–7
  Forest Service 24
  Geological Survey 16, 161
  National Aeronautics and Space Administration (NASA) 200
  National Science Foundation 55
  Office of Foreign Disaster Assistance 271
  Regional Office of Foreign Disaster Assistance, San Jose 135
University of Colorado *see* Colorado, University of
University of Pittsburgh *see* Pittsburgh, University of

Valle de Estrella, Costa Rica 134, 136
Verba, S. 1994 41
Vertinsky, I. 1977 20

Village Councils, India 193
Volcan Reventador, Ecuador 93
voluntary associations, India 182
voluntary co-ordination 19
volunteer action 4
vulnerability 11, 12, 44, 274

water supply, Mexico City 56
Watsonville, CA 171, 172
Weick, K.E. 1990 27, 29, 31–2
Weiner, S.S. 1976 26
Weiss, Carol H. 1977 27, 28
Weiss, Carol H. 1998 27, 28
Weiss, Carol H. (with Bucuvalas) 1984 27
Weissinger-Baylon, R. 1986 24
White, Randy A.1986 82
Whittier, CA 162, 163
  Emergency Coordinator 163
Whittier Fault 161
Whittier Narrows, CA earthquake 1987 8, 13, 45, 74, 161–70
  adaptive behavior 167
  communication and coordination 168–9
  damage and losses 161
  data collection 48
  funding sources 165–67
  information exchange 164
  information search 161–64
  nonlinear logistic regression analysis 168–9
  organizational learning 165
  phase planes 167–68
  response system evaluation 72, 75–6, 194, 231–76
  Richter measurement 161
  sample characteristics 300
Wildavsky, A.1974 19
Wildavsky, A.1988 21

Yin, Robert K. 1993 42